Lithium and Nickel Contamination in Plants and the Environment

World Scientific Series on Advances in Environmental Pollution Management

Series Editors: Dharmendra Kumar Gupta
(*Ministry of Environment, Forest and Climate Change, India*)
Clemens Walther
(*Leibniz University Hannover, Germany*)

Published

Vol. 1 *Lithium and Nickel Contamination in Plants and the Environment*
edited by Masahiro Inouhe, Clemens Walther & Dharmendra K Gupta

World Scientific Series on Advances in Environmental Pollution Management – Volume 1

Lithium and Nickel Contamination in Plants and the Environment

Editors

Masahiro Inouhe
Ehime University, Japan

Clemens Walther
Gottfried Wilhelm Leibniz University Hannover, Germany

Dharmendra K Gupta
Ministry of Environment, Forest and Climate Change, India

World Scientific

NEW JERSEY • LONDON • SINGAPORE • BEIJING • SHANGHAI • HONG KONG • TAIPEI • CHENNAI • TOKYO

Published by

World Scientific Publishing Co. Pte. Ltd.

5 Toh Tuck Link, Singapore 596224

USA office: 27 Warren Street, Suite 401-402, Hackensack, NJ 07601

UK office: 57 Shelton Street, Covent Garden, London WC2H 9HE

Library of Congress Cataloging-in-Publication Data
Names: Inouhe, Masahiro, editor. | Walther, Clemens, editor. |
Gupta, Dharmendra Kumar (Phytoremediation researcher) editor.
Title: Lithium and nickel contamination in plants and the environment /
editors: Masahiro Inouhe, Ehime University, Japan,
Clemens Walther, Gottfried Wilhelm Leibniz University Hannover, Germany,
Dharmendra K Gupta, Ministry of Environment, Forest and Climate Change, India.
Other titles: World Scientific series on advances in environmental pollution management ; v. 1.
Description: Hackensack, NJ : World Scientific, 2024 |
Series: World Scientific series on advances in environmental pollution management ;
Volume 1 | Includes bibliographical references and index.
Identifiers: LCCN 2023053646 | ISBN 9789811283116 (hardcover) |
ISBN 9789811283123 (ebook) | ISBN 9789811283130 (ebook other)
Subjects: LCSH: Plants--Effect of trace elements on. | Lithium--Environmental aspects. |
Nickel--Environmental aspects. | Soils--Lithium content. | Soils--Nickel content.
Classification: LCC QK753.T7 L58 2024 | DDC 571.9/452381--dc23/eng/20231222
LC record available at https://lccn.loc.gov/2023053646

British Library Cataloguing-in-Publication Data
A catalogue record for this book is available from the British Library.

For any available supplementary material, please visit
https://www.worldscientific.com/worldscibooks/10.1142/13588#t=suppl

Desk Editors: Aanand Jayaraman/Amanda Yun

Typeset by Stallion Press
Email: enquiries@stallionpress.com

https://doi.org/10.1142/9789811283123_fmatter

Preface

Lithium (Li) is a chemical element with the atomic number 3; it is the 27th most abundant element and covers about 0.006% (wt.) of the Earth's crust. Li does not occur in metallic form in nature but is found in minerals in small amounts, in nearly all igneous rocks and in the waters of many mineral springs. Li is also found in trace amounts in plants, plankton and invertebrates at concentration levels of 69–5760 ppb. In plant tissue, Li concentration is typically around 1 ppm, with some plant families bioaccumulating more Li than others; Li accumulation does not appear to affect the essential nutrient composition of plants. It is reported that Li at a 5 ppm concentration reduces seed germination in some plant species and can also permeate through ligand-gated ion channels as well as cross both nuclear and mitochondrial membranes. Similar to sodium, lithium can enter and partially block (although not permeate) potassium and calcium channels. Aquatic organisms tend to bioaccumulate Li more than land-dwelling organisms. The biological effects of Li are many and diverse; however, its mechanisms of action are only partially understood.

Nickel (Ni) is an important trace metal element with the atomic number 28. It is a hard and ductile transition metal and is the 22nd most abundant element in the Earth's crust (twice as abundant as Cu). Ni is known to play an important role in the biology of some plants, bacteria, archaea and fungi. Ni is an essential microelement for normal plant growth and development and is also useful for several biological functions. Ni is an essential part of several enzymes, such as glyoxalase-I and urease, required for nitrogen metabolism/assimilation in higher plants as well as helping plants against several biotic/abiotic stresses. Hence, Ni deficiency in plants may lead to stunted plant growth, senescence, reduced N metabolism and reduced Fe uptake. There are two modes of Ni uptake via roots into plants, i.e., passive diffusion and active transport. Higher concentrations of Ni in plants may lead to several deleterious alterations. The major source of Ni exposure is oral consumption, as Ni is essential to plants. Dietary Ni may affect human health through infections by Ni-dependent bacteria, although Ni may also be an essential nutrient for bacteria living in the large intestine, in effect functioning as a prebiotic. Most Ni absorbed by humans is removed by the kidneys and passed out of the body through urine. Ni is not a cumulative poison; however, larger doses or chronic inhalation exposure may be toxic, even carcinogenic, and constitute an occupational hazard.

This book, *Lithium and Nickel in Plants and the Environment*, presents the most recent research on Li and Ni chemistry, sources, acute and long-term impacts on the environment and human health, as well as remediation measures. This book comprises a thorough examination of the effects of Li and Ni exposure, including toxicity and its prevention, so that readers can efficiently evaluate the associated hazards. Recent methods in Li and Ni detoxification, speciation and molecular mechanisms are included, and it offers all the knowledge required for efficient risk assessment, prevention and countermeasures. This edited volume brings together a diverse group of environmental science, sustainability and health researchers to address the challenges posed by the global increasing use of Ni and, particularly, Li, in the context of the transition of our energy systems. The background information provided in this book may help assess and minimise possible negative effects and avoid mass poisoning caused by Li and Ni contamination of soil and plants.

The editors (Prof. Masahiro Inouhe, Prof. Clemens Walther and Dr. Dharmendra K. Gupta) personally acknowledge all the contributing authors for their valuable time, knowledge and eagerness to bring this book to fruition.

https://doi.org/10.1142/9789811283123_fmatter

About the Editors

Dharmendra K. Gupta is a director (S) at the Ministry of Environment, Forest and Climate Change, Government of India. His fields of research include abiotic stress by radionuclides/heavy metals and xenobiotics in plants, antioxidative systems in plants, and environmental pollution (radionuclides/heavy metals) remediation through plants (phytoremediation). Dr. Gupta has been awarded several international awards by entities such as the Japan Society for the Promotion of Science (JSPS), the Belgian Science Policy Office (BELSPO), JAE-Doc. (Spain), TWAS-CNPq (Italy), Royal Society India Fellow (U.K.) and MASAV (Israel). Dr. Gupta has published more than 125 international manuscripts and review articles in international and reputed journals, edited 24 books and served as an editor for two book series from Springer, Switzerland, and one from Scientific World, Singapore.

Masahiro Inouhe is an emeritus professor and a research scientist at the Faculty of Science, Ehime University, Japan. His research focuses on plants' abilities for growth and adaptation to environmental stresses, which includes a range of approaches involving plant physiology, biochemistry and nutritional biotechnology. He has been involved in the development of two major projects: one on cell wall metabolisms associated with plant cell growth and the other on heavy-metal adaptation mechanisms by the formation of metal-binding complexes in plant cells under various environmental conditions. His most prominent achievements consist of novel findings about cell wall proteins controlling cell elongation growth and cadmium-binding polypeptides from various sources. Prof. Inouhe has published more than 100 international manuscripts and review articles in international and reputed journals and edited four books for Plant Physiology and Plant Biochemistry.

Clemens Walther is a professor at the Leibniz Universität Hannover, Germany, and the head of the Institute of Radioecology and Radiation Protection. He has been the chair of the Nuclear Chemistry Section of the German Chemical Society and the head of the German Steering Board of the Competence Center Radiation Research (KVSF) since 2017. He has also been serving as a member of the German Commission for Radiation Protection since 2015 and the vice president of the German Swiss Society for Radiation Protection since 2023. Prof. Walther's past appointments include being dean of the Faculty of Mathematics and Physics (2019–2021), head of the European Network on Nuclear and Radiochemistry Education and Training (2016–2022), member of the extended governing board of the German Society for Mass Spectrometry (DGMS) (2012–2015) and head of the Mass Spectrometry Division of the German Physical Society (DPG) (2012–2015). Prof. Walther has published more than 175 international manuscripts and review articles in international and reputed journals and edited seven books from Springer.

https://doi.org/10.1142/9789811283123_fmatter

Contents

https://doi.org/10.1142/9789811283123_fmatter

Abbreviations

AA	Amino acid
ABA	Abscisic acid
ACC	1-Aminocyclopropane-1-carboxylic acid
AMF	Arbuscular mycorrhizal fungus
APX	Ascorbate peroxidase
AsA	Ascorbic acid
BC	Biochar
BCF	Bioconcentration factor
Br	Bioaccumulation rate
BRs	Brassinosteroids
Ca	Calcium
CAT	Catalase
CiA	Citric acid (Citrate)
DAO	Diamino oxidase
DAS	Days after sowing
DFC	Donnan free space
DHAR	Dehydroascorbate reductase
EDDS	Ethylene diamine disuccinic acid
EDTA	Ethylene diamine tetra acetic acid
EL	Electrolyte leakage
EMF	Ectomycorrhizal fungus
EPS	Extracellular polymeric substance
Fe	Iron
GA_3	Gibberellic acid
GB	Gobindobhog

GPX	Guaiacol peroxidase
GR	Glutathione reductase
GSH	Glutathione (reduced form)
GSNO	S-Nitroso glutathione
GSNOR	S-Nitroso glutathione reductase
GSSG	Glutathione disulphide (oxidised form)
GST	Glutathione-s transferase
H_2O_2	Hydrogen peroxide
HM	Heavy metal
HRT	Hydraulic retention time
IRT	Iron-regulated transporter
IRT1	Iron-regulated transported 1
JA	Jasmonic acid
JAs	Jasmonates
K	Potassium
Li	Lithium
Li_2O	Lithium oxide
LiCl	Lithium chloride
LPS	Lipopolysaccharide
MaA	Malic acid (Malate)
MDA	Malondialdehyde
MDHAR	Monodehydroascorbate reductase
MeJA	Methyl jasmonate
Mg	Magnesium
MG	Methylglyoxal
MT	Metallothionein
MTP	HM transporting
MTPs	Metal tolerance proteins
MuA	Mugineic acid
NA	Nicotianamine
Na	Sodium
NAS	Nicotianamine synthase
Ni	Nickel
NiA	Nicotianamine
NRAMP	Natural resistance-associated macrophage protein
$^1O_2$	Singlet oxygen
$O_2\cdot^-$	Superoxide radical
OA	Organic acid
P	Phosphorus

PAO	Polyamine oxidase
PAs	Polyamines
PC	Phytochelatin
PCS	Phytochelatin synthases
PGA	Polygalacturonic acid
PGPR	Plant growth promoting rhizobacteria
PNUE	Photosynthetic nitrogen use efficiency
POD	Peroxidase
PROX	Proline oxidase
RBOH	Respiratory burst oxidase
REE Y	Rare earth element yttrium
RNS	Reactive nitrogen species
ROS	Reactive oxygen species
RSS	Reactive sulphur species
SAMDC	S-adenosylmethionine decarboxylase
SOD	Superoxide dismutase
TBARS	Thiobarbituric acid
VOCs	Volatile organic compounds
WFS	Water free space

https://doi.org/10.1142/9789811283123_0001

Chapter 1

Distribution of Lithium/Nickel in Soil and Its Uptake by Plants

Pavel Otřísal[*,‡] and Jozef Sabol[†]

[*]*Faculty of Physical Culture, Palacky University Olomouc, Tr. Míru 117, 771 11 Olomouc, Czech Republic*

[†]*Faculty of Security Management, Police Academy of the Czech Republic in Prague, Lhotecká 559/7, 14301 Prague, Czech Republic*

[‡]*pavel.otrisal@upol.cz*

Abstract

The chapter deals with some important aspects of the relationship of lithium and nickel with the ecosystem, which consists mainly of soil, water, plants and air. Some aspects of lithium and nickel use in the energy industry are also mentioned. We begin by considering the fact that the metallic elements lithium and nickel, either alone or in the form of their chemical compounds, are currently considered potential energy materials whose applicability is increasing with the transition to the mass use of electricity and batteries for powering motor vehicles. Both lithium and nickel are commonly found in nature. Even in relatively low concentrations, their presence is very dangerous or even toxic to some animals and biological organisms. On the other hand, certain plants and animals are a natural part of ecosystems and are unable to survive without their presence because they are vital to them. This contradiction and its implications form the main content of this chapter. The most significant effects of lithium and nickel in the environment, particularly in soil, water and plant systems, are presented. The interconnectedness between soil, water and plants is shown in relation to each other. Some of the analytical methods used for the detection of lithium and nickel are also given. In addition, some specific results are presented, which are not intended to specify particular locations in the field, but rather to highlight the ability of researchers to monitor the presence of lithium and nickel in the environment and to create conditions for their removal and possible reuse.

Keywords: Lithium, nickel, risk element, heavy metal, contamination, ecosystem, soil, water, plant, metal, EDTA, DTPA, accumulation, hyperaccumulator

1. Introduction

With the development of modern technologies, the production and consumption of metals and metalloids, and thus their input into ecosystems, are increasing very rapidly. The discussion about the transition to new and highly efficient energy materials is a phenomenon of our time. The world's largest mining companies have already started to prepare in advance for the expected steep rise in demand for materials required for the production of batteries for electric vehicles. However, they currently disagree on what goods will be most in demand in the near, medium and distant futures. While some miners are betting on lithium, others predict that the growth in interest will be more likely to be directed towards nickel, cobalt or platinum (E15.cz, 2019). However, it is essential to note that, at present, batteries for electric vehicles can pose as much of a risk to the automotive industry as chips. Experts warn that while there is a huge capacity for battery production, this is not matched by the supply of materials. In the coming years, producers are likely to face shortages of metals such as cobalt, nickel and lithium. This could slow down or even suspend not only the grandiose plans of carmakers to switch to electric cars but also the European Union's environmental plans (E15.cz, 2021). So, it appears that lithium, nickel, cobalt, copper and rare earth metals will be among the basic raw materials that humanity will desperately need for energy transformation. However, the issues related to the toxicology of lithium and nickel are also growing in importance in connection with the discussions among those involved in research on trace metals and their environmental toxicity to plants.

As a result of human economic activity, the natural landscape is gradually transforming into an anthropogenic landscape, i.e., a landscape dominated by humans. In relation to metals, metal compounds and their toxicity, one of the most significant changes is the underground mining of metal ores. The centuries of mining and metallurgical activity that have accompanied humankind virtually everywhere in the world have also affected the distribution and behaviour of metals and their compounds. The alkaline rare metal lithium and one of the representatives of a large ecologically important group of heavy metals, nickel, are transported into groundwater, mine water and seepage water as a result of mining activities. These metals and their compounds have varying mobility. The contamination of these metals in the vicinity of mining areas is indicated by the abundance of secondary minerals, a significant part of which is unstable and thus soluble in water. These secondary minerals are found both on the surfaces of ore bodies and in the heaps of waste products generated after mining (Andráš, 2014).

In developed countries, water has ceased to be a source of serious bacterial and viral diseases. A greater risk is posed by contamination from chemicals, whose use has been increasing since the middle of the 20th century. The elements at risk, very often referred to in ecotoxicology as environmentally harmful metals or, more simply, heavy metals, originate from anthropogenic activities. These elements (metals) are among the most common and longest-lasting environmental contaminants. If they enter the soil, they persist for thousands of years, and it is very difficult to eliminate their effects on plants and soil fertility. Permissible limits for hazardous elements in soils are set in national legislation. The levels of hazardous elements in soils that have been exposed to high loads over a long period of time often exceed the established limits and can have negative effects on biological and physico-chemical processes. Additionally, these elements can enter the food chain and adversely affect human health. Hazardous elements are among the longest-known toxic substances. In ecosystems, they move along specific pathways of their biological cycles. They can emerge from these cycles and accumulate, for example, in soil or living organisms. The mobility of hazardous elements is influenced by their solubility and the solubility of their compounds in water. There is a direct proportionality which states that the more soluble a compound is in water, the higher the mobility of the element at risk. Also important is the solubility of the hazardous elements in acids, especially sulphuric and nitric acids, which are often present in the environment. Strongly acidic aqueous precipitation can increase the solubility of compounds of risk elements. Hazardous elements can be readily taken up by plants from the soil solution or leached into surface water or drinking water sources. The toxicity of hazardous elements is related to their concentration in the environment. In trace concentrations, many elements are essential for organisms, and their deficiency can cause metabolic disorders in plants and animals. In low concentrations, these metals are a natural part of the Earth's crust. Locally, as part of ores, their natural occurrence can be many times higher (Pavlíková, 2006, 2008).

The content of risk elements in aquatic and soil ecosystems is significantly influenced by microorganisms. Microorganisms are involved in altering the solubility, bioavailability and mortality of these elements. Their content affects the active transport of elements, oxidation–reduction reactions and form extracellular complexes; some compounds transform into volatile or less toxic forms. The mobilisation or immobilisation of hazardous elements follows one or more mechanisms (Pavlíková, 2008):

- oxidisation causing a reduction of risk elements, affecting their oxidation degree;
- pH changes leading to a change in solubility;
- dissolution and extraction of hazardous elements by microbial metabolites and degradation products;
- volatilisation by alkylation reactions (e.g., biomethylation);
- immobilisation leading to the formation of stable minerals or bioaccumulation by organic matter;
- transformation of organometallic complexes.

At the very beginning, it should be stated that, accepting a certain degree of general understanding of the toxicology of lithium and nickel, it is possible to conclude that lithium is not of major toxicological significance in natural concentrations. However, its salts, which may be present in the soil, can undergo hydrolysis, leading to a significant reduction in soil pH. Such soil may be too alkaline for plant growth and unsuitable for certain types of plants. Nickel, on the other hand, is considered an essential element at very low concentrations; however, this is not the case for some types of bacteria.

2. Ecosystem Functions of Soil and Water

2.1. *Soil ecosystem functions*

Soil is the topmost layer of the Earth's crust. It serves as a repository for substances, both natural and harmful, and provides a habitat for plants, animals and humans. It is of great importance for the stability of ecosystems. Soil pollution affects the quality of other environmental components and food (Panáček and Balzerová, 2013).

Soil is an important part of the biosphere and contributes to the capture, immobilisation and decontamination of incoming elements. It also significantly co-determines the direction of their subsequent movement, whether into the atmosphere, hydrosphere or biosphere. Geogenic and anthropogenic factors can influence the total elemental content of the soil. In particular, the parent substrate determines the natural background content of the elements of concern in the soil, whereas anthropogenic activity determines their changes. For example, serpentinite is a metamorphosed grey-green rock formed by the alteration of strongly alkaline igneous rocks. These rocks are characterised by high nickel and chromium contents, and elevated levels of these elements will also be found in basalts, amphiboles and syenites. Weathering residues of limestones and some shales are characterised by elevated cadmium content. Gravel and gravel sands, including nickel, are characterised by significantly lower concentrations of these risk elements. The nickel content of soils generally increases with increasing clay content (Pavlíková, 2008).

When nickel and other elements enter the soil, a number of interactions occur between the soil phases and the elements themselves. These reactions are influenced by a wide range of physico-chemical properties of the soil as well as by the properties of the elements or their compounds. The decisive factor for the uptake of elements by plants is their concentration in the soil solution. The release of elements into the soil solution can be influenced not only by the total content of elements in the soil and their forms but also by other factors such as soil type, sorption, soil reaction, organic matter content, soil redox potential and the action of other ions. The most important processes affecting the behaviour of hazardous elements are those involved in their sorption from the liquid to the solid phase. Several basic mechanisms are involved in the sorption of ions, as shown in the schematic diagram in Fig. 1.

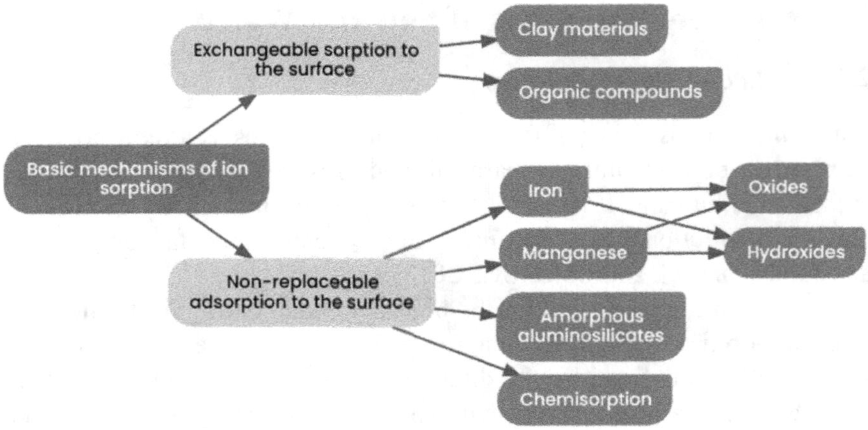

Figure 1. Basic mechanisms of ion sorption.

Soil has a wide range of functions in the environment. From a human perspective, only its reproductive function in agriculture and forestry has long been of interest. It has also been considered a source of building materials, an area for development or other uses. In the ecosystem, however, the soil is a filtering, storage and transport medium for water and solutions. It is also a very good buffering, stabilising and transforming medium. Soil is the medium for energy exchange between the atmosphere, hydrosphere and geosphere. Soil is a necessary condition for the balance of different ecosystems and can therefore be seen as a necessary and absolutely essential condition for life on land. For ecological biotests, the so-called soil ageing is an important feature. In real conditions of "living" soil, natural processes occur that affect not only the soil itself but also the toxicity and bioavailability of substances present in the soil. The bioavailability of rare and heavy metals and their compounds decreases with time due to their sorption into soil particles and their entrapment within soil micropores. The toxicity of the substances present in the soil usually decreases with time. As a consequence of soil ageing, substances freshly injected into artefactual soil usually have a higher toxicity than the same amount under realistic conditions. Soil ageing, and thus the reduction of lithium and nickel toxicity, is enhanced by soil drying and wetting (Kočí and Mocová, 2009).

Lithium, nickel and most other elements of concern are present in soil solution, usually in the form of cations. Their adsorption depends on the density of negative charges on the surface of soil colloids. This reversible sorption occurs on the surface of both permanently charged aluminosilicates and humic polymers and oxyhydroxides, whose charge depends on the pH of the environment. Heavy metals compete for the occupation of binding sites with calcium and magnesium ions, which are present in greater abundance. For this reason, the sorption process is not predominant among other hazardous elements in the soil.

Specific adsorption occurs by exchanging cations and anions with surface ligands, usually by forming covalent bonds. This sorption occurs on the surface of several compounds, including quartz, amorphous aluminosilicates and aluminium hydroxides. However, aluminium and iron oxides are the most important, while amorphous silicon and aluminium oxides are less important. Chemical sorption involves reactions that produce new compounds that reduce the solubility of elements, both on the sorbent surface and through the formation of a solid phase in the environment. Precipitation reactions can only take place on contaminated soils.

2.2. *Uptake of hazardous elements by plants*

The movement of hazardous elements to plant roots occurs through diffusion and mass soil flow. In the immediate vicinity of the roots, the reaction of the elements of concern with organic acids secreted by the plant results in the formation of chelates, increasing the diffusion gradient and accelerating the uptake of elements. The transport of an element from an external solution through the cell wall is a passive process, in which ions are transported via diffusion. The plant significantly influences the uptake of the element of concern into the cell interior, its transport and distribution into different parts of the plant, and through its exudates, it participates in the solubilisation and precipitation of elements found mainly in the rhizosphere of roots. Hazardous elements penetrate the roots through the epidermis and the cortex through the apoplast or symplast into the xylem. A part of them is bound irreversibly by non-metabolic fixation in cell walls (to negative charges of carboxyl groups of polygalacturonic acid, which is part of pectins in cell walls), or passes through the cell wall and is further transported into cells, and the apoplast directly transports the part. The transport of an element or its binding is determined by the interaction of the plant genotype and the element itself (Pavlíková, 2008).

2.3. *The fate of the toxicant in the aquatic environment*

Most of the naturally occurring metals and semi-metals of the periodic table can be found in water. When assessing metal pollution, a separate group of metals called heavy or toxic metals is often discussed. Heavy metals, which include nickel, are defined chemically by a bulk density greater than 5,000 kg m^{-3} or by the precipitation of their salts with sodium sulphide to form sparingly soluble sulphides. The term 'toxic metal' can be more easily defined, whereas the term 'heavy metal' is often used as a synonym for toxic metal, which can lead to inconsistencies. For example, lithium or beryllium are toxic metals but not heavy metals. Therefore, it is not recommended to use heavy metals as a synonym for poisonous metals (Kopp, 2015).

Water erosion preferentially carries away lighter particles with higher sorption capacities than other soil components. Most of the foreign substances (mainly from emissions) are bound to these particles. Sedimentation occurs mainly in large reservoirs. Preferably, reducible ions, which include nickel ions, are deposited in sediments (Kalač *et al.*, 2015).

In terms of the overall flux of toxicants in the aquatic environment, including lithium, nickel and their compounds, microbial degradation plays the most important role. Although microorganisms decompose most organic substances, we cannot exclude the fact that inorganic compounds are also subject to partial microbial decomposition. The course and intensity of biodegradation are influenced by a number of factors, including the concentration of the toxicant, the amount of organic substrate and abiotic and biotic environmental factors. Increases in temperature, sufficient oxygen and nutrient supply lead to an acceleration of the process. The entry of the toxicant into the aquatic and, consequently, soil ecosystems takes place through different pathways. These include intentional water intoxication, thus the planned activity of applying toxic substances to water and subsequently to the soil, and the unintentional introduction of toxic substances into aquatic and soil systems (Kopp, 2015).

3. Selected Basic Factors Affecting Plant and Mushroom Growth

Pollutants or contaminants can be products of natural processes independent of human actions. However, pollutants primarily originate from areas of human activity, affecting the surrounding air, water and soil. Water and air transport toxic substances (pollutants) into the environment very quickly. In contrast, the transport of substances in soil is gradual, with soil acting more as a reservoir for toxic substances. Due to the interconnection and continuity of processes in the environment as a whole, the toxic substance will reach a different environment than the one it originally contaminated. The interrelationship of these physical, chemical and biological mechanisms is not yet fully understood.

For this reason, the distribution of toxic substances in soil and plants is a matter that has not been fully and sufficiently investigated and clarified. Although the toxicokinetics and toxicodynamics of lithium and nickel and their distribution in nature are not fully understood, it is important to recognise that these hazardous elements and especially their compounds can cause health problems in humans, animals and plants. The degree of their harmful effects depends primarily on their concentration. Toxicity is mainly caused by heavy metals, which have a proclivity to accumulate in plant and animal organisms and thus contribute to the bioaccumulation of harmful substances in them (Linhart, 2019). Bioaccumulation is a lethal hazard for top predators because they consume organisms that may have already absorbed significant amounts of toxic substances, and thus, the cumulative concentration of chemical and especially harmful substances in top predators can increase alarmingly (Borkovcová and Žáková, 2015).

The ability of plants to grow is influenced by a number of factors, which can be broadly divided into internal and external factors. In studying the effect of toxic substances on plant growth or, more generally, the interrelationship between soil and plants, it is necessary to consider a number of interacting parameters and factors broadly classified as external growth factors. Their subdivision is shown in the diagram in Fig. 2, from which it is apparent that there are many factors affecting plant growth. The presence of metals involved in plant growth is essential and irreplaceable. However, the degree to which they are beneficial depends on their concentrations in the soil. Some metals are vital for plants, whereas some are toxic or significantly affect plant growth and yield.

A special and very interesting category to consider in terms of the accumulation of harmful metallic compounds is mushrooms. It is well known that mushrooms are neither plants nor animals but form a separate kingdom of organisms. Unlike plants, mushrooms do not contain chlorophyll, which releases oxygen or energy through photosynthesis. Mushrooms form a large network of cellular filaments (hyphae) called

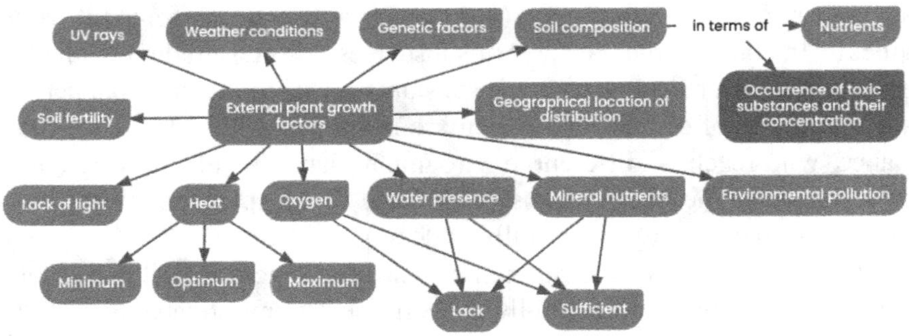

Figure 2. External plant growth factor.

mycelium underground. The accumulative capacity of mushrooms is a property that allows them to accumulate heavy metals from the environment, i.e., from soil, water and air. As a suitable bioindicator, the common loosestrife (*Lycoperdon perlatum*) is used. Although mushrooms are capable of accumulating nickel and other heavy metals, this does not mean that they must always contain critical amounts of these metals and that we must, therefore, strictly avoid eating them. Mushrooms are an integral part of living nature, and their cultivation has become an important sector of agricultural production in recent years. People have been using many of them not only as food but also for medicinal purposes since ancient times in many regions of the world.

From the point of view of the distribution of rare and heavy metals, it should be noted that soil is the largest abiotic reservoir of these elements, from which they enter the plant cycle. Nutrients are returned to the soil through decomposition processes. Since nutrients serve as limiting elements in both terrestrial and aquatic ecosystems, their cycling is very important. A general model of nutrient cycling is shown in Fig. 3.

Soil is one of the most essential components of the environment. Together with other components, such as the geosphere, atmosphere, hydrosphere and biosphere, it forms a functional unit called an ecosystem. For individual ecosystems, soil contamination is a threat, particularly in terms of disrupting their functions and possibly the health of plants, animals and humans. Furthermore, ecosystem contamination can lead to a loss of soil fertility and thus contribute to yield losses. This process can also be expressed schematically, as shown in Fig. 4.

Soil contamination is always associated with the need for soil removal to prevent contaminants from entering plants and other living organisms.

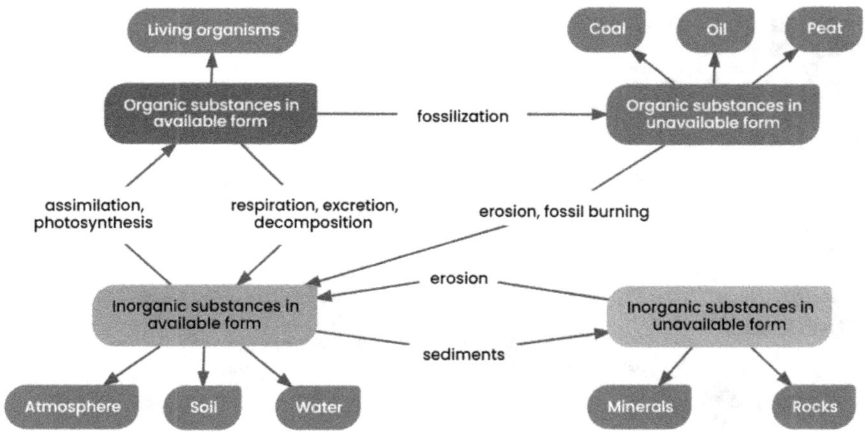

Figure 3. General model of the substance cycle. Adopted from Borkovcová and Žáková (2015).

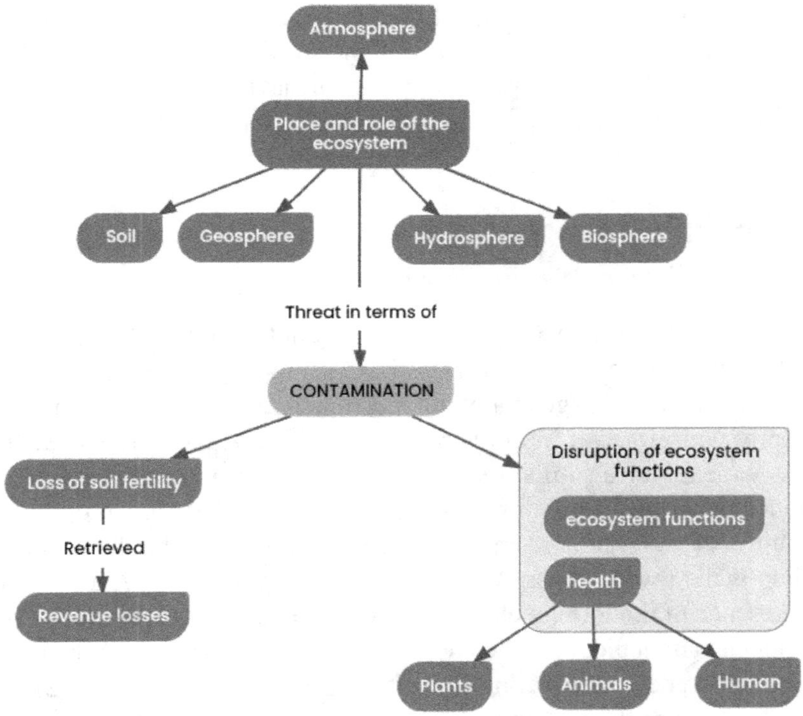

Figure 4. Place and role of the ecosystem.

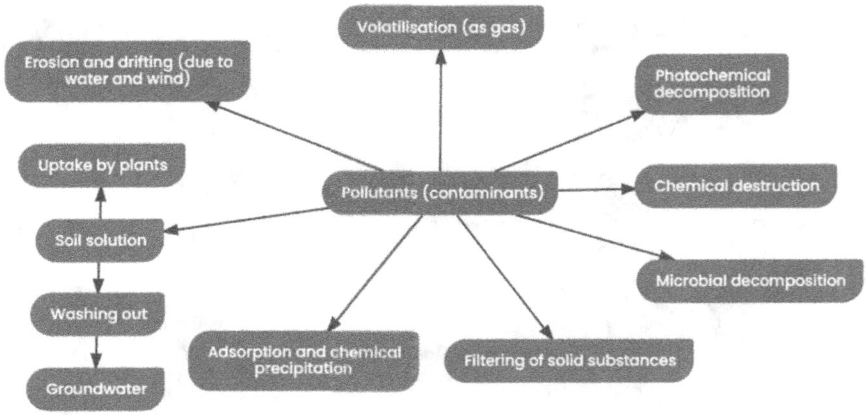

Figure 5. Behaviour of pollutants in soil.

In order to effectively prevent and combat soil contamination, it is imperative to understand how contaminants enter the soil and how much of them are present. The information must therefore be available and known not only on the mode of contamination and the specific contaminant but also on its quantity, thus its concentration. The behaviour of contaminants in soil can be explained by referring to the diagram shown in Fig. 5.

4. Lithium

4.1. *General properties and history*

Lithium is the lightest metal. Although it may not seem like it at first sight, it is lighter than water. However, testing for this physical property is not very convenient and is generally not recommended because lithium reacts violently with water to form hydrogen and LiOH. Although this reaction is less violent than a similar reaction of sodium, it is not recommended to conduct experiments of this type. Another very important property of lithium is that it is the most toxic of the alkali metals. The discoverer of lithium is the Swede Johan August Arfwedson, who, in 1817, discovered the presence of the foreign element in petalite and soon afterwards also in spodumene and lepidolite when he analysed mineral finds from the island of Utö in Sweden (Das Institut für seltene Erden und Metalle, 2022).

The study of the negative effects of lithium is linked to its presumed or actual toxicity. In particular, its occurrence in food and living organisms, where it is considered to affect a wide range of metabolic processes in the human body, has created a need to study its toxicology and the responses to its effects. It has become apparent that there is a continuing need to collect and evaluate relevant toxicological information on the occurrence of lithium in plants and animals, thereby contributing to the understanding of the processes involved in the mechanism of its effects and, last but not least, to obtain information relevant, for example, to the pharmacological industry engaged in the research, development and production of pharmaceuticals. The use of lithium in medicine dates back to the middle of the 19th century. The first physician to make a contribution to its medical application was the English physician Alfred Garrod. He saw the potential for its use in the treatment of gout and kidney stones without achieving the desired therapeutic effect. Further, lithium was used in its early medical applications in the form of lithium bromide, which was believed to have a significant effect on the treatment of epilepsy. However, even in this case, it was not shown to be more effective than other bromides (Diding *et al.*, 1969).

It should be noted that real applications for lithium did not appear until shortly after the Second World War. Until then, it had been used virtually only in applications related to lubricants (mineral oils thickened with lithium stearate) and in the glass industry (as lithium carbonate or lithium oxide). This situation changed when, in the United States, the requirement for tritium, which can be obtained from lithium, arose for the construction of hydrogen bombs. This began with widespread mining of lithium, especially at Kings Mountain in North Carolina. Due to the short half-life of lithium, large quantities of the element were required, especially between 1953 and 1963. Significant stocks of lithium appeared on the market after the end of the Cold War, around 1993. In addition to the standard technologies associated with its extraction by mining, brine extraction, a cheaper method of extracting lithium, became increasingly important. Currently, relatively significant amounts of lithium are used for battery production, elastomer polymerisation, construction and the organic synthesis of pharmaceuticals and agrochemicals. In those industries involved in the production of batteries and accumulators, lithium is the most important production and technological input (Nayak *et al.*, 2017).

Lithium is an element that occurs naturally in nature. However, lithium occurs in relatively small amounts throughout the Earth's crust. Lithium is classified as an intrinsically non-hazardous metal. As in other cases in toxicology, the relationship between the dose of a toxic substance containing lithium, the recipient of the dose and the length of time that the recipient is exposed to the dose must be carefully monitored. It is therefore very important to consider aspects of toxicokinetics, which is one of the key areas of toxicology and their medical applications and deals with the monitoring of the fate of a toxic substance in the body from its administration to its elimination, i.e., from its entry into the body to its metabolic transformation or elimination.

Lithium is one of the metals that is not essential for life. The effects of lithium toxicity on plants are still unclear due to the limited information available on its mobility from soil to plants. Lithium, as a pure element, is highly reactive. Therefore, its existence in its elemental form is rather unexpected and very unlikely. Because of its high reactivity, it reacts with other elements and thus occurs in nature in the form of salts or minerals.

Lithium is a light metal and has the lowest density of solid elements under standard conditions. At room temperature, it is only stable for long periods in completely dry air, but even so, it reacts slowly to form lithium nitride. In moist air, a dull grey layer of lithium hydroxide quickly forms on the surface. Like all alkali metals, elemental lithium reacts with skin moisture and causes serious burns and skin damage. Many lithium compounds that form lithium ions in an aqueous solution are characterised as harmful to health, unlike the corresponding sodium and potassium compounds. As a trace element, lithium is a common component of mineral water in the form of its salts.

4.2. *Distribution of lithium in soil*

Lithium represents about 0.006% of the Earth's crust. Thus, it is less abundant in the Earth's crust than zinc, copper and wolfram but slightly more abundant than cobalt, tin and lead. Although lithium is more abundant in nature than lead, it is more difficult to obtain because of its lower local availability and confinement and, therefore, its greater distribution in the environment. We consider it important to note that lithium is present in drinking water and some foods. Typical foods that contain lithium are meat, fish, eggs and dairy products. For example, meat weighing 100 g contains about 100 μg of Li. Various plants, such as tobacco or resin, collect and accumulate lithium compounds from the soil. The average plant dry matter content is between 0.5 and 3 ppm of lithium. In ocean waters, the average lithium concentration is about 180 ppb but only about 3 ppb

Table 1. Lithium concentrations in different soil types. Based on Groot (2009).

Experimental location	Soil type/Soil series	Li concentration (mg kg^{-1})/(mg L^{-1})
USA	Cecil (Clayey, kaolinitic, thermic Typic Hapludult)	11.49
USA	Iredell (Fine, mont morillonitic, thermic Typic Hapludalf)	25.38
USA	Madison (Clayey, kaolinitic, thermic Typic Hapludult)	11.93
USA	Louisa (Fine, mont morillonitic, thermic Ruptic Ultic Dystrochrept)	33.29
USA	Wakulla (Sandy, siliceous, thermic Psammentic Hapludult)	5.82
USA	Bonifay (Loamy, siliceous, thermic Grossarenic Plinthic Paleudult)	3.74
Jordan Valley	Ustochreptic and ustollic camborthids and calciorthids, ustic torriorthents	1.04–1.56
Northern Jordan Valley		1.05–1.06
Middle Jordan Valley		1.19–2.68
Southern Jordan Valley		0.95–1.48

in river water. Mineral waters are also a source of lithium (Das Institut für seltene Erden und Metalle, 2022).

Shahzad (2016), referring to Bradford (1966), reports that lithium concentrations were recorded in 100 of 400 water samples, with some of these samples being monitored at concentrations greater than 0.05 mg L^{-1}. This concentration is already high enough to cause toxic and other adverse effects on citrus plants. Lithium concentrations in various soil series have been documented to ultimately reduce food quality (Table 1).

4.3. *Distribution of lithium in plants*

Lithium compounds are widely distributed in nature. However, they occur relatively rarely, thus in low to negligible concentrations. The average concentrations of lithium in its different sources are 30 mg kg^{-1} in the Earth's crust, 25 mg kg^{-1} in soil, 2 mg kg^{-1} in drinking water, 170–190 μg L^{-1} in seawater and 2 ng m^{-3} in the atmosphere (Groot, 2009). Lithium concentrations vary in different types of water. However, the importance of studying Li in plants is increasingly emphasised. For the sake of completeness, Li concentrations in foods have ranged widely, from 0.01 ppm in bananas to 55 ppm in oats (Shacklette *et al.*, 1978).

In industry, lithium is mainly used in the production of dyes and batteries. It is also used in producing metal alloys and, perhaps unexpectedly, in manufacturing chemicals for cleaning swimming pools. The water handled as pool cleaning waste then passes into the environment, where it is a direct source of lithium toxicity for plants. Research carried out in the US as early as 1963 has shown the toxic effect on citrus plant growth and stunting in southern California. In this context, Shahzad (2016) reported that moderate to high lithium concentrations of 4–40 mg Li kg^{-1} were found in citrus leaves. The researchers believe that the effect on citrus growth was due to irrigation with water that was contaminated with lithium.

Kent (2022) states that more data collection regarding the ability of wheat plants to transport lithium through them is necessary and highly desirable in the long term. In particular, he considers it necessary to study the conditions under which lithium deposited in old leaves is able to move into new leaves and affect their sensitivity and growth. In his experiments, he investigated the mechanisms of uptake translocation and the ultimate fate of lithium in wheat.

Some plant species are sources of lithium. However, its content varies in different parts of the plant, and the time of harvesting has a major influence on its quantity (IS MUNI, 2022). The importance of lithium in plants has not yet been sufficiently explored. However, despite this fact, many studies have revealed its various effects on plant physiology. It is not clear whether lithium is essential for plant growth. In relation to this issue, there is inconsistent information specifying different, critically essential properties of lithium in different living organisms. It is not without interest that various plant species can absorb significant amounts of lithium, even though it does not appear to be essential for their growth and development. On the other hand, there is also information that lithium can support some important functions in the metabolic changes of halophytes, i.e., plants that can tolerate high soil or water salinity, with salinity levels most commonly in the range of 3–4% in the case of the sea (Wikipedia, 2022). However, it has been confirmed that lithium is toxic to plants when present in high concentrations in soil (Kabata-Pendias and Mukherjee, 2016). Shahzad (2016) refers to the author Borovik-Romanova (1965), who analysed 138 plants from eight soil types and found that plants generally have an average content of 0.15–0.3 mg Li kg^{-1}, with those rooted in saline soils being found to contain 2–10 times more than this value. Based on the information and data presented above, it can be concluded that threshold lithium concentrations in plants vary widely.

Lithium in higher concentrations has been shown to alter the growth of corn roots and the ability of corn to orient itself according to the direction of gravity, which is called 'gravitropism'. In contrast, low levels of lithium have, in some cases, improved plant productivity, increased yield, accelerated ripening and increased disease resistance. For example, lithium at low concentrations (5 mg dm^{-3}) effectively increases maize biomass production by 15%. Thus, there is no clear evidence yet to support the essential or beneficial role of lithium for plants, leaving the hypothesis of its essentiality questionable. Nonetheless, it can be assumed that lithium plays an essential role in the biochemistry of some plants.

Many of the biological functions of lithium are very likely to be affected by its small ionic radius. This fact produces strong polarising effects and may displace some monovalent and divalent cations (e.g., K^+, Mg^{2+} or Ca^{2+}) at various binding sites of enzymes and biological membranes. The Li^+ cation probably plays an important role in the inhibition of Mg^{2+}-containing proteins in both plant and mammalian systems (Bach, 1990). Furthermore, it can be assumed that lithium will perform similar functions in plants as magnesium or calcium, or may completely replace them in normal metabolic processes. However, these considerations may be partly influenced by the fact that the basic nature of lithium has not yet been fully explored. There is also the question of whether plants are able to use all the lithium present in their surroundings, especially in the soil. It is equally debatable what the optimal conditions are that will facilitate the movement of lithium molecules into a particular plant. It appears that the problems associated with the study of the life cycle of lithium in nature, specifically in relation to plants, need to be further addressed. Lithium enters plants during their growth, i.e., when they are drawing nutrients from the soil. There is no doubt that it remains in them for the duration of their lives and passes back into the soil after they die. It is also unclear to what extent plants are able to accumulate lithium and its compounds in their structure and change their chemical properties in a manner typical of specific lithium compounds. There is a lack of relevant scientific information on the relationship between the amount of lithium that can be extracted from soil and the amount taken up by plants. Similarly, it has not been sufficiently demonstrated whether and in what quantities lithium is transferred to the soil and subsequently to plants as a waste product of chemical production, which treats lithium as a potentially usable energy material. This trend of increased lithium use may lead to the release of lithium into the environment and potentially contaminate the surrounding air, soil and water. Despite the many issues that have been raised and debated in connection with the occurrence of lithium in the environment, particularly in terms of its possible transfer to plants, it must be stated that the transfer of lithium from natural sources to humans has been determined to pose virtually no risk (U.S. Environmental Protection Agency, 2022). On the other hand, plants that are capable of retaining lithium have been studied, though the extent of this ability varies. This will undoubtedly create a further need for more in-depth studies of the relationship between the uptake of lithium and its compounds, the soil and its quality and composition, and plants. The continuation of the research efforts is exemplified by the work of Anderson (1990).

Although standard criteria for assessing water quality are currently available, they still do not include consideration of lithium concentrations or amounts in all cases. As noted above, lithium may pose a hazard to certain plants. Several concentrations of lithium were used during the measurements, with the ranges shown in Table 2. The following

Table 2. Tolerance of plants to soil addition of lithium sulphate. Based on California Agriculture (1964).

Plant tested	**Lithium range tested (ppm)**	**Lithium range producing 25% growth depression (ppm)**	**Lithium content in tissue at 25% growth depression (ppm)**		
			Blade	**Petiole**	**Root**
Avocado	0–6	6	120	400	60
Soybean	0–20	7	40	40	80
Sour orange	0–6	8	180	ND*	50
Grape	0–16	12	300	450	400
Tomato	0–40	12	ND*	380	100
Red kidney bean	0–20	12	140	140	520
Cotton	0–40	25	1100	350	250
Red beet	0–80	35	5500	2200	90
Sweet corn	0–120	70	160	340	180

*Not determined.
Source: Bingham *et al.*, 1964.

summarises the results for those lithium concentrations that produce mild toxicity symptoms.

The potential hazard of lithium to crops undoubtedly deserves continued attention and further study in all irrigated areas. At the same time, however, it appears that there is no real reason for the elevated lithium concentrations in some crops to become a cause for panic and alarmist messages.

4.4. *Lithium and environmental protection*

In connection with the issue of lithium distribution in the environment and its potential transfer to plants and animals, it is also necessary to mention the issue of environmental protection. One of the basic principles of environmental protection, and therefore of limiting the distribution of lithium in the environment, is to prevent contamination of surface water, groundwater and soil. It is necessary to take all organisational and technical measures aimed at preventing it from entering sewers and drains. Based on the facts stated above and in conjunction with the requirement to comply with emission limits, it is therefore necessary to limit environmental exposure as much as possible by preventing contamination of surface water, groundwater and soil (Marhold, 1980; Riedl and Vondráček, 1980).

5. Nickel

5.1. *General properties and history*

Nickel is a chemical element with the symbol Ni and a proton number of 28. It is a transition metal that occurs at a concentration of 0.0008% in the Earth's crust. Nickel is most commonly found in compounds in oxidation state II, though oxidation states III and IV are also known but are unstable.

Metals, especially heavy metals, including nickel, are considered major toxicants. When they are monitored analytically in the environment, the results are commonly presented as the concentration of the metal in the substrate being evaluated. This form of expression leads to the assumption that the metals are present in one standard form and that the same concentration at different locations will always mean the same risk when assessing environmental risk. However, this is a misconception because the behaviour of metals in the environment and their effects on the body are fundamentally dependent on the chemical and physical forms in which the metal occurs (Anděl, 2011).

Nickel is always found in the wild together with cobalt. Nickel also occurs in minerals together with sulphur, arsenic and antimony. Nickel tetracarbonyl [$Ni(CO)_4$], which is a component of cigarette smoke, has the most harmful effects (Borkovcova, 2015). Nickel is a trace element and belongs to a group that may be considered essential. In terms of the distribution of this metal in soil and its ability to be transferred to plant organisms, it should be noted that nickel is highly stable at room temperature in air (slow oxidation), water, hydrochloric acid and hydroxides. Dilute acids, which can occur in nature, for example, as a result of water falling on the Earth's surface in the form of so-called acid rain, react very slowly with nickel.

Nickel compounds form what is known as 'boundary cations' with a covalent index of 1.75–3.4. The boundary index expresses the product of the radius of the ion and its electronegativity. The mechanism underlying the toxicity of nickel compounds usually involves their ability to displace cations belonging to group A, which has a covalent index of ≤1.75, or other boundary cations (zinc) from metalloproteins. These cations are usually less toxic than group B cations with a covalent index of ≥3.4, but much more toxic than group A cations with a covalent index of ≤1.75. The toxicity ranking of metal cations to algae and flowering plants is given in Table 3. Thus, it can be stated that the order of toxicity of the above groups of cations in plant organisms generally decreases in the following order:

$$B > \text{boundary} > A$$

Nickel is classified under a fairly large group of metals called 'heavy metals'. These metals are extremely undesirable for the human body and

Table 3. Order of toxicity of elements in plant organisms. Based on Kalač *et al.* (2015).

Organism	Order of toxicity
Algae	Hg > Cu > Cd > Fe > Cr > Zn > Ni > Co > Mn B B boundary
Flowering plants	Hg > Pb > Ag > Cu > Cd > Ni > Co > Mn > Zn B B B B boundary

can have severe effects on the health of the population. Nickel enters the environment mainly as part of the effluent from electroplating plants (Kalač *et al.*, 2015).

The technological processing of nickel and its use are often associated with the ability of this element to provide very high-quality alloys. It is also widely used as a battery material and, last but not least, as a catalyst for a wide range of chemical reactions. It is also used as a material for the construction of electrical resistors.

5.2. *Distribution of nickel in soil*

As already mentioned, nickel occurs naturally and relatively abundantly in the environment. It can be found in soil, water and food. Relatively significant amounts of nickel are released, for example, during volcanic eruptions. In nature, nickel is mostly associated with the presence of sulphur and iron. It can also often be found in combination with arsenic and antimony. Most nickel is thought to occur in the Earth's inner and outer cores. Large reserves of nickel are stored on the ocean floor. Pure nickel can also be found in combination with iron in meteorites. The Earth's core is also composed largely of nickel and iron (NiFe). Nickel is extracted from its ores, in which it is usually found as an oxide mixed with iron (laterite and garnierite) or as a NiFe sulphide or pentlandite. Volcanic activity, fires, soil erosion and evaporation from seawater are also natural sources of nickel in the environment. Humans increase the amount of nickel in the environment mainly through mining, metallurgy and the burning of fossil fuels, including waste. It is, therefore, evident that nickel is ubiquitous in the environment. Due to its easy binding with iron and magnesium atoms, most nickel is found in the soil profile (Arnica, 2022). Nickel is therefore found in the Earth's crust, where it is present in amounts of only about 0.008%. Research shows that most nickel is found on Earth and other planets in the solar system in the core, where it forms an alloy with iron and some other elements. In the Earth's core, its mass fraction is about 5.2%, according to the latest models (Das Institut für seltene Erden und Metalle, 2022). However, nickel is virtually absent in its elemental form.

The fine particles of nickel and its compounds that are contained in the dust are carried by the airflow. In conjunction with the processes outlined above, nickel-containing particles are released into the atmosphere from both natural and anthropogenic sources. Through dry and wet deposition, nickel and its compounds are released into other environmental compartments, such as water and soil. Nickel contained in soil can leach into groundwater.

Nickel can occur naturally in surface waters through the process of weathering of bedrock that contains its deposits. The nickel contained in the water then undergoes various physical and chemical processes that affect its continued persistence in the environment. Under normal conditions, nickel combines very well chemically with abundant iron or magnesium particles. For this reason, large amounts of nickel can be found in soils and sediments.

The anthropogenic source of nickel is mainly wastewater from the metal finishing, ceramic and glass industries. Other sources may be nickel-plated parts of equipment in contact with water, for example, in water distribution networks. Nickel is present in water in oxidation state II and predominantly in its inorganic forms. The toxicity of nickel compounds to aquatic organisms, similar to that of lithium, is strongly influenced by the physical and chemical properties of the water. In waters with low concentrations of calcium and magnesium, a concentration of around 10 mg L^{-1} was found to be lethal for some aquatic organisms, for example, coleus. Thus, nickel is considered a lethally toxic substance in this case and at this concentration (Kopp *et al.*, 2015).

The natural environment associated with the soil is very diverse. One important characteristic that can be used to describe soil is its pH. It has already been stated that nickel is resistant to corrosion. Its occurrence in nature, and therefore in soil, is linked to its compounds, which are characterised by their existence and their relationship to a particular pH. At low pH, it takes the form of the nickel ion Ni^{+2}, whereas at neutral to alkaline pH, it is in the stable form of nickel hydroxide $Ni(OH)_2$, and in a strongly alkaline environment in the presence of oxygen, it forms nickel oxides, which react with acids to produce nickel salts and hydroxides.

The average nickel content in soils is 40 mg kg^{-1}. Its content is significantly influenced by the nature of the underlying rocks. Soils with serpentinite in the bedrock may contain nickel in a 100–7,000 mg kg^{-1} concentration range. Nickel occurs in soils mainly as Ni^{2+}. Although serpentine soils are often weakly acidic, groundwater flowing from ultramafic rocks and serpentinites is strongly alkaline. The buffering capacity and geochemistry of serpentine soils are influenced not only by the high basic saturation but also by the soil organic component, which includes biomass containing living organisms, such as bacteria, detritus, i.e., dead plant remains, and humus containing an amorphous and colloidal mixture of organic matter. Nickel is soluble in acidic environments. Ni^{2+} and $Ni(OH)^+$ ions are mainly found in soils with a pH > 7.8. In acidic soils, nickel occurs as Ni^{2+}, $NiSO_4$ and $NiHPO_4$, depending on the ratio of phosphates and sulphates in the soil. The sequential leaching method shows that approximately 10% of nickel is bound to carbonates, about 20% to FeMn oxides and 50% to silicates. Nickel bound loosely or by adsorption forces is only minimally present.

The concentration of nickel in soil depends not only on its pH but also on the quantities of clay particles and organic matter. Organic matter has a strong ability to sorb nickel and its compounds. Nickel is characterised by its increased mobility but also by its limited transfer factors and toxicity, in particular phytotoxicity, which is very often associated with the use of pesticides. Phytotoxicity expresses the degree of toxicity of a given product intended for plant treatment and is frequently used for the assessment of herbicides (Petr, 2016).

The fate of nickel and its compounds in soil processes is relatively straightforward: soil elements adsorb most nickel compounds. Subsequently, the element becomes immobile and poses no further threat to the environment. However, if the surrounding environment is acidic, the permeation of soil horizons into groundwater occurs, which can retard the growth of microorganisms. Most microorganisms are however able to develop their own resistance to nickel. The ability of nickel to accumulate in plants or animals has not been sufficiently demonstrated. It can therefore be reasonably assumed that it does not directly affect the food chain. On the other hand, animals living near refineries have been found to have an increased incidence of all types of cancer, probably related to elevated concentrations of nickel in the air (Schindlerová, 2020).

For example, diethylenetriaminepentaacetic acid (DTPA) is used to determine the bioavailability of nickel and its compounds at a concentration of 5×10^{-3} M. This method was originally developed for calcareous and neutral soils with low metal concentrations. However, it has been verified that this method can also be highly effective for the determination of nickel content (concentration) in soils and soil profiles. Ethylenediaminetetraacetic acid (EDTA) can be used in the same way.

For soil profiles (layers) that do not need to be specified in detail for the purposes of this communication, the bioavailability of nickel and its compounds was found to decrease with depth. In the upper part of the forest soil profile, its concentration values reach 19.5% and 23.8% when determined using DTPA and EDTA, respectively. These values refer to extractable nickel relative to its total concentration. The values gradually decrease toward the lower horizon, where they reach 2.31% for the DTPA method and 2.74% for the EDTA method in relation to extractable nickel. The trend is similar for agricultural land, with values of 1.62% DTPA-extractable nickel and 2.46% EDTA-extractable nickel found for the 2 cm soil horizon and 0.46% DTPA-extractable nickel and 0.73% EDTA-extractable nickel in the lower horizon. In the case of a minefield, the average leachability was found to be 2.04% nickel by DTPA and 3.04% by EDTA. It is important to note that nickel also tends to accumulate in plants, and its concentrations may be higher than indicated by the EDTA and DTPA leachates used. The values of total nickel measured in winter grain reach values of 86 mg kg^{-1}, but the values of nickel in soil were measured at 14.8 mg kg^{-1} in leachate using DTPA and 22.6 mg kg^{-1} in leachate using EDTA at a soil depth of 2 cm (Pipkova, 2010).

The availability of nickel compared to chromium and cobalt varies depending on many conditions. These include, for example, the mineralogical composition of the soil and the different binding affinities of these metals to different minerals. The vertical distribution of these metals in the soil profile may also play a role. Compared to chromium, nickel is relatively easy to release and is more readily available to plants. In the forest soil profile, the values of 19.5% of the total nickel content in the top horizon in the DTPA and 23.8% in the EDTA leachates. Its availability decreases with increasing depth, with 2.31% of the total nickel content in the DTPA leaching and 2.74% in the EDTA leaching at depth. In contrast, in agricultural soils, the values of releasable nickel are lower by about one order of magnitude. In the top horizon (up to a thickness of 2 cm), a value of 1.62% of the total nickel content was obtained in the DTPA and 2.46% in the EDTA leachates. In this profile, availability also decreases with increasing depth. At the lower horizon, 0.46% was released by DTPA leach and 0.73% by EDTA leach. From the values measured in the downhole field, the average recoveries of 2.04% of the total nickel content from the DTPA leach and 3.04% from the EDTA leach are based on the average recoveries. In contrast, chromium is only released from the soil at low concentrations compared to nickel. In the forest soil profile, values of one-tenth of a percent are reached in almost all measured values in both DTPA and EDTA leachates. In the agricultural soil profile, the values are a further order of magnitude lower. In the minefield, the average is 0.03% by DTPA leachate and 0.08% by EDTA leachate of the total chromium content in the soil. It was also found that the bioavailability of nickel decreases with depth in the soil profiles. This phenomenon can be explained by a lower rate of weathering and, therefore, the formation of stronger bonds in the deeper horizons. Concentrations in winter cereals (86 mg kg^{-1}) indicate the ability of metals, especially nickel, to accumulate in plants. Analyses show that the concentration of phytodegradable nickel, predicted by extraction of DTPA (14.8 mg kg^{-1}) and EDTA (22.6 mg kg^{-1}) in the 2 cm soil horizon, is less in the soil than in the grain, a possible indicator of this very accumulation phenomenon and the ineffectiveness of these reagents to accurately determine the availability (Pipkova, 2010).

5.3. *Distribution of nickel in plants*

The primary routes of nickel exposure in the body are through the consumption of plant-derived foods. Foods with a high nickel content include nuts, legumes, cereals, liquorice and chocolate. For example, the nickel contents in cocoa beans, soybeans, walnuts, peanuts and fruit flakes are 9.8, 5.2, 3.6, 2.8 and 1.8 $\mu g\ g^{-1}$, respectively. Although nickel concentrations vary by food type, average levels generally range from 0.01 to 0.1 $\mu g\ g^{-1}$ (McGachy, 2021).

In general, there is relatively little information on nickel uptake by plants. In prokaryotes, evolutionarily very old organisms that evolved 3–3.5 billion years ago, uptake is mediated by nickel/cobalt transporter proteins (Andresen *et al.*, 2018). Nickel is quite controversial despite the existence of several enzymes that usually contain it. The controversy arises from the fact that they are not dependent on nickel and that its role as a positive ion, or cation, has been found to be easily fulfilled by other divalent cations without significant problems (Xu, 2022).

At higher concentrations, nickel is toxic to plants and, secondarily, to plant consumers. In essence, its mechanism of action is mostly based on the inhibition of the -SH group in enzymes, regardless of the form in which nickel and its compounds enter the human body. Some plants can accumulate nickel and then, depending on its concentration in the soil, become toxic when consumed chronically.

Brooks (1987) suggests that nickel is toxic to non-adapted plants. Some plants may be stressed by its presence. Its critical limit for plant growth is set at about 0.2 mg L^{-1} in soil solution.

The availability of nickel, chromium and cobalt varies depending on many conditions, such as the mineralogical composition of the soil, the different binding of these metals to different minerals or the vertical distribution of these metals in the soil profile. Nickel is relatively easy to release compared to chromium and is, therefore, more readily available to plants.

It has already been stated that nickel is one of the most significantly toxic heavy metals. Atmospheric deposition can release nickel from the air into soil or water. Nickel binds to particles containing iron and manganese, which are often found in soil and sediment. Therefore, most of the nickel in the environment occurs here. In natural water at a pH of 5–9, Ni^{2+} is the dominant form of Ni occurrence. In this pH range, nickel can sorb to iron and manganese oxides or form complex compounds with inorganic ligands. The toxicity of nickel to some aquatic organisms is relatively high; therefore, its permissible concentration in watercourses is more strictly limited than in drinking water. Plants take up nickel from the soil mainly through their roots and are able to accumulate it. A decrease in pH increases nickel mobility and thus its uptake by plants (Integrated Pollution Register — IRZ, 2022).

It may be of some interest that nickel is toxicologically classified as a major poison, particularly in the form of the compounds NiS, NiO and nickel tetracarbonyl. It is highly allergenic and a human carcinogen (lung, nasal septum and, more rarely, pharyngeal cancer). Nickel is also hazardous to aquatic organisms, including plants. It is therefore more strictly limited in surface waters than in drinking water.

It is not without interest that nature itself is capable of reducing the concentration of nickel in the soil. An interesting alternative for reducing nickel concentrations in soils is therefore the use of plants that can tolerate elevated concentrations of metals in the soil and have the ability to accumulate these elements in their tissues. This ability is called 'hyperaccumulation'. Plants that have this ability are called 'hyperaccumulators'. For some plants, metals such as zinc, manganese, nickel and copper are even essential parts of their life cycles and categorised under essential micronutrients. The mechanism of phytoremediation of soils with elevated metal content can be simplistically described as the removal of metals from the soil through their transport into roots, stems and leaves. This method is currently used in various remediation applications. Plants are planted in the contaminated area, subsequently harvested and removed, and the area is replanted until the concentration of metals in the soil is

reduced to an acceptable level. Hyperaccumulators are usually defined as species capable of accumulating metals at levels 100 times higher than the concentrations found in normal, non-accumulating plants. One approach to phytoremediation is phytoaccumulation, which is based on the absorption of contaminants by the roots of the plant with subsequent accumulation in the above-ground part of the plant. Harvested plants must be treated as waste. A prerequisite for the method to work is the hyperaccumulation property of the plant species towards the remediated contaminant. Good accumulators of nickel include *Alyssum murale* (tarewort), which is shown in Fig. 6.

It is possible to extract these metals from plants that are contaminated with nickel through a process in which suitable plants are sown, harvested, dried and burned. The combustion extracts energy from the plants, and the remaining ash can be refined or extracted to obtain the captured heavy metals, including nickel.

The theoretical findings described in the preceding paragraphs have been confirmed by botanists at the University of the Philippines. In fact, botanists have discovered a new species of plant with unusual abilities. These are characterised by the ability to accumulate nickel in amounts

Figure 6. *Alyssum murale*. Photo courtesy of Dana Holubová, Botanical Garden, Faculty of Science, Masaryk University.

hundreds to thousands of times greater than most other plants (Český rozhlas, 2014).

The ability to accumulate nickel without poisoning the plant is a relatively rare phenomenon. Worldwide, it is found in about 450 species, a mere fraction of the estimated 300,000 vascular plant species in existence. It is relatively abundant in the *Violaceae* family.

The hyperaccumulation of nickel has been described by scientists in some representatives of three genera:

- *Agatea* (one species);
- *Hybanthus* (at least six species);
- *Rinorea* (Figs. 7 and 8).

The last named is a pantropical genus of forest shrubs and trees, which includes 225–275 species. At least three of these are capable of hyperaccumulating nickel. This ability was first recorded in 1977 in museum specimens of *Rinorea bengalensis*, which grows throughout Southeast Asia, and then in *Rinorea javanica* from Indonesia, Thailand

Figure 7. Leaves of *Rinorea niccolifera*. Photo courtesy of Dr. Edwino S. Fernando.

Figure 8. *Rinorea niccolifera* in its natural habitat. Photo courtesy of Dr. Edwino S. Fernando.

and Vietnam. In 1994, scientists also described it in a previously unnamed species on the Indonesian island of Sulawesi. A fourth species has recently been discovered in the Philippines.

Rinorea niccolifera is native to Luzon, the largest island in the Philippines. However, this shrub has only been discovered in a small part of the province of Zambales, where it grows in soil rich in metals, especially chromium and nickel. It takes up nickel from the soil in huge quantities, with its leaves containing up to 18,000 $\mu g\ g^{-1}$ (18,000 ppm). This is hundreds to thousands of times more than the vast majority of vascular plants can survive without harm or severe poisoning. *Rinorea niccolifera*'s hyperaccumulation of nickel also far exceeds that of *Rinorea javanica* (up to 2,170 $\mu g\ g^{-1}$) and an unnamed species from Sulawesi (up to 1,830 $\mu g\ g^{-1}$). The discovery of *Rinorea niccolifera* may thus be of practical importance, especially in technologies such as phytoremediation and 'phytodolation'. Phytoremediation uses plants capable of accumulating metals through the treatment of contaminated soils. Phytomining, in turn, is used to extract commercially valuable metals and semi-metals where traditional mining would be impractical or disadvantageous. In fact, extracting them from the leaves or shoots of plants is much easier than extracting them from the ground.

High-resolution secondary ion mass spectrometry has been used with great efficiency to investigate the presence of metal ions in plants.

The mass spectrometry method is used to study the distribution of metals on leaf sections. Using this method, high concentrations of nickel have been observed and elucidated in the region of unicellular star-shaped trichomes (hairs on the leaf surface) in the epidermal layer. However, this technique requires complex sample preparation processes for analysis. The disadvantages of this method are its destructiveness and difficulty quantifying the data (Bahelková, 2021).

5.4. *Nickel and environmental protection*

In connection with the issue of nickel distribution in the environment and its potential transfer to plants and animals, it is also necessary to mention the problem of environmental protection. Nickel is still abundantly used in ore dressing technologies, smelters and refineries. It is also a crucial component of batteries, and it is used extensively for plating and the production of alloys. Somewhat unexpectedly, nickel is also used in the cosmetics industry for the manufacture of cosmetic products, especially shampoos and hairsprays. It is undoubtedly worth noting that smoking tobacco products is a very significant source of indoor nickel contamination. In the case of the long-term effects (several years) of nickel, there is a risk of its permanent introduction into nature, for example, in the form of technological sludge. Another danger with untreated sludge produced through existing technologies is its potential to harbour infectious microorganisms. Based on these facts, it is evident that the need for isolating nickel before it becomes a part of waste and enters the environment will continue to be a very significant technological challenge.

6. Conclusion

Lithium and nickel are metals that are currently considered very important in terms of their processing properties. They share a common potential for use in the energy sector. The trend towards emissions reduction associated with the excessive production of carbon dioxide from internal combustion engines has prompted not only the countries of the European Union but also other major economies in the world to switch to other sources of energy for powering motor vehicles. Even though lithium and nickel occur naturally and do not pose significant toxicological risks, aspects of their distribution in soil and plants need to be addressed in a long-term and systematic manner. Certain enzymatic abilities to inhibit specific functional groups in plant systems may be toxicologically significant, not only from a preventive point of view but also in terms of safeguarding the health of patients. However, it must be stated for the time being that many of the biological processes involved in the transfer of lithium and nickel from ecosystems to soil and plants are not sufficiently understood, elucidated or investigated. It is, however, somewhat optimistic that a number of very modern and sophisticated methods have been developed that can be used to study the lithium and nickel content of soils. It is also positive that the issue of studying the presence of lithium- and nickel-based contaminants in the environment is receiving attention from researchers, with a number of studies focusing on monitoring problem sites, such as landfills.

References

Anděl P. (2011). *Ecotoxicology, Bioindication and Biomonitoring*, 1st edn. Evernia, Liberec, pp. 54–55 (in Czech).

Anderson C.E. (1990). Lithium in plants. In: Bach R.O., Gallicchio V.S. (eds.) *Lithium and Cell Physiology*. Springer.

Andráš P. (2014). *Remnants of Old Mining Activity at Abandoned Cu-deposits*, 1st edn. Vodní zdroje Ekomonitor, Chrudim, pp. 189–190 (in Czech).

Andresen E., *et al.* (2018). Trace metal metabolism in plants. *J. Exp. Bot.* 69: 909–954. doi: 10.1093/jxb/erx465.

Arnika (2022). Nickel (in Czech). https://1url.cz/srhcP. Accessed 19 December 2022.

Bahelková M. (2021). Distribution of elements in plant materials (in Czech). https://1url.cz/8rhYB. Accessed 20 December 2022.

Bach R.O. (1990). Some aspects of lithium in living systems. In: Bach R.O., Gallicchio V.S. (eds.) *Lithium and Cell Physiology*, 1st edn. Springer.

Borkovcová M., Žáková M. (2015). *Biology for the Waste Management Industry*. 1st edn. Mendelova Univerzita v Brně, Brno, pp. 24–27 (in Czech).

Borovik-Romanova T.F. (1965). In: Khitarov N.I. (ed.) *The Content of Lithium in Plants*. Jerusalem. 1969. Trans. USSR: Inst. Geokem. Akad. Nauk.

Bradford V.G.R. (1963). Lithium survey of California's water resources. *Soil Sci.* 96:77–81.

Brooks R.R. (1987). *Serpentine and Its Vegetation*. Dioscorides Press, Portland, OR, pp. 454.

Bingham F.T., Brandford G.R., Page A.L. (1964). Toxicity of lithium. California Agriculture. https://1url.cz/9rhcX. Accessed 20 December 2022.

Český rozhlas (2014). The plant that eats nickel (in Czech). https://1url.cz/arhY2. Accessed 19 December 2022.

Das Institut für seltene Erden und Metalle (2022a). Lithium, Li, atomic number 3 (in Czech). https://1url.cz/orhPa. Accessed 30 November 2022.

Das Institut für seltene Erden und Metalle (2022b). Nickel, Ni, atomic number 28 (in Czech). https://1url.cz/8rhcY. Accessed 5 December 2022.

Diding N., Ottosson J.O., Schou M. (1969). Lithium in psychiatry. In: *Proceedings of a symposium in Lidingö*, March 22, Sweden, Copenhagen, Munksgaard.

E15.cz (2019). The world's miners are betting on electromobility. But they're at a loss as to what to mine (in Czech). https://1url.cz/brhPW. Accessed 12 December 2022.

E15.cz (2021). The transition to electromobility is faltering. The hunger for nickel, cobalt and lithium will outstrip mining capacity (in Czech). https://1url.cz/wrhPo. Accessed 13 Dec 2022.

Groot P.A. (2009). Lithium. https://doi.org/10.1016/B978-0-444-51115-7.00002-4.

Integrated Pollution Register — IRZ (2022). Nickel and compounds such as Ni (in Czech). https://1url.cz/irhYw. Accessed 17 December 2022.

IS MUNI (2022). Metals in biosystems II (in Czech). https://1url.cz/WrhcG. Accessed 17 December 2022.

Kabata-Pendias A., Mukherjee A.B. (2007). *Trace Elements from Soil to Human*. Springer, p. 23. http://dx.doi.org/10.1007/978-3-540-32714-1.

Kalač P., *et al.* (2015). *Environmental Chemistry*. University of South Bohemia, České Budějovice, Czech Republic (in Czech).

Kočí V., Mocová K. (2009). *Ecotoxicology for Chemists*, 1st edn. University of Chemistry and Technology, Prague, pp. 16–19 (in Czech).

Kopp R. (2015). *Basics of Aquatic Ecotoxicology*. Mendel University, Brno, Czech Republic (in Czech).

Linhart I. (2019). *Basic Terms in Toxicology, Ecology and Ecotoxicology*. University of Chemistry and Technology, Prague, Czech Republic, p. 66.

Marhold J. (1980). *Industrial Toxicology Review: Inorganic Substances*, 2nd edn., Avicenum, Prague, p. 528 (in Czech).

Mcgachy L. (2021). *Toxicology and Ecotoxikology I*. University of Chemistry and Technology, Prague, Czech Republic (in Czech).

Nayak P.K., Yang L., Brehm W., Adelhelm P. (2017). From Lithium-Ion to Sodium-Ion Batteries: Advantages, Challenges, and Surprises. https://doi.org/10.1002/anie.201703772.

Panáček A., Balzerová A. (2013). *Fundamentals of Toxicology and Ecotoxicology*. 1st edn. Palacky University in Olomouc, Olomouc, Czech Republic, pp. 38–39 (in Czech).

Pavlíková D. (2006). *Ecotoxicology*. Czech University of Life Sciences, Prague, Czech Republic, pp. 58–59.

Pavlíková D. (2008). *Ecotoxicology*. Czech University of Life Sciences, Prague, Czech Republic. Pp. 66–75 (in Czech).

Petr J. (2016). Geochemical characterization of the soil environment with a focus on contamination of forest soils along selected roads (in Czech). https://1url.cz/0rhcn. Accessed 18 December 2022.

Pipkova Z. (2010). Distribution, binding and mobility of Ni in soils on former Ni ore deposit in the Křemže area (in Czech). https://1url.cz/HrhYu. Accessed 20 December 2022.

Riedl O., Vondráček V. (1980). *Clinical Toxikology*. 5th edn. Avicenum, Praha, p. 820 (in Czech).

Shacklette H.T. (1978). Trace elements in plant foodstuffs. In: Oehme F.W. (ed.) *Toxicity of Heavy Metals in the Environment*, 4th edn. Marcel Dekker, New York, pp. 25–66.

Shahzad B. (2016). Lithium toxicity in plants: Reasons, mechanisms and remediation possibilities — A review. Plant Physiol. Biochem. 107:104–115.

Schindlerová M. (2020). Accumulation of heavy metals in the soil around the landfill Mrsklesy municipal waste dump (in Czech). https://1url.cz/Urhcy. Accessed 18 December 2022.

U.S. Environmental Protection Agency (2022). Test guidelines for pesticides and toxic substances. https://1url.cz/5rhcg. Accessed 11 December 2022.

Wikipedia (2022). Halophilic plants (in Czech). https://1url.cz/7rhcF. Accessed 9 December 2022.

Xu Z. (2022). Influences of lithium on soil properties and enzyme activities. *Chemosphere* 313:137458.

https://doi.org/10.1142/9789811283123_0002

Chapter 2

Contamination of Lithium in Soils and Its Uptake in Plant Species

Anitha Karri, Tamanna Parida, Shaik Riyazuddin, Namuduri Srinivas*

Department of Environmental Science, GITAM School of Science, GITAM (Deemed to Be University), Visakhapatnam 530045, India

**snamudur@gitam.edu*

Abstract

Lithium (Li) is taking frontline in electronic items and vehicles worldwide. Great uses of Li draw attention and are in demand everywhere, hence production also increases daily. The major problem with Li is its toxicity, which eventually enters human and animal life via the food chain. It is non-degradable and has a long shelf-life — it can sustain anywhere for a more extended period but causes much toxicity to the host. Removal of Li is a big challenge, and expensive methods need to be employed. Thus, environmentalists came up with a sustainable approach called phytoremediation — the use of plants to clean up hazardous contaminants. In this method, the plant, or hyperaccumulator, grows on Li-contaminated soils and absorbs a higher concentration of Li without compromising its own immunity. In this chapter, we reviewed the source, pathway of contamination and remediation of Li using plant species.

Keywords: Lithium, hyperaccumulator, phytoremediation, Li-batteries, electronic vehicle

1. Introduction

Improper disposal and aggressive uses of lithium (Li) are emerging problems facing the world. Production in Li has grown three times more since 2000, which approximates 77,000 tons per year (USGS, 2020; Kelly *et al.*, 2010). Li plays a vital role in electronic products and vehicle rechargeable Li batteries, leading to global Li demand (Mohr *et al.*, 2012). E-waste dumped in soil likely leads to Li contamination (Li ions and Li_2O) due to the degradation of electronic products (Avila-Arias *et al.*, 2019; Li and Achal, 2020) and vehicular grease accumulated with 1% Li that enters the environment through road runoffs (Zeng and Li, 2014). The primary reason for the world's shift to electric vehicles is the increase in the prices of fossil fuels, environmental issues related to climate change and decreased production as well as the enormous consumption. In the USA alone, transport accounts for 28% of the total greenhouse gas (GHG) emissions, and in India, the transport sector is responsible for 14% of energy-related CO_2 emissions (Godinho and Coetzee, 2020; International Energy Agency, 2020a). Recently, international policies have also stressed reducing carbon emissions. This has ultimately resulted in the world embracing Electric vehicles, which run on Li-ion batteries. It has been analysed that from 2014 to 2021, the cost of imported Li-ion cells has risen from \$180 million to over \$1.2 billion, marking a sevenfold increase (GoI, 2023).

Another everyday use of Li is in ceramics and as a catalyst in producing synthetic rubber and plastics (Aral and Vecchio-Sadus, 2011). The increasing usage of Li has resulted in the metal finding its way into the biological environment. Li at low levels benefits living organisms as it helps in neurotransmission (Jakobsson *et al.*, 2017). Mood-stabilising drugs and ceramics possess Li, and are one of the significant sources of Li entering water and soil through the leaching process of landfills, sewage, and storm water (Aral and Vecchio-Sadus, 2008; Hao *et al.*, 2017). Li is considered a nonessential nutrient as its absence causes no deficiency symptoms in humans, although intake of minimum Li regulates neurotransmission, which indicates a positive impact on living organisms (Jakobsson *et al.*, 2017). On the other hand, in excess Li reduces cell growth and affects the reproductive organ in humans (Sobolev *et al.*, 2019). It is clear that high doses (15–20 mg l^{-1} blood concentrations) damage the kidney and cause visual impairment. nausea, cardiac arrest and coma. Like humans, animals are also affected by Li, presenting kidney damage, etc. (Bradford, 1973; Robinson *et al.*, 2018). The main aim of this chapter is to review the source, contamination and sustainable treatment method of Li.

2. Sources of Lithium and Ways of Contamination

We need to know the sources of Li and its types, which will lead us to the source of contamination. These are divided into three categories: Brines, Li available as pegmatite and sediment-hosted deposits. Li is reactive in nature and occasionally found in a free state. It is mostly clubbed with mineral components such as aluminum silicide or apatite and in salts like Li hydroxide (LiOH) and lithium chloride (LiCl), and it regressively takes part in contamination (Dessemond *et al.*, 2019; Kushnir, 2013; Bleiwas and Coffman, 1986; Speirs *et al.*, 2014). Li concentration in granite, diabase, shales, sandstones and carbonates is 22, 15, 66, 15 and 5 mg kg^{-1}, respectively. As mentioned earlier, the universal demand for Li has drastically increased in the past few years, and anthropogenic sources of lithium pollution are a phenomenon of the recent past. It has also been estimated that by 2025 more than 80% of the market will be dominated by Li batteries (Harper *et al.*, 2019). Li-ion batteries also contain cobalt, manganese, nickel, iron, copper, etc. (Sun *et al.*, 2017; Siqi *et al.*, 2019). One major concern with Li batteries is that they have to be replaced in 2–3 years (Harper *et al.*, 2019). Most of them end up in landfills and recycling facilities (Bernardes *et al.*, 2004) and Li could consequently reach the human body. It is estimated that 42.50% of Li can leach into the environment if a battery containing Li nickel manganese cobalt (NMC) oxide is disposed into a landfill (Richa, 2016).

3. Lithium Soil and Its Accumulation

The existence of Li can be noticed in nature albeit in minimum concentrations. Li concentrations were reported in the Earth's crust (30 mg kg^{-1}), drinking water (2 mg kg^{-1}), soil (25 mg kg^{-1}), atmosphere (2 ng m^{-3}) and seawater (170–190 $\mu g\ l^{-1}$) by earlier researchers (Birch, 1976; Thomas and Merian, 1991). It is noticed that Li widely varies in foods; for instance, oats possess 55 parts-per million (ppm) of Li, whereas in bananas it is reportedly present at the rate of 0.01 ppm (Shacklette *et al.*, 1978). The point source of Li contamination in water and soil are the metal alloys and batteries industry, colour production and Li-based cleaning agents for pools, leading to environmental pollution (Schrauzer, 2002). Bradford (1973) observed that the growth of citrus plants in southern California was affected by highly contaminated Li water. The concentration of Li was recorded at 0.05 mg l^{-1} in 100 samples out of 400, indicating the toxicity level of water.

Borovik-Romanova *et al.* (1965) experimented with 138 plants by using eight different soil types where they observed plants from saline soil contained 2–10 times more Li concentrations than the normal type (0.15–0.3 mg Li kg^{-1} in dry weight) and a correlation has been seen in the availability of Li on the surface soil and the soil unit (Kabata-Pendias and Pendias, 1992). It even varies the texture of the soil from 56 mg kg^{-1} (calcareous soils) to 1.3 mg kg^{-1} (light organic soil) (Baber *et al.*, 2016). The accumulation of Li in soils mainly depends on the type of mineral and composition of the rock, as well as the soil pH, and it is evaluated that acidic soils promote the solubility of lithium (Sobolev *et al.*, 2019; Aral and Vecchio-Sadus, 2011). Physical characteristics such as texture, permeability and bulk density determine the dynamics of Lithium accumulation. Clayey regions behold more lithium when compared to the non-clayey regions, and similarly, top soils behold less Li than the beneath-level soils (Schrauzer, 2002; Török *et al.*, 2021). Climate factors also influence Li accumulation. In arid areas, Li present in soils moves upward and may precipitate with chlorides and sulfates. In humid areas, however, the movement of Li is more towards the deeper layers of the soil (Ammari *et al.*, 2011).

Thus, the reclamation of polluted soils has become one of the major concerns among the scientific community across the globe. Among various methodologies used for pollutant removal, phytoremediation has been gaining huge prominence recently due to its low cost and high efficiency. Phytoremediation involves using plants to extract and remove elemental pollutants from soil or lower their bioavailability in soil (Berti and Cunningham, 2000). The significant advantages of using phytoremediation are that it is economically feasible, environmentally friendly, can be applied on a large scale, improves soil fertility and reduces erosion (Aken *et al.*, 2009; Jacob *et al.*, 2018).

4. Lithium Uptake and Remediation

The primary route of lithium uptake by plants is through the root system and it is transported through the cell walls to the xylem via the Casparian strip. The translocation of lithium to the leaves and shoots by the xylem or phloem is still unknown (Shakoor *et al.*, 2023; Tanveer *et al.*, 2019). After translocation to the shoots, Lithium is stored in various cell organelles, thereby demonstrating its tolerance in relation to uptake. *Apocynum venetum* (*A. venetum*) can tolerate up to 25 mg l^{-1} of Lithium chloride and is commonly used for studying the Li uptake of plant species. Sword-leaf dogbane *A. venetum* can tolerate up to 25 mg l^{-1} LiCl, corresponding to 72% of the total Li content in this species (Qiao *et al.*, 2018; Shakoor *et al.*, 2022a, 2022b) (Fig. 1). Study related to the impact of lithium on plant growth parameters reported by various researchers are given in Table 1.

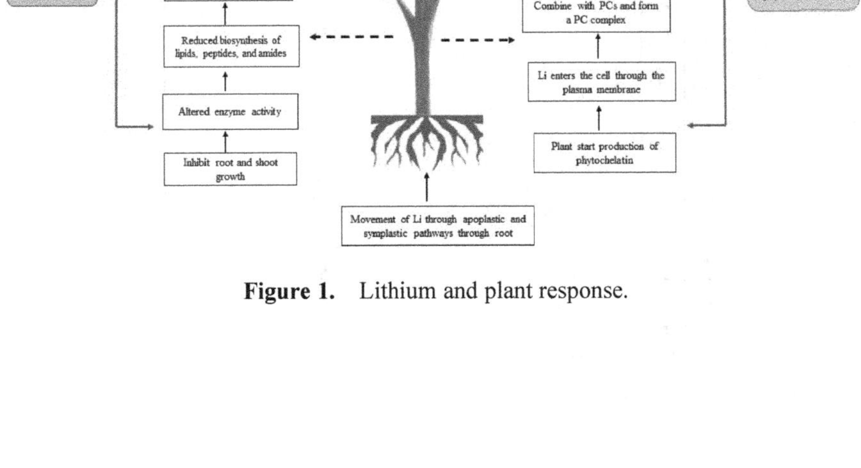

Figure 1. Lithium and plant response.

Table 1. Impacts of lithium on plant development and some of the studies have been highlighted in the table.

Lithium source	Concentration (mg L^{-1})	Plant species	Response	References
Lithium chloride	0, 5, 25, 50	*Zea mays* L. var. saccharata Kcke & *Helianththus annuus L.*	Fresh shoots and roots biomass decreased, leaf areas decreased, and photosynthetic pigment decreased	Hawrylak-Nowak *et al.* (2012)
Lithium chloride, lithium hydroxide	0, 5, 25, 50, 100	*Lactuca sativa var. capitata*	Observed necrotic spots on older leaves, L-Ascorbic acid decreased, and water content reduced.	Kalinowska *et al.* (2013)
Lithium chloride	0, 50, 200, 400	*Apocynum venetum*	Dry biomass decreased, and photosynthetic pigments and parameters decreased.	Jiang *et al.* (2014)
Lithium sulfate	0, 10, 20, 30, 40	*Lactuca sativa L.*	Stem diameter increased, shoot/root ratio increased	da Silva (2012)

5. A Survey on Remediation of Lithium-Contaminated Soils

Scientific communities across the globe have realised the significance of the remediation of Li-contaminated soils and have identified proper techniques to immobilise Li from reaching organisms. The general phenomena in the remediation of metals such as extraction, reducing toxicity and mobility were categorised by Wuana and Okieimen (2011). Li, as mentioned earlier, is exceedingly mobile and depends on soil type and climate type to disperse. As a result, phytoremediation is considered to be the most cost-effective and eco-friendly process. Muthusaravanan *et al.* (2018) reviewed the sources of heavy metal (HM) pollution, soil contamination by HMs and their effects on the environment and health of organisms as well as the types of phytoremediation techniques and plants utilised in detail. They concluded that among the various processes of phytoremediation, i.e., phytodegradation, phytovolatilisation, phytoaccumulation, and phytoextraction, data on phytodegradation and hydraulic control are rare and that these are less explored techniques. In this regard, they concluded that the impetus must be placed on biological mechanisms controlling pollution mitigation. Various scientists across the globe have utilised phytoremediation for Li immobilisation. Some trials were conducted on Li-contaminated soil with or without any chelating agent and it was observed that some species recorded a maximum absorbance level below 4000 mg kg^{-1}. Jiang *et al.* (2019) experimented on two different types of Li-containing soil with the plant *Apocynum venetum* and observed that germination growth was higher in LiCl than the Li_2CO_3 under the same condition. For 0–150 mmol Li_2CO_3, germination was 3–91%, whereas 4–90% was seen in 0–400 mmol l^{-1} LiCl, and thus they concluded that venetum seeds are highly tolerating of Li, especially LiCl.

Bakhat *et al.* (2019) revealed that in a minimum concentration of Li (20 mg Li kg^{-1}), spinach showed a good response, whereas, at a higher level, pigmentation occurred with the interference of calcium and potassium uptake in plants and enhancement of the antioxidant enzymes in the shoot. This study indicates that uptake and accumulation of Li have mostly been seen in leaves and shoot zones, which can cause health hazards to human health after consumption. Hawrylak-Nowak *et al.* (2012) put sunflower and maize plants into the Li-contaminated soil (0–50 mg Li dm^{-3}) to study the leaf area, biomass and photosynthesis, and they observed that at 50 mg Li dm^{-3}, both the species got sensitive and

Table 2. Research related to lithium immobilisation by plant species.

Plant species	Lithium species	Concentration	Observations	References
Brassica oleracea var. capitata, Brassica napus, ***Helianthus annuus, Spinacia oleracea, Solanum lycopersicum***	Lithium chloride	20, 50, 100, 150, 200, 300, 400, 500 and **1000 mg kg^{-1}**	*C. hirsute, H. annuus, B oleracea var. capitata* **accumulated more amount of lithium when compared to the other plant species.**	Kavanagh *et al.* (2018)
Apocynum venetum	Lithium chloride	50, 200, and 400 mg kg^{-1}	Plant survived even at 400 mg/kg, about 1800 mg/kg of lithium was accumulated in the leaves	Jiang *et al.* (2014)
Brassica juncea	Lithium chloride	380 mg kg^{-1}	About 23% of lithium was accumulated in leaves and 70% was imbibed in the root ecosphere	Elektorowicz and Keropian (2015)

symptoms were plant specific. In the case of the sunflower plant at 50 mg Li dm^{-3}, chlorophylls a, b, and carotenoids were decreased, which was absent in the case of maize. The photosynthetic pigment of maize was healthy at that concentration. The research related to lithium immobilisation by different plant species are given in Table 2.

6. Route of Lithium from Plants to Human Beings

The uptake of Li differs due to uneven distribution on the Earth's crust. Li can be absorbed entirely by small intensities and get disturbed throughout the body when consumed as a form of salt or bioavailable element through plants. This depends much on the composition and the absorption rate also changes due to the variations in the different components of the diet (Schrauzer, 2002). Li at higher levels is considered toxic to humans as it may result in hyperchloremic acidosis, muscle seizures, renal acidosis, coma, renal disorders, diarrhoea, vomiting, nausea, cardiac arrests, development of hyperparathyroidism, etc. (Shahzad *et al.*, 2017; Malhi *et al.*, 2020; Gong *et al.*, 2016). Li concentrations above 110 mg l^{-1} are reported to cause nausea, vomiting, lethargy, tremor and fatigue, and concentrations above 250 mg l^{-1} can cause coma, seizures, hyperthermia, hypotension and even death (Boltan and Fenves, 2008; Dunne, 2010).

7. Conclusion

Due to the high demand for Li for batteries and electronics, proper disposal and management must be made mandatory to minimise the ill impacts of Li on the biological environment, especially in a country like India, where the population is large and demand for space is enormous. In target (III) of Schedule II of the Battery Waste Management Rules (2020), set forth by the Ministry of Environment, Forest and Climate Change of the Government of India, as a target for the extended producer responsibilities it has been envisaged that *The recycling of Waste Battery means recycling of Battery materials such as lead, nickel, lithium, cobalt, plastics, rubber, glass, etc*. Thus, companies that manufacture Li-ion batteries must have a clear-cut management plan for collecting and recycling this soft, silvery, light-weight metal. At the same time, funding must be allocated for researching phytoremediation techniques, which have emerged as the most promising phenomenon.

References

Aken B.V., Correa P.A., Schnoor J.L. (2009). Phytoremediation of polychlorinated biphenyls: New trends and promises. *Environ. Sci. Technol.* 44:2767–2776.

Ammari T.G., Al-Zu'bi Y., Abu-Baker S., Dababneh B., Gnemat W., Tahboub A. (2011). The occurrence of lithium in the environment of the Jordan Valley and its transfer into the food chain. *Environ. Geochem. Heal.* 33: 427–437.

Aral H., Vecchio-Sadus A. (2008). Toxicity of lithium to humans and the environment — A literature review. *Ecotoxicol. Environ. Saf.* 70:349–356.

Aral H., Vecchio-Sadus A. (2011). Lithium: Environmental pollution and health effects.

Avila-Arias H., Nies L.F., Gray M.B., Turco R.F. (2019). Impacts of molybdenum, nickel, and lithium-oxide nanomaterials on soil activity and microbial community structure. *Sci. Total Environ.* 652:202–211.

Bakhat H.F., Rasul K., Farooq A.B.U., Zia Z., Fahad S., Abbas S., Hammad H.M. (2020). Growth and physiological response of spinach to various lithium concentrations in soil. *Environ. Sci. Pollut. Res.* 27:39717–39725.

Berti W.R., Cunningham S.D. (2000). Phytostabilization of metals. In: Raskin I., Ensley B.D. (eds.) *Phytoremediation of Toxic Metals: Using Plants to Clean-up the Environment*. John Wiley & Sons Inc., New York, pp. 71–88.

Birch N.J. (1976). Possible mechanism for biological action of lithium. *Nature* 264:681.

Bleiwas D.I., Coffman J.S. (1986). Lithium availability-market economy countries. Bureau of Mines Information Circular 9102:23.

Bolan N., Hoang S.A., Tanveer M., Wang L., Bolan S., Sooriyakumar P., Rinklebe J. (2021). From mine to mind and mobiles–Lithium contamination and its risk management. *Environ. Pollut.* 290:118067.

Boltan D.D., Fenves A.Z. (2008). Effectiveness of normal saline diuresis in treating lithium overdose. In: *Baylor University Medical Center Proceedings*, vol. 21. Taylor & Francis, pp. 261–263.

Bradford G.R. (1973). Lithium. In: Chapman H.D. (ed.) *Diagnostic Criteria for Plants and Soils*, 2nd Printing. Quality Printing Co., Abilene, Texas, pp. 218–224.

da Silva R.R., de Faria A.J.G., Alexandrino G.D.C., Ribeiro E.A., dos Santos A.C.M. (2019). Enrichment of lithium in lettuce plants through agronomic biofortification. *J. Plant. Nutrit.* 42:2102–2113.

Dessemond C., Lajoie-Leroux F., Soucy G., Laroche N., Magnan J.F. (2019). Spodumene: The lithium market, resources and processes. *Minerals* 9:334.

Dunne F.J. (2010). Lithium toxicity: The importance of clinical signs. *Briti. J. Hosp. Med.* 71:206–210.

Elektorowicz M., Keropian Z. (2015). Lithium, vanadium and chromium uptake ability of *Brassica juncea* from lithium mine tailings. *Internat. J. Phytorem.* 17:521–528.

Export Import Data Bank, Department of Commerce, GoI, Ministry of Commerce and Industry. https://tradestat.commerce.gov.in/eidb/Icomq.asp?hs=85076000. Accessed April 2023.

Garbisu C., Alkorta I. (2001). Phytoextraction: A cost-effective plant-based technology for the removal of metals from the environment. *Biores. Technol.* 77:229–236.

Godinho C., Coetzee K. (2020). Climate transparency report. Retrieved from the International Climate Transparency Partnership. https://www.climate-transparency.org/wp-content/uploads/2020/11/Climate-Transparency-Report-2020.pdf.

Gong R., Wang P., Dworkin L. (2016). What we need to know about the effect of lithium on the kidney. *Am. J. Physiol. Renal. Physiol.* 311: F1168–F1171.

Habashi F. (1997). *Handbook of Extractive Metallurgy*. Wiley-VCH, New York.

Harper G., Sommerville R., Kendrick E., Driscoll L., Slater P., Stolkin R., Anderson P. (2019). Recycling lithium-ion batteries from electric vehicles. *Nature* 575:75–86.

Hawrylak-Nowak B., Kalinowska M., Szymańska M. (2012). A study on selected physiological parameters of plants grown under lithium supplementation. *Biol. Trace Elem. Res.* 149:425–430.

https://www.epa.gov/transportation-air-pollution-and-climate-change/carbon-pollutiontransportation#:~:text=%E2%80%8BGreenhouse%20gas%20(GHG)%20emissions,terms%20than%20any%20other%20sector. https://www.rsc.org/periodic-table/element/3/lithium. Accessed 8 April 2022.

https://www.weforum.org/agenda/2023/01/chart-countries-produce-lithium-world/.

International Energy Agency. (2020a). Data and Statistics (India CO_2 emissions by sector). https://www.iea.org/data-and-statistics/data-browser?country=INDIA&fuel=CO2%20emissions&indicator=CO2BySector.

Jacob J.M., Karthik C., Saratale R.G., Kumar S.S., Prabakar D., Kadirvelu K. (2018). Biological approaches to tackle heavy metal pollution: A Survey of literature. *J. Environ. Manag.* 217:56–70.

Jakobsson E., Argüello-Miranda O., Chiu S.W., Fazal Z., Kruczek J., Nunez Corrales S. (2017). Towards a unified understanding of lithium action in basic biology and its significance for applied biology. *J. Memb. Biol.* 250:587–604.

Jiang L., Wang L., Mu S.Y., *et al.* (2014). Apocynum venetum: A newly found lithium accumulator. *Flora-Morphol. Distribut. Funct. Ecol. Plant.* 209:285–289.

Kabata-Pendias A., Pendias H. (1992). *Trace Elements in Soils and Plants*. CRC Press, Boca Raton, FL.

Kalinowska M., Hawrylak-Nowak B., Szymańska M. (2013). The influence of two lithium forms on the growth, L-ascorbic acid content and lithium accumulation in lettuce plants. *Biol. Trace Elem. Res.* 152:251–257.

Kavanagh L., Keohane J., Cabellos G.G., Lloyd A., Cleary J. (2018). Induced plant accumulation of lithium. *Geosciences* 8:5.

Kelly T., Matos G.R., Buckingham D.A., DiFrancesco C.A., Porter K.E., Berry C. (2005). Historical statistics for mineral and material commodities in the United States (No. 140). US Geological Survey.

Koval'skiĭ V.V., Borovik-Romanova T.F., Letunova S.V., Ginzburg E.O. (1965). Some data on the trace element content of microorganisms. *Mikrobiologiia* 34:403–406.

Kushnir D. (2013). Lithium resource flows over time: Implications and challenges. *Abstr. Pap. Am. Chem. Soc.* 245.

Li W., Achal V. (2020). Environmental and health impacts due to e-waste disposal in China–A review. *Sci. Total Environ.* 737:139745.

Malhi G.S., Bell E., Porter R.J., Boyce P., Mulder R., Hopwood M., Berk M. (2020). Lithium should be borne in mind: Five key reasons. *Aust. New Zeal. J. Psychiat.* 54:659–663.

Marques A.P., Rangel A.O., Castro P.M. (2009). Remediation of heavy metal contaminated soils: Phytoremediation as a potentially promising clean-up technology. *Crit. Rev. Env. Sci. Technol.* 39:622–654.

Mesjasz-Przybyłowicz J., Nakonieczny M., Migula P., Augustyniak M., Tarnawska M. (2004). Uptake of cadmium, lead nickel and zinc from soil and water solutions by the nickel hyperaccumulator *Berkheya coddii*. *Acta Biol. Cracoviensia Ser. Bot.* 46:75–85.

Mohr S.H., Mudd G.M., Giurco D. (2012) Lithium resources and production: Critical assessment and global projections. *Minerals* 2:65–84.

Muthusaravanan S., Sivarajasekar N., Vivek, J.S., Paramasivan T., Naushad M., Prakashmaran J., Al-Duaij O.K. (2018). Phytoremediation of heavy metals: Mechanisms, methods and enhancements. *Environ. Chem. Lett.* 16:1339–1359.

Qiao L., Tanveer M., Wang L., Tian C. (2018). Subcellular distribution and chemical forms of lithium in Li-accumulator *Apocynum venetum*. *Plant Physiol. Biochem.* 132:341–344.

Richa K. (2016). Sustainable management of lithium-ion batteries after use in electric vehicles. Rochester Institute of Technology.

Robinson B.H., Yalamanchali R., Reiser R., Dickinson N.M. (2018). Lithium as an emerging environmental contaminant: Mobility in the soil-plant system. *Chemosphere* 197:1–6.

Schrauzer G.N. (2002) Lithium: Occurrence, dietary intakes, nutritional essentiality. *J. Am. Coll. Nutrit.* 21:14–21.

Scott A.D., Smith S.J. (1987). Sources, amounts, and forms of alkali elements in soils. In: Advances in Soil Science. Springer, New York, pp. 101–147.

Shacklette H.T., Erdman J.A., Harms T.F., Papp C.S. (1978). Trace elements in plant foodstuffs. *Toxicity of Heavy Metals in the Environment. Part I.* Marcel Dekker Inc., New York, p. 25.

Shahzad B., Tanvee M., Hassan W., Shah A.N., Anjum S.A., Cheema S.A., Ali I. (2016). Lithium toxicity in plants: Reasons, mechanisms and remediation possibilities — A review. *Plant Physiol. Biochem.* 107:104–115.

Shakoor N., Adeel M., Azeem I., Ahmad M.A., Zain M., Abbas A., Rui Y. (2022a). Interplay of higher plants with lithium pollution: Global trends, meta-analysis, and perspectives. *Chemosphere* 310:136663.

Shakoor N., Adeel M., Zain M., Zhang P., Ahmad M.A., Farooq T., Rui Y. (2022b). Exposure of cherry radish (Raphanus sativus L. var. Radculus Pers) to iron-based nanoparticles enhance its nutritional quality by trigging the essential elements. *NanoImpact* 25:100388.

Shakoor N., Adeel M., Ahmad M.A., Zain M., Waheed U., Javaid R.A., Rui Y. (2023). Reimagining safe lithium applications in the living environment and its impacts on human, animal, and plant system. *Environ. Sci. Ecotechnol.* 100252.

Siqi Z., Guangming L., Wenzhi H., Juwen H., Haochen Z. (2019). Recovery methods and regulation status of waste lithium-ion batteries in China: A mini review. *Waste Manag. Res.* 37:1142–1152.

Sobolev O.I., Gutyj B.V., Darmohray L.M., Sobolieva S.V., Ivanina V.V., Kuzmenko. (2019). Lithium in the natural environment and its migration in the trophic chain. *Ukrain. J. Ecol.* 9:195–203.

Speirs J., Contestabile M., Houari Y., *et al.* (2014). The future of lithium availability for electric vehicle batteries. *Renew. Sustain. Ener. Rev.* 35:183–193.

Sun X., Hao H., Zhao F., Liu Z. (2017). Tracing global lithium flow: A trade-linked material flow analysis. *Res. Conserv. Recycl.* 124:50–61.

Tanveer M., Hasanuzzaman M., Wang L. (2019). Lithium in environment and potential targets to reduce lithium toxicity in plants. *J. Plant Growth Regul.* 38:1574–1586.

Thomas D.W., Merian E. (1991). *Metals and Their Compounds in the Environment.* In: Merian E. (ed.). VCH, Weinheim, pp. 789–801.

Török A.I., Moldovan A., Levei E.A., Cadar O., Tănăselia C., Moldovan O.T. (2021). Assessment of lithium, macro-and microelements in water, soil and plant samples from karst areas in Romania. *Materials* 14:4002.

US Department of Energy “2019 US Energy and Employment Report (USEER),” January 2019. https://www.usenergyjobs.org/2019-report. Accessed May 6, 2022.

US Geological Survey US Department of the Interior. (2020). Mineral Commodity Summaries. 2015-05-14[2015-09-29]. http://minerals.usgs.gov/minerals/pubs/commodity/manganese/mcs-2015-manga.pdf.

Wuana R.A., Okieimen F.E. (2011). Heavy metals in contaminated soils: A review of sources, chemistry, risks and best available strategies for remediation. International Scholarly Research Notices, 2011:20 pages. Article ID 402647, doi:10.5402/2011/402647.

Zeng X., Li J. (2014). Spent rechargeable lithium batteries in e-waste: Composition and its implications. *Front. Environ. Sci. Eng.* 8:792–796.

https://doi.org/10.1142/9789811283123_0003

Chapter 3

Physiological Responses of Plants to Environmental Lithium Pollution

David W. M. Leung* **and Peter J. Prendergast**

School of Biological Sciences, University of Canterbury, Private Bag 4800, Christchurch 8140, New Zealand

**david.leung@canterbury.ac.nz*

Abstract

Lithium (Li) is considered an emerging metal contaminant in the environment. To better prepare for the challenges to crop productivity from Li pollution, it is necessary to have a better understanding of the physiological basis/mechanisms of the responses of plants to Li stress. The overview presented here of the toxic effects of Li shows that different plants may exhibit variations in the relationship between Li bioaccumulation and the severity of growth inhibition and reduction in the contents of the different photosynthetic pigments. To deepen our understanding of how different plants respond to Li exposure, many studies have focussed on the increase in oxidative stress, which could cause damage to biomolecules, such as membrane lipids and proteins. These damages could be translated into growth inhibition effects and the loss of photosynthetic pigments. In addition, since various mineral nutrients, such as K, Ca and Mg, play vital roles in various metabolic reactions and pathways, alterations in mineral contents in plants exposed to Li could result in metabolic reprogramming leading to growth alterations. It is proposed here that gene expression changes, particularly those involved in plant hormone metabolism and transport as well as autophagy, could also underlie the physiological responses of plants to Li stress.

Keywords: Abiotic stress, autophagy, calcium signalling, jasmonic acid, metabolic reprogramming, mineral nutrients, oxidative stress, photosynthetic pigments, phytohormones

1. Introduction

Investigations into the possible impact of lithium (Li) on plants have never been more prevalent than they are today. This is largely due to the growing interest in renewable energy sources as a result of the rise in fossil oil and gas costs, as well as the global quest for a sustainable economy in response to the challenges posed by climate change. An exponential surge in demand is anticipated for Li battery-powered electric vehicles as well as other Li battery-powered technologies and products (Al-Thyabat *et al.*, 2013). This growing demand is likely to be associated with some environmental consequences, including higher environmental exposure of plants to Li due to an increase in traditional Li mining-related activities as well as the eventual disposal by end users of electronic devices and batteries with Li components. In light of Li being an emerging environmental contaminant (Chow, 2022), it is crucial to comprehend the physiological effects and toxicity of Li pollution on plants. In particular, it is urgent that we gain a better understanding of the abilities of different plant species to accumulate and cope with (or react to) Li in their different tissues. This knowledge base has the potential to aid in the development and identification of plant species that could be used for biotechnology-based solutions, including phytomining of Li and phytoremediation of Li-contaminated soil (da Silva *et al.*, 2019; Shahzad *et al.*, 2016). These new plant-based biotechnologies could, in turn, provide eco-friendly alternative methods of obtaining Li and biological alternatives to counteract the projected increased environmental exposure to Li on plants that might be susceptible to excess levels of environmental Li (Shakoor *et al.*, 2023).

The overall goal here is to highlight some of the knowledge gaps in the physiology of plants that interact with Li in the environment. First, we provide an overview of some of the well-known changes, ranging from no change to severe growth retardation, in plants following exposure to Li. Although Li, particularly at low doses, could appear to stimulate some

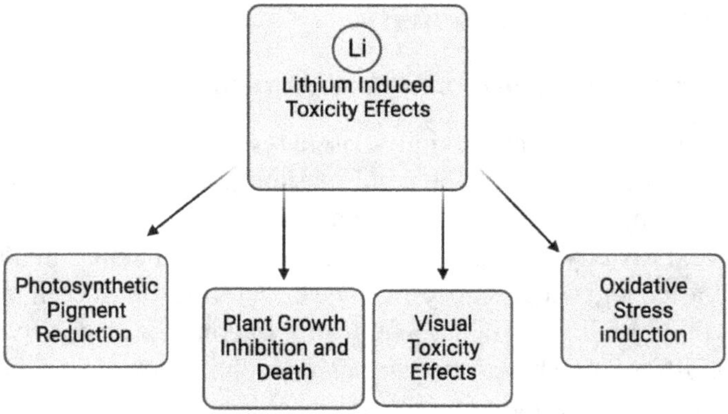

Figure 1. Some common changes that have been linked to Li stress-induced toxicity in plants. (Created with BioRender.com).

physiological changes associated with the growth of some plants, in light of emerging excess Li in the environment, the objectives of most studies have largely been investigations into Li stress and the more well-known physiological/biochemical markers of stress or toxicity problems (Fig. 1). In the second part of this chapter, some of the physiological/ biochemical/molecular mechanisms are discussed, including a cellular process that may underpin the physiological changes in plants following exposure to Li.

2. Significance of Li supply to plants

Li is often regarded as inessential for plant growth and development. However, the exact nature of the relationship between Li supply and plants is still unclear. Li is commonly regarded as highly toxic and is known to induce detrimental growth defects (Mulkey, 2005; Hawrylak *et al.*, 2012). For example, citrus plants have shown severe toxic effects in response to 4–40 mg Li kg^{-1} soil, as evident from necrotic and chlorotic spots on their leaves (Kabata-Pendias, 2010). In contrast, low levels of Li have been associated with favourable growth effects. For example, the fresh biomass of maize was increased by 15% following treatment with 5 mg L^{-1} of Li (Hawrylak-Nowak *et al.*, 2012). The overall role and importance of Li in plants remain mostly unknown. Thus, it is clear that more information is needed to assess the physiological significance of Li in plants.

3. Li-induced plant toxicity

3.1. *Adverse plant growth and development effects of Li*

Specific elements generate toxicity-based responses when their concentration exceeds an optimal level, resulting in decreased plant productivity and growth (Anjum *et al.*, 2015, 2016a, 2016b). The optimal level of Li tolerance and toxicity for a plant has to be determined experimentally and is highly varied and species-specific. The effect of Li on overall plant size and growth is highly variable, with both tolerant and sensitive plants exhibiting different responses to varying concentrations of Li. There have been reports of favourable growth benefits at low concentrations, while at high concentrations (>200 mg Li kg^{-1} soil), most plant species exhibit detrimental responses. These include biomass and root reduction, as well as complete growth inhibition for particularly sensitive species.

There are few reports of plants that can tolerate high concentrations of Li. The Asteraceae and Solanaceae families of plants are among the best-known Li-tolerant species, showing only minor development defects over a broad range of Li concentrations (Schrauzer, 2002; Kabata-Pendias and Mukherjee, 2007). Jiang *et al.* (2014) reported that certain plants in the *Apocynum* genus seem to be extremely resistant to Li exposure. Both *Apocynum pictum* and *A. venetum* had no adverse effects on shoots, roots or leaves when grown in 0–200 mg Li kg^{-1} soil. However, in response to higher concentrations, there was a marked reduction in their total biomass and root dry weight. Li *et al.* (2009) found that *Brassica carrinata* was able to tolerate up to 200 mg L^{-1} Li; however, in response to higher concentrations, growth inhibition and even severe growth retardation could also occur. Although *Helianthus annuus* (sunflower) and *Zea mays* (maize) have also been regarded as tolerant to Li, both plants exhibited a similar reaction to Li, with a 10% and 15% rise in shoot biomass at 5 mg Li^{-1} Li, respectively. They showed, however, a 27% and 32% reduction in shoot biomass at 50 mg L^{-1} Li (Hawrylak *et al.*, 2012). Similar to sunflower and maize plants, lettuce plants responded to both low and high doses of Li. There was a large rise in its root growth at 2.5 mg L^{-1} Li and a significant decrease in its root growth at 50 mg L^{-1} Li (Kalinowska *et al.*, 2013).

Contrastingly, citrus plants are highly intolerant and vulnerable to Li, even at low concentrations (Kabata-Pendias, 2010). However, certain plant species exhibit growth inhibition at substantially lower Li concentrations and might be considered very sensitive to Li. Bingham *et al.* (1964) found a 25% reduction in the growth of avocado, soybean and sour orange at 6, 7 and 8 mg Li kg^{-1} soil, respectively, while Li concentrations greater than 100 mg kg^{-1} completely inhibited growth.

As already noted, identifying Li toxicity-based symptoms is not always straightforward, as they can be variable across species. Apart from growth parameters, such as reduction in shoot or root biomass, shoot height or root length, the other most commonly identified symptoms include necrosis at leaf margins and roots, leaf chlorosis and abscission of leaves (Shahzad *et al.*, 2016). Necrosis is also a commonly observed symptom following incomplete pathogen–plant interactions. Li-induced necrotic spots have been demonstrated to be mediated via ethylene (Naranjo *et al.*, 2003). Such spots have been observed on leaves of lettuce and sunflower grown at 100 and 50 mg Li kg^{-1} soil, respectively (Kalinowska *et al.*, 2013; Hawrylak *et al.*, 2012). Necrotic spots tend to appear on older leaves, correlated with Li accumulation. For example, Naranjo *et al.* (2003) found this in *Nicotiana tabacum* (tobacco), a member of the Solanaceae family, as 50 mM LiCl brought about severe leaf curling and necrotic spot formation on the older leaves. However, to a lesser extent, similar symptoms were also seen when the tobacco plants were grown at lower Li concentrations (15 and 30 mM), and their onset was markedly delayed. The root systems of plants can also display necrosis and chlorosis. For example, corn and citrus plants have shown severe formation of necrotic spots in their roots, which is associated with injured root development (Kabata-Pendias, 2010; Kabata-Pendias and Mukherjee, 2007). The response-based variability to Li exposure is not well understood and is related to the current limited understanding of the role and function of Li in plant physiology and biochemical processes.

3.2. *Toxic effects of Li on photosynthetic pigments*

Li has also been shown to affect the contents of photosynthetic pigments as well as carbon fixation in leaves. Hawrylak-Nowak *et al.* (2012) reported that there was no decrease in chlorophyll a and b contents of the leaves of sunflower seedlings after transfer to a nutrient solution supplemented with different Li concentrations (between 5 and 50 mg L^{-1}) for 14 days. There was, however, a 16% decrease in the carotenoid content at 25 mg L^{-1} Li.

Similarly, there was no effect of 5–25 mg L^{-1} Li on the photosynthetic pigments in the leaves of maize seedlings after two weeks of growth in nutrient solutions with different added Li concentrations. However, compared to the sunflower seedlings in the same study, in response to 50 mg L^{-1} Li in the nutrient solution, there were relatively greater losses in the photosynthetic pigment contents (45% and 67% decreases in chlorophylls a and b and carotenoids, respectively).

Although Li bioaccumulation in the shoots of both sunflower and maize seedlings after two weeks of growth in a nutrient solution was dose-dependent, more Li was taken up by sunflower shoots than maize shoots in response to 5–50 mg L^{-1} Li. In response to 50 mg L^{-1} Li in the nutrient solution, the sunflower shoots accumulated about five times more Li than maize shoots, or 3,292 and 695 mg Li kg^{-1} dry shoot biomass, respectively (Hawrylak-Nowak *et al.*, 2012). In light of the Li contents in the shoots of both plants when grown at 5–50 mg L^{-1} Li, the lack of effect of Li on chlorophyll a and b contents in sunflower leaves compared to that in maize seedlings seems to support the hypothesis that sunflower is more tolerant than maize plants to Li exposure and bioaccumulation. This suggestion is also consistent with the reduction in root dry biomass observed in maize plants with increasing concentrations of Li in the nutrient solution but not in the sunflower plant. In contrast, the shoot dry biomass of both plants seems to be of less indicative value of their different responses to the growth-inhibiting effect of Li, as only at the highest Li concentration tested (50 mg L^{-1}) was a reduction observed in shoot dry biomass. Similarly, if the finding of the effect of Li on leaf area were considered in isolation, it would seem intriguing that Li had a slight but significantly promotive effect of 5 or 25 mg L^{-1} Li on the leaf area of maize plants but not on that of sunflower. Clearly, there is an incomplete understanding of the physiological basis of the interrelationships between the effects of Li on the contents of photosynthetic pigment contents, Li bioaccumulation and different growth parameters.

Li *et al.* (2009) found that *Brassica carrinata* showed a reduction in chlorophyll content at concentrations greater than 200 mg L^{-1} Li. A surface-floating aquatic plant, *Salvinia natans*, showed a decrease in chlorophyll a and b contents, but not carotenoid content, when grown in a nutrient solution supplemented with 10 mg L^{-1} Li (Torok *et al.*, 2022). Interestingly, the plants grown in a nutrient solution supplemented with 20 mg L^{-1} Li had accumulated twice the amount of Li than those grown in 10 mg L^{-1} Li. However, they did not show further reductions in photosynthetic pigment contents. While it is known that for each plant exposed to Li solution, the photosynthetic pigment contents could be adversely affected, the dose–response relationship is not clearly understood. This might be due to the lack of more precise investigations into this relationship. For example, since experiments were not conducted with Li concentrations below 10 mg L^{-1} other than the control (0 mg L^{-1} Li), it is not possible to determine if there is a dose–response relationship between a decrease in photosynthetic pigments and the uptake of increasing Li concentrations. Alternatively, there might be a threshold level of tolerance to Li bioaccumulation, which could vary among plant species. Also, it is possible that reductions in photosynthetic pigment contents might not be the primary target of the onset of Li-induced toxicity. This suggestion is consistent with the finding of a recent study showing that the cell wall and vacuole are the major subcellular compartments for Li bioaccumulation rather than other vital organelles, including the nucleus and chloroplasts, in soybean shoots (Shakoor *et al.*, 2023).

The formation of necrotic spots and a reduction in chlorophyll content could be interlinked with a steady breakdown of photosynthetic pigments. This could also be explained by the enhanced catalytic activity of chlorophyll degradation and bleaching enzymes, such as chlorophyllase and Mg-dechelatase (Harpaz-Saad *et al.*, 2007). The potential mechanisms of photosynthetic inhibition by Li remain largely unclear. However, Li could be causing structural damage to chloroplasts or interfering with biochemical processes required for pigmentation production. These may result in the Calvin cycle and electron chain reaction inhibition. The interaction pathway between Li and photosynthetic pigment reduction represents an area that needs additional future research.

4. Biochemical basis underpinning Li toxicity on plant growth

4.1. *Oxidative damages*

It is important to identify the key biochemical changes associated with plant growth inhibition or even death at species-specific high Li concentrations, although our current understanding of this relationship is still poor. One proposed mechanism is that excessive Li exposure could favour the onset of oxidative stress by affecting the regulation of reactive oxygen species (ROS) production and antioxidant defence enzymes

(Saad *et al.*, 2017). Li has been shown to increase lipid peroxidation and malondialdehyde (MDA, a lipid peroxidation index), while also down-regulating key antioxidant defence enzymes, such as superoxide dismutase and catalase (Bueso *et al.*, 2007; Oktem *et al.*, 2005). Interestingly, Liu *et al.* (2018) found increased production of hydrogen peroxide (H_2O_2), superoxide anion ($O_2^{\bullet-}$), and hydroxy radical (·OH) in zebrafish in response to 25 and 50 mg L^{-1} Li. Jing *et al.* (2022) also found an increase in ROS production as well as a reduction in antioxidant enzyme capacity in mice exposed to 20 mg L^{-1} of lithium carbonate. This enhanced ROS production can lead to increased lipid peroxidation owing to the downregulation of antioxidant defence and Fenton-based processes mediated by $O_2^{\bullet-}$ and OH^- (Clemens, 2006). A similar phenomenon was observed in plants; for example, a recent work by Kuloğlu *et al.* (2022) on *Allium cepa* (onion) found a 59.1% increase in MDA content and a decrease of 60.8% in glutathione (GSH). Furthermore, they discovered significant changes in the activity of superoxide dismutase and catalase in response to 100 mg L^{-1} of lithium carbonate. Lipid peroxidation is a well-studied biochemical process that has been demonstrated to induce membrane rupture and damage. This occurs by stealing electrons from lipids in the cell membrane, resulting in a significant reduction in plant biomass (Ayala *et al.*, 2014; Shahzad *et al.*, 2016). An increase in the rate of this reaction to the point where basal defence regulation is overwhelmed could explain the reported plant growth inhibitory effects at high Li concentrations. In addition to this, the observed downregulation of antioxidant defence enzymes by Li suggests enhanced transcriptional control of genes related to oxidative defence (Singh *et al.*, 2010). ROS have been shown to increase in response to Li exposure and inhibit antioxidant production (Erakovic *et al.*, 2000; Tanveer and Wang, 2020). Large-scale ROS generation has been found to have permanent impacts on plant growth and development, which can result in growth inhibition or death (Huang *et al.*, 2019). ROS have been shown to interact with numerous membranes, lipids, nucleic acids, photosynthetic pigments and many types of biomolecules. It is therefore possible that the mediation of growth inhibition and plant death occurs when high ROS levels disrupt vital biochemical processes as well as the core plant metabolism (Mittler *et al.*, 2022). Li-induced oxidative stress seems to be a potential pathway leading to the inhibition and stagnation of plant growth and development. More research into the oxidative disruption caused by Li in diverse plant-based systems is, however, required for more definite results.

4.2. *Effects on mineral contents in plants*

Halophytes have been shown to elicit a metabolic response to Li but still show signs of toxicity at high concentrations (Hawrylak-Nowak *et al.*, 2012). The biological activity of Li is frequently attributed to its short ionic radius, which causes the displacement of different ions at derived membranes and enzyme active sites. Duff *et al.* (2014) reported that Li could displace Mg ions in plant and mammalian proteins, which might influence the biological functions of the affected proteins. In soybean plants grown in different concentrations of lithium oxide (Li_2O), the contents of P, K, Ca, Mg and Na were found to be altered in the shoots and roots (Shakoor *et al.*, 2023). Under Li stress (200 mg Li_2O kg^{-1} soil), the contents of all the inorganic nutrients studied, such as P, K, Ca, Mg, and Na, were reduced. Since these nutrients are essential to many metabolic processes associated with plant growth and development, it is possible that Li could bring about metabolic reprogramming by influencing the mineral contents in plant cells (Shakoor *et al.*, 2023).

4.3. *Calcium signalling*

Li has also been shown to inhibit the inositol cycle as well as disrupt calcium signalling in plants (Berridge, 1993; Naranjo *et al.*, 2003). Allender *et al.* (1997) discovered that Li ions could be producing negative effects on plant growth by interfering with Ca metabolism. Stevenson *et al.* (2000) hypothesised that Li might interact via the inositol signalling pathway, which also involves the glutamate receptor and the Ca-dependent pathway. These processes, together with H^+-ATPase (proton pump), could therefore impact plant growth (Allender *et al.*, 1997; Stevenson *et al.*, 2000). Li has also been shown to cause metabolic changes through ion-based substitutions, which may manifest as a form of metabolic regulation that might affect plant growth. However, the precise nature of this is still not completely known (Kabata-Pendias and Mukherjee, 2007).

4.4. *Gene expression changes underlying physiological responses*

It was found that the brown seed line of *Brassica carinata* was more tolerant to Li bioaccumulation than the yellow seed line (Li *et al.*, 2009).

While the yellow seed line died after exposure to 90 mM Li, the brown seed line managed to survive and accumulate over 60 mg g^{-1} dry biomass. The transcriptional changes of about 1,000 genes in the brown and yellow seed lines of *B. carinata* in response to 60 mM LiCl were found (Li *et al.*, 2009). More genes involved in cell wall, hormone and stress-related metabolism, secondary metabolism, metal transport and cellular detoxification were upregulated more than fivefold in the brown seed line than in the yellow seed line (Das *et al.*, 2012; Eraković *et al.*, 2020). The altered gene expression profiles are consistent with the differential tolerance levels of the brown and yellow seed lines to Li exposure. The changes in plant hormone contents were not quantified in this study; however, the elevated expressions of metal transport genes (for example, heavy-metal-associated domain protein) and stress-associated genes (for example, chitinase) were validated (Li *et al.*, 2019).

There is evidence that not only the expression profiles of the genes associated with plant hormone metabolism are altered upon exposure to different doses of Li (Li *et al.*, 2019), but the concentrations of plant hormones could also vary (Shakoor *et al.*, 2023). In soybean roots, the contents of abscisic acid (ABA), salicylic acid (SA) and gibberellic acid (GA) were lowered under Li stress (200 mg Li_2O kg^{-1} soil) compared to the control (plants not exposed to Li) (Shakoor *et al.*, 2023). In contrast, the level of jasmonic acid (JA) in the roots of soybeans under Li stress was higher than the control. When soybean plants were grown at a lower Li concentration (50 mg Li_2O kg^{-1} soil), the ABA content in the roots was higher than that in the control, while the contents of JA, SA and GA in the roots were the same as those of the control. Since plant hormones play critical roles in growth and development processes, changes in plant hormone metabolism are likely to underpin Li-induced physiological/biochemical responses associated with the regulation of these processes.

4.5. *LiCl is an activator of autophagy: Implications*

Autophagy is a unique eukaryotic cellular process that removes damaged proteins or organelles or recycles nutrients associated with the regulation of cell survival (Das *et al.*, 2012). LiCl at 20 mM is an activator of autophagy, for example, in cultured human corneal fibroblasts (Nie *et al.*, 2018). In plants, there is a basal level of autophagic activity associated with normal plant growth, while some stress-associated plant hormones, such as brassinosteroids, ethylene and methyl jasmonate (MeJA), can increase the activity of autophagy in plants under a variety of abiotic and biotic stress conditions (Wang *et al.*, 2021).

There do not seem to be any investigations into whether LiCl application can directly enhance autophagy in plants. However, in post-harvest storage of tomato fruits, it was found that 50 mM LiCl enhanced the protection against *Botrytis cinerea*–induced fruit decay through MeJA treatment, which induced autophagy in the fruit during storage, while 5 mM hydroxychloroquine (an autophagy inhibitor) reduced the protective effect of JA (Min *et al.*, 2022). In this study, there was no treatment of LiCl alone or with the hydroxychloroquine treatment on the disease severity, nor was there any information on autophagic activity under LiCl treatment alone or with the hydroxychloroquine treatment. It is, therefore, not clear whether or not the effect of LiCl on tomato fruit (Min *et al.*, 2022) might be related to an increase in autophagic activity.

Interestingly, the concentration of LiCl (50 mM) that might activate autophagy was also used in previous studies, for example, LiCl-induced necrotic spot formation on tobacco leaves (Naranjo *et al.*, 2003). Although in many other studies, lower concentrations of LiCl were often used to study their effects on plants, it seems worthwhile to investigate the possibility that some of the effects observed might be associated with altered autophagic activity. It is proposed here that autophagy, as a cellular process/mechanism, might be linked with plant hormones and plant cell survival/adaptation under adverse plant growth conditions, including Li stress. This line of investigation would potentially add new knowledge about the mechanisms underpinning the physiological response of plants to Li exposure.

5. Conclusion and future perspectives

There are many avenues of research that can lead to a more comprehensive understanding of the physiological basis of the response of plants to Li exposure. Some of them have been outlined in this mini-review. Li is considered an emerging environmental contaminant with impacts on plants. Moreover, with climate change expected to bring about changes to the physiology of plants, future investigations must take into account the interactions of Li exposure and climate change factors, including drought stress, high temperature stress and disease susceptibility, with plants.

References

Al-Thyabat S., Nakamura T., Shibata E., Iizuka A. (2013). Adaptation of minerals processing operations for lithium-ion (LiBs) and nickel metal hydride (NiMH) batteries recycling: Critical review. *Miner. Engg.* 45:4–17.

Allender W.J., Cresswell G.C., Kaldor J., Kennedy I.R. (1997). Effect of lithium and lanthanum on herbicide induced hormesis in hydroponically-grown cotton and corn. *J. Plant Nutr.* 20:81–95.

Anjum S.A., Tanveer M., Hussain S., Bao M., Wang L., Khan I., Ullah E., Tung S.A., Samad R.A., Shahzad B. (2015). Cadmium toxicity in maize (*Zea mays* L.): Consequences on antioxidative systems, reactive oxygen species and cadmium accumulation. *Environ. Sci. Pollut. Res.* 22:17022–17030.

Anjum S.A., Ashraf U., Khan I., Tanveer M., Ali M., Hussain I., Wang L.C. (2016a). Chromium and aluminum phytotoxicity in maize: Morpho-physiological responses and metal uptake. *CLEAN–Soil Air Water* 44:1075–1084.

Anjum S.A., Tanveer M., Hussain S., Shahzad B., Ashraf U., Fahad S., Hassan W., Jan S., Khan I., Saleem M.F. (2016b). Osmoregulation and antioxidant production in maize under combined cadmium and arsenic stress. *Environ. Sci. Pollut. Res.* 23:11864–11875.

Ayala A., Muñoz M.F., Argüelles S. (2014). Lipid peroxidation: Production, metabolism, and signaling mechanisms of malondialdehyde and 4-hydroxy-2-nonenal. *Oxid. Med. Cell Longev.* 2014:360438.

Berridge M.J. (1993). Inositol trisphosphate and calcium signalling. *Nature* 361:315–325.

Bingham F., Bradford G., Page A. (1964). Toxicity of lithium to plants. *Calif Agric.* 18:6–7.

Bueso E., Alejandro S., Carbonell P., Perez-Amador M.A., Fayos J., Bellés J.M., Rodriguez P.L., Serrano R. (2007). The lithium tolerance of the Arabidopsis cat2 mutant reveals a cross-talk between oxidative stress and ethylene. *Plant J.* 52:1052–1065.

Chow A.T. (2022). Proactive approach to minimise lithium pollution. *J. Enviorn. Qual.* 51:872–876.

Clemens S. (2006). Toxic metal accumulation, responses to exposure and mechanisms of tolerance in plants. *Biochimie* 88:1707–1719.

da Silva R.R., de Faria A.J.G., Alexandrino G.D.C., Ribeiro E.A., dos Santos A.C.M., Deusdara T.T., do Nascimento I.R., Nascimento V.L. (2019). Enrichment of lithium in lettuce plants through agronomic biofortification. *J. Plant Nutr.* 42:2102–2113.

Das G., Shravage B.V., Baehrecke E.H. (2012). Regulation and function of autophagy during cell survival and cell death. *Cold Spring Harb. Persp. Biol.* 4:a008813.

Duff M., Kuhne W., Halverson N., Chan C.S., Kitamura E., Hawthorn M.N., Stafford C., Milliken C., Caldwell E. (2014). mRNA Transcript abundance during plant growth and the influence of Li^+ exposure. *Plant Sci.* 229:262–279.

Eraković V., Župan G., Varljen J., Laginja J., Simonić A. (2000). Lithium plus pilocarpine induced status epilepticus-biochemical changes. *Neurosci. Res.* 36:157–166.

Harpaz-Saad S., Azoulay T., Arazi T., Ben-Yaakov E., Mett A., Shiboleth Y.M., Hörtensteiner S., Gidoni D., Gal-On A., Goldschmidt E.E. (2007). Chlorophyllase is a rate-limiting enzyme in chlorophyll catabolism and is posttranslationally regulated. *Plant Cell* 19:1007–1022.

Hawrylak-Nowak B., Kalinowska M., Szymańska M. (2012). A study on selected physiological parameters of plants grown under lithium supplementation. *Biol. Trace Elem. Res.* 149:425–430.

Huang H., Ullah F., Zhou D.X., Yi M., Zhao Y. (2019). Mechanisms of ROS regulation of plant development and stress responses. *Front. Plant Sci.* 10:800.

Jiang L., Wang L., Mu S.Y., Tian C.Y. (2014). *Apocynum venetum*: A newly found lithium accumulator. *Flora-Morphol. Distrib. Funct. Ecol. Plant* 209:285–289.

Jing H., Wang F., Gao X.J. (2022) Lithium intoxication induced pyroptosis via ROS/NF-κB/NLRP3 inflammasome regulatory networks in kidney of mice. *Environ. Toxicol.* 37:825–835.

Kabata-Pendias A. (2000). *Trace Elements in Soils and Plants*. 3rd Edition, Boca Raton, CRC Press.

Kabata-Pendias A. (2010). *Trace Elements in Soils and Plants*. 4th Edition. Boca Raton, CRC Press.

Kabata-Pendias A., Mukherjee A.B. (2007). *Trace Elements from Soil to Human.* Springer-Verlag Berlin, Heidelberg.

Kalinowska M., Hawrylak-Nowak B., Szymańska M. (2013). The influence of two lithium forms on the growth, l-ascorbic acid content and lithium accumulation in lettuce Plants. *Biol. Trace Elem. Res.* 152:251–257.

Kuloğlu S.S., Yalçin E., Çavuşoğlu K., Acar A. (2022) Dose-dependent toxicity profile and genotoxicity mechanism of lithium carbonate. *Scient Rep.* 12:13504.

Li X., Gao P., Gjetvaj B., Westcott N., Gruber M.Y. (2009). Analysis of the metabolome and transcriptome of *Brassica carinata* seedlings after lithium chloride exposure. *Plant Sci.* 177:68–80.

Liu D., Gao L., Zhang Z., Tao S., Pang Q., Li A., Deng H., Yu H. (2018). Lithium promotes the production of reactive oxygen species via GSK-3β/TSC2/TOR signaling in the gill of zebrafish (*Danio rerio*). *Chemosphere* 195:854–863.

Min D., Li Z., Fu X., Li F., Li X., Zuo J., Zhang X. (2022). Autophagy is involved in methyl jasmonate-mediated resistance against *Botrytis cinerea* in postharvest tomato fruit by regulating jasmonate signaling and reactive oxygen species homeostasis. *Sci. Hort.* 305:111361.

Mittler R., Zandalinas S.I., Fichman Y., Van Breusegem F. (2022). Reactive oxygen species signalling in plant stress responses. *Nat. Rev. Mol. Cell Biol.* 23:663–679.

Mulkey T.J. (2005) Alteration of growth and gravitropic response of maize roots by lithium. *Gravit. Space Res.* 18:119–120.

Naranjo M.A., Romero C., Bellés J.M., Montesinos C., Vicente O., Serrano R. (2003). Lithium treatment induces a hypersensitive-like response in tobacco. *Planta* 217:417–424.

Nie D., Peng Y., Li M., Liu X., Zhu M., Ye L. (2018). Lithium chloride (LiCl) induced autophagy and downregulated expression of transforming growth factor β-induced protein (TGFBI) in granular corneal dystrophy. *Exp. Eye Res.* 173:44–50.

Oktem F., Ozguner F., Sulak O., Olgar Ş., Akturk O., Yilmaz H.R., Altuntas I. (2005) Lithium-induced renal toxicity in rats: Protection by a novel antioxidant caffeic acid phenethyl ester. *Mol. Cell Biochem.* 277:109–115.

Saad A.B., Rjeibi I., Alimi H., Ncib S., Smida A., Zouari N., Zourgui L. (2017) Lithium induced, oxidative stress and related damages in testes and heart in male rats: The protective effects of *Malva sylvestris* extract. *Biomed. Pharma.* 86:127–135.

Schrauzer G.N. (2002) Lithium: Occurrence, dietary intakes, nutritional essentiality. *J. Am. Coll. Nutr.* 21:14–21.

Shahzad B., Tanveer M., Hassan W., Shah A., Anjum S.A., Cheema S., Ali I. (2016). Lithium toxicity in plants: Reasons, mechanisms and remediation possibilities — A review. *Plant Physiol. Biochem.* 107:104–115.

Shakoor N., Adeel M., Ahmad M.A., Husaain M., Azeem I., Zain M., Zhou P., Li Y., Xu M., Rui Y. (2023). Environment relevant concentrations of lithium influence soybean development via metabolic reprogramming. *J. Hazard Mater.* 441:129898.

Singh R., Tripathi R., Dwivedi S., Kumar A., Trivedi P., Chakrabarty D. (2010). Lead bioaccumulation potential of an aquatic macrophyte *Najas indica* are related to antioxidant system. *Biores. Technol.* 101:3025–3032.

Stevenson J.M., Perera I.Y., Heilmann I., Persson S., Boss W.F. (2000). Inositol signaling and plant growth. *Trend Plant Sci.* 5:252–258.

Tanveer M., Wang L. (2020). Lithium: Is it good or bad for plant growth? In: *Metal Toxicology Handbook*. Boca Raton, CRC Press, pp, 455–464.

Torok A.I., Moldovan A., Kovacs E., Cadar O., Becze A., Levei E.A., Neag E. (2022). Lithium accumulation in *Salvinia natans* free-floating aquatic plant. *Materials* 15:7243.

Wang P., Wang T., Han J., Li M., Zhao Y., Su T., Ma C. (2021). Plant autophagy: An intricate process controlled by various signaling pathways. *Front. Plant Sci.* 12:754982.

https://doi.org/10.1142/9789811283123_0004

Chapter 4

Physiological and Biochemical Responses of Crop Plants to Nickel Toxicity

Kiran Saroy, Aditi Bisht, and Neera Garg*

Department of Botany, Panjab University, Chandigarh 160014, India

**gargneera@gmail.com; garg_neera@yahoo.com*

Abstract

Nickel (Ni) is an essential, naturally occurring micronutrient required for various metabolic processes in plants, such as ureolysis, hydrogen metabolism, methane biogenesis and acetogenesis, since it is an important component of the enzymes glyoxalases, ureases, superoxide dismutases, hydrogenases, etc. Ni constitutes the active site of two metalloenzymes, i.e., urease and hydrogenase, in plants, which are involved in nitrogen (N) metabolism. Urease is responsible for the breakdown of urea into ammonia and carbon dioxide, while hydrogenase catalyses the oxidation of molecular hydrogen and recovers the energy for the reduction of enzymes. Elevated levels of Ni ($\geq$20 mg kg^{-1}) in the soil pose a significant threat to the agricultural productivity of crop plants worldwide owing to their extreme toxicity. Consequently, Ni has now been ranked 57th in priority hazardous substances by the American Agency for Toxic Substances and Disease Registry (ATSDR, 2022). The common symptoms of Ni toxicity in plants include reduced germination, declined shoot and root growth, membrane damage, reduced nutrient absorption and abnormal flower shape. All these toxic effects result in reduced photosynthesis and respiration rates, loss of key osmolytes, etc., which cause oxidative stress and ultimately reduce the yield of crops. In addition, excess Ni has a negative impact on legume–rhizobia symbiosis by affecting nodulation potential and N_2-fixation processes, as well as decreasing ureides biosynthesis. However, significant variations in terms of Ni metal-induced sensitivities among various crop plants have been observed at various levels. Therefore, this chapter reviews the updated information regarding Ni stress-induced physiological, biochemical and molecular responses in crop plants, especially legumes.

Keywords: Antioxidant defence, growth, legumes, nickel toxicity, nitrogen metabolism, ROS

1. Introduction

Heavy metal (HM) toxicity is one of the major environmental constraints that negatively influences the growth and productivity of crop plants (Gupta *et al.*, 2014; Prajapati *et al.*, 2019; Vischetti *et al.*, 2022). Among HMs, nickel (Ni) is an essential, naturally occurring micronutrient required for plant growth and development (Bhalerao *et al.*, 2015; Shahzad *et al.*, 2018). It plays diverse roles in the metabolism of plants, including ureolysis, methane biogenesis, hydrogen metabolism and acetogenesis (Vatansever *et al.*, 2017; Barcelos *et al.*, 2018). Moreover, deficiency of Ni leads to reduced urease activity, which results in delayed nodulation and reduced nitrogen (N_2) fixation in legumes (Küpper and Kroneck, 2007; Afzal *et al.*, 2015). Although Ni is essential for several metabolic phenomena, its presence at elevated levels in the soil is extremely toxic to crop plants and affects their physiological and metabolic processes, which ultimately reduce their yield (Vatansever *et al.*, 2017; Haseeb *et al.*, 2022).

Leguminous crops require more Ni than other crops since it plays a vital role in nodulation, N_2 fixation and ureides (allantoin, ALN, and allantoic acid, ALA) transport to other plant organs (González-Guerrero *et al.*, 2014; Freitas *et al.*, 2018). However, high Ni concentrations in the growing media reduce nodule formation and impair nodule functioning in legumes (Zobiole *et al.*, 2010; Garg and Saroy, 2019). Numerous studies have also reported that high Ni concentrations have negative impacts on the structural and anatomical dynamics of plants, which reduces photosynthesis, the uptake of water and nutrients and final productivity (Aqeel *et al.*, 2021; Einhardt *et al.*, 2021). Moreover, Ni at high concentrations interacts with NADPH and induces oxidative damage to plant cells (Tanveer, 2020). Therefore, this chapter focuses on updated information related to the adverse impacts of Ni toxicity on the metabolic profile of agricultural crops.

2. Nickel Toxicity in Soil and Its Uptake Mechanisms by Plants

Ni is one of the ubiquitous trace metals and the 24th most abundant element in the Earth's crust. It belongs to Group VIII B of the periodic table with an atomic number of 28 (Sachan and Lal, 2017; Naz *et al.*, 2022). It was discovered and extracted from the niccolite ore, which is composed of nickel arsenide, by Baron Axel Fredrik Cronstedt (a Swedish mineralogist and chemist) in 1750. Ni has a variety of oxidation states ranging from 1 to 4, with Ni (II) being the most common in biological systems (Sreekanth *et al.*, 2013). It enters the environment from various natural and anthropogenic sources. Natural sources of increased Ni include dusts from volcanic emissions, smoke particles from forest fires, windblown soil dusts, the weathering of rocks and soils and aerosols from oceanic dusts (Liao and Selim, 2018). The anthropogenic activities may involve mining works, smelters, burning of coal, oil combustion sewage, alloy manufacturing, electroplating and batteries, phosphate fertilisers and pesticides (Bhalerao *et al.*, 2015; Klein and Costa, 2015). In soil, Ni may be present as either soluble compounds, such as chlorides and nitrates, or insoluble compounds, such as oxides and sulphides. Ni plays important roles in various physiological processes, starting from seed germination to yield production (Poonkothai and Vijayavathi, 2012; Shahzad *et al.*, 2018). Although it is an essential micronutrient for normal plant growth and development (Patra *et al.*, 2019), its presence at elevated levels is extremely toxic to plants and has negative impacts on their structural and anatomical dynamics, which reduces crop yield by interfering with many physiological and biochemical processes (photosynthesis, nutrient and water uptake, enzyme activities, etc.) of plants (Jan *et al.*, 2019; Hassan *et al.*, 2019; Ahmed *et al.*, 2023).

2.1. *Nickel in soil*

Although Ni is uniformly distributed through the soil profile, its concentration is typically high at the soil surface due to the deposition of industrial waste and agricultural activities (Dhaliwal *et al.*, 2019; Ahmed *et al.*, 2023). Ni is found in a variety of inorganic and organic forms, such as crystalline minerals or precipitates, as well as water soluble and free ions. Soil properties, such as soil composition, texture, surface area, free Fe and Mn oxides, organic matter contents, soil sorption potential and pH, influence the phyto-availability of Ni (Laghlimi *et al.*, 2015; Dhaliwal *et al.*, 2019). Molas and Baran (2004) found that soluble Ni concentrations were higher at pH levels ranging from 5.1 to 5.6, implying that acidic soils have higher Ni solubility. Therefore, Ni availability in the soil is influenced by soil pH, base cation saturation (BCS) and soil inherent properties. Since certain organic acids, such as histidine, increase Ni phyto-assimilation while decreasing its phytotoxicity, parental material and organic matter have a major impact on Ni solubility (Krämer *et al.*, 1996; Iyaka, 2011). In general, soil enzymatic activities may vary under Ni stress because Ni can react with sulphydryl (–SH) groups of enzymes, causing inhibition or inactivation of the enzymatic activities (Shaw and Raval, 1961). In a study, soil enzymatic activity diminished with increasing concentration of Ni, and the order of sensitivity for the three enzymes was urease > β-glucosidase > protease activity (Moreno *et al.*, 2003). Moreover, Kucharski *et al.* (2009) found that the sensitivity of the analysed enzymes varied in the following sequence: urease > dehydrogenases > alkaline phosphatase > phosphatase > catalase > aryl-sulphatase > β-glucosidase.

2.2. *Nickel uptake from soil to plants*

Ni is mostly absorbed by plants from their roots, either by passive diffusion or active transport (Harasim and Filipek, 2015). Ni transport is influenced by a number of factors, including plant types, soil pH, concentration in the substratum and organic content in the soil (Bhalerao *et al.*, 2015). Ni uptake commonly declines at higher pH values in the soil solution due to the formation of less soluble complexes. For instance, it has been reported that Ni uptake increased as pH increased up to 5 but decreased as pH increased to 8 in *Lathyrus sativus* (Panda *et al.*, 2007). Soluble Ni compounds (such as chlorides, sulphate and nitrates) are preferentially absorbed passively through the cation transport system;

however, they could also be absorbed via the ABC, ZIP and Mg ion transport systems because of the similar charge/size ratio of the essential metal ions (Fe^{2+}, Mg^{2+}, Zn^{2+} and Cu^{2+}) (Demidchik and Maathuis, 2007; Shahzad *et al.*, 2018). Conversely, insoluble Ni compounds (oxides and sulphides) enter the plant via endocytosis (Ahmad and Ashraf, 2011). Moreover, low-affinity Zn transporters are known to play a vital role in Ni uptake; these are present at the cell membranes of the epidermal root cells of *Thlaspi arvense* and *Alyssum murale* (Deng *et al.*, 2014). Instead, Ni uptake by roots appears to be catalysed by poorly selective cation transporters, notably the members of the ZRT/IRT-like protein (ZIP) family (Van der Pas and Ingle, 2019). Nkongolo *et al.* (2018b) observed differential gene expression in red maple plants when treated with Ni, with more upregulated than downregulated genes responsible for the encoding of proteins related to transport and translation in plants. Moreover, Ni toxicity led to the downregulation of transcripts that encode the Ni transporter family: natural resistance-associated macrophage proteins (NRAMP3) and AT2G16800 in white pine and jack pine (Moarefi and Nkongolo, 2022).

Ni may be chelated by histidine (His) and peptides (Vacchina *et al.*, 2003) because Ni has a high affinity for imidazole ring (Kasprzak *et al.*, 1987). Due to the presence of carboxyl, amino and imidazole groups, His plays a key role as a Ni chelator, which can enhance Ni transport in plants (Callahan *et al.*, 2007). Ni metal–chelate complexes are instantly absorbed by roots, and the complex is transported via plasmalemma in the roots (Dalir and Khoshgoftarmanesh, 2014, 2015). Dalir and Khoshgoftarmanesh (2015) proposed that Ni uptake occurs in plants with the help of xylem via the symplastic pathway. After uptake, Ni may be deposited in the roots or transferred to the shoots and aboveground plant organs through the transpiration stream via stelar tissue, such as xylem and phloem, and finally accumulate in cellular parts (Page and Feller, 2015; Hassan *et al.*, 2019).

3. Ni toxicity Induce Physiological and Biochemical Constraints

Higher dosages of Ni could lead to several deleterious alterations in plants (Fig. 1), which appear in the form of reduced seed germination, stunted plant growth and development, the induction of oxidative stress, altered metabolic processes including photosynthetic and non-photosynthetic

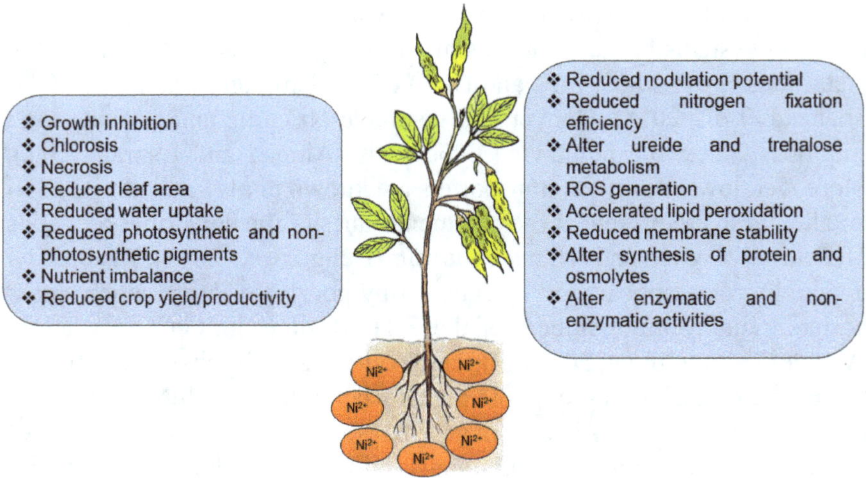

Figure 1. An overview of nickel (Ni) toxicity in plants.

efficiency, mineral nutrient disorders and reduced reproductive potential leading to ultimate death (Fabiano *et al.*, 2015; Rizwan *et al.*, 2019; Ahmad *et al.*, 2023).

3.1. *Growth and productivity*

Excess concentrations of Ni in soil induce a number of physiological alterations and severe toxicity symptoms, e.g., reduced vegetative growth, including plant height and total biomass production, as observed in various agricultural crops, such as sunflower, cabbage, maize and barley (Bhalerao *et al.*, 2015; Sachan and Lal, 2017; Duan *et al.*, 2020; Badawy *et al.*, 2022). Ozdener and Kutbay (2009) and Singh *et al.* (2012) reported that excess Ni induces changes in all energy-driven processes during germination and plant development, thus delaying the emergence of embryonic shoots in *Eruca sativa* and *Vigna mungo.* Sheoran *et al.* (1990) found that *Cajanus cajan* plants grown in sand under $NiCl_2$ (1 mM) had reduced leaf area. According to Molas (1998), excess Ni reduced the amount of intercellular spaces, palisade and mesophyll sponge in *Brassica oleracea*. Similarly, according to Seregin and Kozhevnikova (2006), Ni toxicity had detrimental effects on the ultrastructure of wheat leaves, including the thickness of mesophyll cells, vascular bundles, vessel diameter and the

width of leaf epidermal cells. Ni-induced decreases in plant growth parameters have also been reported in tomato (Mosa *et al.*, 2016; Badawy *et al.*, 2022), cotton (Khaliq *et al.*, 2016), *Brassica* (Ansari *et al.*, 2015) and *Glycine max* plants (Sirhindi *et al.*, 2015). Mosa *et al.* (2016) reported that Ni stress caused distortion in the cell nucleus and nucleolus of tomato root tips. In addition, Rehman *et al.* (2016) reported that Ni stress decreased maize growth and biomass, which might be due to a reduction in nutrient uptake and induced oxidative stress. Ni stress reduced root and shoot growth (Mir *et al.*, 2018) in soybean, which could be attributed to a reduction in mitotic activities and degraded root architecture, which restricted the uptake of water and minerals by the plant (Fashola *et al.*, 2016). Moreover, Jan *et al.* (2019) revealed that Ni reduced growth parameters and dry biomass in *Oryza sativa*, similar to the results for Glycine max (Yasin *et al.*, 2019).

Roots are the main plant organ that comes into direct contact with 50% of the Ni accumulated in roots, with over 80% found in the vacuoles and 20% in the cortex (Sreekanth *et al.*, 2013; Jan *et al.*, 2019). Amjad *et al.* (2020) revealed that Ni concentration was higher in the roots than in the leaf tissues in maize hybrids (Pioneer-32F10 and Syngenta-8441). Ni-polluted soils result in reduced yield and poor seed quality in soybean legumes (Mir *et al.*, 2018). The reduced yield in crop plants is mostly associated with the Ni concentration that accumulates inside the plant in various organs. The inhibition of root growth simultaneously affects the aboveground plant parts, leading to reduced plant biomass as well as reproductive potential. Mujeeb *et al.* (2019) reported that cowpea seedlings exposed to Ni toxicity exhibited reduced seedling growth, accompanied by a constant decline in the dry weight of roots, shoots, leaves and seedlings.

3.2. *Photosynthesis and mineral nutrition*

Elevated Ni levels inhibit photosynthesis through disruption of chloroplast structure, blockage of chlorophyll (Chl) biogenesis and electron transport reactions, reduced activities of enzymes taking part in the Calvin cycle, etc. (Bhalerao *et al.*, 2015; Khaliq *et al.*, 2016; Mujeeb *et al.*, 2019). In addition, a deficiency of CO_2 due to stomatal closure is a key factor contributing to reduced photosynthetic activity. A reduced number and size of chloroplasts and the disorganisation of ultrastructure were reported under Ni stress in *Spirodela polyrhiza* and *Lemna minor*

(Appenroth *et al.*, 2010). Ni stress led to declined Chl contents in *Eruca sativa* leaves through both Mg and Fe deficiencies (Kamran *et al.*, 2016). Since the primary effect of Ni toxicity is the displacement of essential mineral ions (such as Fe and Mg), Ni stress can alter the structure of the Chl molecule and the activity of ribulose-1,5-bisphosphate carboxylase oxygenase (RuBisCo) (Soliman *et al.*, 2019; Nabi *et al.*, 2020). Ni interferes with the photosynthetic electron transport chain (ETC) (Drążkiewicz and Baszyński *et al.*, 2010) and the availability of cytochromes *b6f* and *b559* in *Zea mays* leaves (Koch and Schneider, 2016). In a study, higher Ni levels in soil decreased the amount of Chls and carotenoids in *Triticum aestivum* L. Bhakar-02 due to the disruption of Chl synthesis and absorption of Mg (Shafeeq *et al.*, 2012). Beri and Sharma (2016) reported that the leaves of Ni-treated lentil plants had a reduced amount of Chl and declined Chl a / Chl b ratio. Inhibition of photosynthesis also resulted from reductions in the number and size of stomata and limited plant CO_2 uptake in Ni-treated leaves. Moreover, some *in vivo* studies on photosynthetic protein complexes demonstrated that Ni primarily inactivates photosystem I (PS I) (Rokka *et al.*, 2005), whereas *in vitro* studies indicated photosystem II (PS II) as its primary target (Tripathy *et al.*, 1981). Sheoran *et al.*, (1990) noted photosynthetic inhibition in pigeon pea leaves treated with 1 mM $NiCl_2$ solution; they suggested that Ni stress might have targeted the key enzymes of photosynthesis, such as RuBisCo (Alam *et al.*, 2007). In some cases, the inhibition of Calvin cycle reactions results in the aggregation of ATP and NADPH, which then induces a pH gradient through the thylakoid membranes, thereby blocking PS II activity (Fisher *et al.*, 2019). This, in turn, further triggers stomatal closure, limiting photosynthesis through blockage of CO_2 uptake (Sharma *et al.*, 2019). Excessive production of ATP and NADPH can result in overall photosynthetic inhibition due to inefficient biochemical pathway reactions (Shahzad *et al.*, 2018).

Ni competes with other nutrients for uptake and translocation, and it can reduce the concentration of many nutrients up to deficiency levels (Shahzad *et al.*, 2018). Therefore, Ni toxicity reduces the absorption of essential minerals in plants through direct or indirect means, which is also responsible for reduced photosynthesis. The direct effect is caused by high Ni affinity for functional -SH groups of enzymes, resulting in protein conformational changes and inactivation as well as metabolic disorders. Conversely, the indirect effect of Ni on enzyme activity results from ion-induced imbalances caused by competitive inhibition of macro- and micronutrient absorption and transport (Mir *et al.*, 2018; Hassan *et al.*, 2019). Ni interferes with essential mineral (micro or macro) uptake, either by affecting the root cell wall permeability, thus altering the activity of nutrient transporters, or by competing for common binding sites due to the nearly identical ionic radii of Ni^{2+} with Mg^{2+}, Fe^{2+} and Zn^{2+} (Mitra, 2018; Rehman *et al.*, 2020). Ni toxicity decreased the concentrations of Cu, Mn, Zn and Fe in both shoots and roots of the maize plant (Rehman *et al.*, 2016); it also decreased macronutrient (N, P, K, Ca, Mg and S) contents in wheat (Matraszek *et al.*, 2016) due to low-energy status, which decreases active uptake and nutrient transport. In conclusion, elevated Ni concentrations disrupt the uptake and availability of nutrients, resulting in a severe decline in plant growth and developmental processes.

3.3. *Nitrogen fixation and ureide biosynthesis*

Legumes are able to form a symbiotic relationship with rhizobia to fix substantial amounts of atmospheric nitrogen (N_2) and convert it into ammonia (NH_3) in their root nodules. It is believed that Ni plays a crucial role in biological nitrogen fixation (BNF) in legumes (Vijayakumar, 2018; Zheng *et al.*, 2019), i.e., nodulation and N_2 fixation, as well as ureides (ALN and ALA) transport to other plant organs (Freitas *et al.*, 2018). Ni acts as a cofactor for urease and catalyses the conversion of urea into ammonium ions (NH_4^+), which are further used by plants as a source of N (Barcelos *et al.*, 2018). In addition, Ni is the functional part of hydrogenase (NiFe), a crucial enzyme involved in the recycling of H_2, a product of N_2 reduction (Alfano and Cavazza, 2020). The presence of high concentrations of Ni affects the infection process by reducing root hair formation, thereby limiting the number of active nodules in legumes (de Macedo *et al.*, 2016). In addition, the functional efficiency of nodules (ARA) declines under Ni toxicity, which could be directly correlated to reduced levels of LHb protein. Poor root nodulation in soybean plants caused direct toxic effects on rhizobia and LHb synthesis, thereby interfering with the N_2-fixing system under Ni stress (de Macedo *et al.*, 2016). The inhibition of nodulation and nitrogenase activity under Ni toxicity has also been reported in *Frankia-Alnus* (Wheeler *et al.*, 2001) and *Lens culinaris – Rhizobium leguminosarum* symbiosis (Saad *et al.*, 2016). Haddad *et al.* (2015) reported a significant decline in nodulation, N_2 fixation and N uptake under HM stresses (Ni, Co, Cr, Cd, Cu and Pb) on *Vicia faba*, *Trifolium alexandrinum* and *Glycine max*. This could be attributed to the reduction in O_2 flux into the nodule under stress and impaired carbon © metabolism, which could limit bacteroid respiration (González *et al.*, 2001). This fact supports the hypothesis that decreased atmospheric N_2 fixation is due to the low symbiotic activity of microbes at high Ni concentrations in soil.

Ureides (ALN and ALA) have been known as important N-rich compounds, playing a crucial role in the assimilation, metabolism, transport and storage of N in several plant species, such as *Vigna unguiculata* and *V. radiata* (Todd *et al.*, 2006; Freitas *et al.*, 2018). Alamillo *et al.* (2010) reported that the induction of ureides (ALN and ALA) in nodules mitigated the negative impacts of drought stress and conferred resistance to *Phaseolus vulgaris* plants. To elucidate this, researchers proposed that ureide accumulation might play a regulatory role and result in N-feedback inhibition of the nitrogenase complex (King and Purcell, 2005). Ultimately, ureides get translocated to the leaves as urea, which is an important source of NH_3, and the efficiency of urea decreases after its hydrolysis with the enzyme urease, thereby reducing NH_3 emissions under stress, leading to large N losses (Sanz-Cobena *et al.*, 2008). Legumes require N in large quantities, and its limited availability negatively effects their growth (Takagi *et al.*, 2018). Some studies reported that Ni application augmented urease activity, preventing urea accumulation in the leaflets of cucumber and soybean plants (Khoshgoftarmanesh *et al.*, 2014; Ojeda-Barrios *et al.*, 2016), while others observed a decline in urease activity in soybean leaves (de Queiroz Barcelos *et al.*, 2017). Moreover, Garg and Saroy (2019) observed significantly decreased total ureides (TUs) in nodules as well as urea in pigeon pea genotypes, along with reduced activity of the enzyme urease under Ni stress.

3.4. *Oxidative stress*

Ni has a direct impact on plant metabolism and can induce the generation of reactive oxygen species (ROS), which can lead to oxidative damage to the plants (Bhalerao *et al.*, 2015; Abd-Allah *et al.*, 2019), for example in *Hyoscyamus niger* (Nasibi *et al.*, 2013), cotton (Khaliq *et al.*, 2016), *Glycine max* (Sirhindi *et al.*, 2016; Mir *et al.* 2018), mustard (Abd-Allah *et al.*, 2019) and tomato (Jahan *et al.*, 2020). ROS, including superoxide anion radical ($O_2^{\bullet-}$) and hydrogen peroxide (H_2O_2), are continuously produced in plant tissues, such as mitochondria, peroxisomes and chloroplast

(Tripathy and Mohanty, 2017; Snezhkina *et al.*, 2019). Increased ROS has damaging effects on cell membrane stability, proteins, respiration, lipids, photosynthesis, DNA through lipid peroxidation, etc. (Kashyap and Garg, 2018; García-Caparrós *et al.*, 2020). Ni is not a redox-active metal and does not seem to be an effective catalyst due to its relatively high oxidation/reduction potential. It cannot directly produce ROS but can produce $OH^{\bullet}$ through a Fenton/Haber–Weiss reaction (Amari *et al.*, 2020). Moreover, it has been suggested that Ni stress enhances the generation of ROS, such as H_2O_2 and $O_2^{\bullet -}$, that are responsible for the production of LPO (Janků *et al.*, 2019; Amjad *et al.*, 2020). H_2O_2 is an important ROS that increases under high concentrations of Ni in leaves and roots of *Triticum aestivum*, as reported by Hao *et al.* (2006) and Gajewska and Sklodowska (2007), and in *Glycine max* (Sirhindi *et al.*, 2016; Mir *et al.*, 2018; Einhardt *et al.*, 2021). Pandey and Rajeev (2010) observed an increase in MDA under Ni stress in eggplant and confirmed the adverse impact of Ni on lipid composition and the H-ATPase activity of the plasma membrane. Ni stress (0.5–1.5 mM) significantly enhanced MDA contents in pigeon pea plants, which is the main indicator of ROS damage, leading to destabilisation of the membrane integrity, functionality and ion balance in the cytoplasm (Shahzad *et al.*, 2018).

4. Adaptive tolerance mechanisms

Under Ni stress, the level of ROS rises inside the plant organs, which damages the cellular components. Therefore, in order to ameliorate the damaging effects triggered by ROS, plants need to be harmonically activated (García-Caparrós *et al.*, 2020; Saroy and Garg, 2021). ROS effects are alleviated with the help of the plant's inherent antioxidative enzymatic as well as non-enzymatic mechanisms (Table 1). The different mechanisms involved in the scavenging of ROS work in a coordinated manner (Biswas *et al.*, 2020).

4.1. *Antioxidants machinery*

ROS production in plant cells is regulated by a complex antioxidant mechanism that includes both enzymatic and non-enzymatic antioxidants, which are efficient enough to play a vital role in the detoxification of excess ROS (Sirhindi *et al.*, 2016; Jan *et al.*, 2019; Nabi *et al.*, 2020).

Table 1. Different antioxidants and their catalysed reactions.

Enzymatic antioxidants	Reactions catalysed/Functions
Superoxide dismutase (SOD)	$2O_2^{\bullet-} + 2H^+ \rightarrow H_2O_2 + O_2$
Catalase (CAT)	$H_2O_2 \rightarrow H_2O + {}^1/_2\, O_2$
Guaiacol peroxidase (GPX)	$2GSH + H_2O_2 \rightarrow GSSG + 2H_2O$
Antioxidants of ascorbate-glutathione (AsA–GSH) cycle	
Ascorbate peroxidase (APOX)	$H_2O_2 + 2AsA \rightarrow H_2O + 2MDHA$
Ascorbate (AsA)	Detoxifies H_2O_2 via action of APX
Monodehydroascorbate reductase (MDHAR)	$MDHA + NADPH \rightarrow AsA + DHA + NADP^+$
Dehydroascorbate reductase (DHAR)	$DHA + 2GSH \rightarrow AsA + GSSG$
Glutathione (GSH)	Helps in the detoxification of DHAR
Glutathione reductase (GR)	$GSSG + 2NADPH \rightarrow 2GSH + 2NADP^+$
Non-protein thiols (NP-SH), phytochelatins (PCs)	Scavenge ROS and sequestration of HMs in the vacuoles

Various investigators confirmed that the enzymatic activities increased in order to combat oxidative stresses, which have been associated with high tolerance levels of plants to HMs (Sirhindi *et al.*, 2016; Jan *et al.*, 2019; Zaid *et al.*, 2019). Under stressful situations, the equilibrium between ROS production and antioxidant machinery is severely disturbed, which inhibits many physiological and biochemical functions essential to the adequate functioning of plants (Al Mahmud *et al.*, 2019). Enzymatic antioxidants include superoxide dismutase (SOD), catalase (CAT), guaiacol peroxidase (GPX) and the enzymes of the ascorbate–glutathione (AsA–GSH) cycle—ascorbate peroxidase (APOX), monodehydroascorbate reductase (MDHAR), dehydroascorbate reductase (DHAR) and glutathione reductase (GR) (Gajewska and Skłodowska, 2007; Bhalerao *et al.*, 2015; Wu *et al.*, 2017). SOD is a ubiquitous metalloenzyme and catalyses the conversion of $O_2^{\bullet-}$ to H_2O_2 and O_2 (Mishra and Sharma, 2019). SOD is associated with metal cofactors in the form of Cu/Zn-SOD, Fe-SOD and Mn-SOD (Garg and Manchanda, 2009). They are distributed in plant cells, with the localisation of Cu/Zn-SOD in chloroplasts and cytosol, Fe-SOD in plastids and Mn-SOD in peroxisomes and mitochondria.

Ni-SOD has also been reported in *Streptomyces* (Kim *et al.*, 1996). However, reactions catalysed by SOD result in the enhanced production of H_2O_2, which, at higher concentrations, can further aggravate oxidative damage. To combat this situation, rapid action of the H_2O_2 scavenging system is needed (Soliman *et al.*, 2019). The increase in H_2O_2 levels stimulates the augmented production of detoxifying enzymes, such as CAT and GPX. CAT is a heme-containing, tetrameric enzyme whose function is to degrade H_2O_2 without involving any electron donor and is mainly found in the cytosol, peroxisomes, mitochondria, glyoxisomes and root nodules (Anjum *et al.*, 2017). CAT dismutates H_2O_2, resulting in the formation of water and O_2 (Garg and Manchanda, 2009). GPX is also a heme-containing enzyme and is involved in the elimination of excess H_2O_2. APOX is a central constituent in the AsA–GSH cycle and performs the same function of scavenging H_2O_2 as CAT (Bielen *et al.* 2013). The most abundant antioxidant molecule found in different ROS generating sites in plants is AsA. In chloroplasts, the AsA is present in its reduced form and protects photosynthetic machinery by eliminating H_2O_2, $O_2^{\bullet-}$ and •OH and quenching $^1O_2$ (Akram *et al.*, 2017). AsA is maintained in a reduced form via MDHAR and DHAR. MDHAR is a flavin adenine dinucleotide (FAD) enzyme that regenerates AsA along with DHA by consuming NADPH from the short-lived molecule MDHA (Sharma *et al.*, 2012). Different isoforms of MDHAR have been found to be confined to the cytosol, mitochondria, peroxisomes, chloroplasts and glyoxisomes (Sharma *et al.*, 2012). DHAR is a thiol enzyme and uses reduced GSH as an electron donor to catalyse the reduction of DHA into AsA. The GSH level in cells is maintained by the activity of GR, which occurs in chloroplasts, cytosol and mitochondria, as well as in plastids and peroxisomes. It mediates the NADPH-dependent catalytic reduction of GSSG into GSH. Besides this, GR is a flavin-containing pyridine nucleotide disulphide oxido-reductase that catalyses the conversion of GSSG to GSH and the oxidation of NADPH (Harshavardhan *et al.*, 2017). It also maintains the reduced status of GSH and AsA pool, which in turn preserves the cellular redox state against reactive oxygen metabolites. Studies on the activation of defence mechanisms under Ni stress have produced diverse results, with some observing an enhancement in SOD and CAT, along with APOX activities and peroxidase in soybean (Sirhindi *et al.*, 2016; Barcelos *et al.*, 2018), while others recorded a slight upsurge in SOD activity with a decrease in CAT activity as well as in the enzymes of the

AsA–GSH cycle in soybean under Ni toxicity (Mir *et al.*, 2018). Moreover, a decrease in SOD and CAT activities has also been observed under Ni stress in *Hydrocharis dubia* leaves (Zhao *et al.*, 2008). In a study, Gajewska and Sklodowska (2007) showed that Ni stress caused significant constriction of antioxidant activities, including those of SOD and CAT, whereas it escalated the activities of GPX and APOX in wheat plants. Similarly, under Ni stress, increased activities of APOX and GPX in *Oryza sativa* (Maheshwari and Dubey 2009) and of CAT and SOD in *Brassica* plants (Sharma *et al.*, 2008) and *Solanum nigrum* (Soares *et al.*, 2016) were detected. At the transcript level, increased expressions of APX, CAT, POD and GR genes were reported in rice and soybean plants grown in Ni-challenged environments (Sirhindi *et al.*, 2016; Rizwan *et al.*, 2018). In addition, the activities of these antioxidant enzymes varied among the genotypes of a plant species (Garg and Bhandari, 2016; Garg and Kashyap, 2019; Saroy and Garg, 2021), with tolerant genotypes displaying stronger antioxidant defence mechanisms under stress than the sensitive ones. However, the activities of these antioxidants can differ with the duration, plant species and plant organs under stress conditions (Anjum *et al.*, 2017).

4.2. *Formation of metal complex by thiol derivatives*

Different thiol derivatives, including GSH and phytochelatins (PCs), contain -SH groups, which are essential for HM chelation in plants (Ulrich and Jakob, 2019; Amari *et al.*, 2020). The most common intracellular mechanism that can be carried out by thiol compounds for the maintenance of low concentrations and detoxification of free metals in plant cytoplasm is chelation (Jan *et al.*, 2019; Mehrandish *et al.*, 2019; Piotrowska-Niczyporuk *et al.*, 2020). GSH is a water-soluble, tripeptide γ-glutamylcysteinyl-Gly and non-protein thiol which is considered the most essential intracellular defence molecule combating ROS-induced oxidative injury in plants (Hasanuzzaman *et al.*, 2017). GSH is responsible for the scavenging of $OH^{\cdot}$ and $^1O_2$ and safeguarding the enzymes of thiol groups. It exists in reduced form (GSH) in plant tissue in almost all cellular partitions, such as mitochondria, cytosol, vacuole, endoplasmic reticulum (ER), chloroplast, peroxisomes and also apoplast (Nahar *et al.*, 2015). GSH performs a vital function in the AsA–GSH cycle, in which it regenerates AsA and the oxidised form glutathione disulphide (GSSG), which is then reprocessed back to GSH through the activity of NADPH-dependent GR. The equilibrium between two forms of GSH is necessary to retain the cellular redox condition (Ahmad *et al.*, 2009). Wang *et al.* (2009) reported high levels of Ni in *Zea mays* L., which enhanced the production of GSH and the GSH/GSSG ratio. Freeman *et al.* (2005) discovered that in *Thlaspi*, a high GSH amount was attained by the initiation of GR, a main enzyme involved in the synthesis of GSH. Increased levels of GSH and NPTs, along with improved GSH/GSSG ratios, were observed in pigeon pea under Cd and As toxicities, which negated the harmful effects of ROS generation (Bisht and Garg, 2022; Bhalla *et al.*, 2023). Besides, under HM stresses, GSH performs a vital function in safeguarding the plants from toxicity through the chelation of HMs for transport and sequestration inside the vacuole or by acting as a precursor for PC biosynthesis, thus maintaining cellular redox status (Hasanuzzaman *et al.*, 2017).

PCs are low-molecular-weight, Cys-rich small polypeptides that share a common structure of (γ-Glu-Cys)nGly ($n = 2$–11) (Hasan *et al.*, 2017). The synthesis of PCs plays a vital role in providing tolerance to HMs in plants (Emamverdian *et al.*, 2015). PCs form complexes with toxic HMs (Cd, Ni, Zn, Pb, Cu, etc.) by attaching to sulphydryl and carboxyl groups in the cytosol and then transporting them into the vacuole for sequestration through transporters (Xiao *et al.*, 2020). Vacuolar sequestration is necessary for maintaining HM equilibrium in plants, which is mediated through ATP-dependent vacuolar pumps (V-ATPase) and tonoplast transporters (Sharma *et al.*, 2016). The production of PCs increased in *Triticum aestivum* with increasing concentrations of Zn and Cd (Sun *et al.*, 2005). Carrier *et al.* (2003) reported the storage of approximately 80% of leaf Cd in the cytosol and vacuoles of *Brassica napus* under Cd toxicity. Various PC synthase (PCS) genes responsible for PC production have been reported in *Arabidopsis thaliana* (*AtPCS1*), *Triticum aestivum* (*TaPCS1*), *Brassica juncea* (*BjPCS1*), *Oryza sativa* (*OsPCS1*) and *Cajanus cajan* (*CcPCS*) (Lee *et al.*, 2003; Jan and Parray, 2016; Bisht *et al.*, 2022) under various stressed conditions. Moreover, enhanced cadmium and arsenite tolerance was noticed in genetically modified tobacco plants because of the overexpression of the *NtPCS1* gene, which played important roles in imparting metal(loid) tolerance, thus improving growth and development in tobacco plants (Lee and Hwang, 2015). Additionally, PCs have an important role in metal ion homeostasis, including the maintenance of the GSH/GSSG ratio and sulphur (S) metabolism in plant cells (Hasan *et al.*, 2015). Compared to free metal ions, the PC–metal complexes are highly consistent and less toxic (Franchi *et al.*, 2014). PCs have been reported to display HM tolerance in *Ceratophyllum demersum* under Pb toxicity, where stress leads to an increase in GSH and Cys contents, with an upsurge in PC levels, which consequently reduces ROS production (Mishra *et al.*, 2006). Similarly, *Avicennia germinans* also displayed tolerance to Cu and Cd toxicities by increasing the transcriptional level and expression of *AvMt2* through sequestration in the vacuoles (Gonzalez-Mendoza *et al.*, 2007). Elevated Ni levels induced PC synthesis from GSH using PCS in *Thlaspi japonicum* (Mizuno *et al.*, 2003) and *Nicotiana tabacum* (Nakazawa *et al.*, 2001). In addition, a vacuolar metal-ion transporter protein (TgMTP1) was reported to play an essential role in the sequestration of Ni into the leaf vacuole of *Thlaspi* sp. and *Alyssum murale* (Persans *et al.*, 2001; Agrawal *et al.*, 2013), which protected the plants from the deleterious effects of HMs. Moreover, Ni stress triggered the expression of 1-aminocyclopropane-1-carboxylic acid (*ACCD*) and *GR* genes in the roots and needles of *Picea glauca*, which imparted metal tolerance to the plants (Boyd and Nkongolo, 2021). Transcriptomic data of Ni-stressed *Medicago sativa* revealed a significant increment in the mRNA levels of *GST*, *PCs* and *Prx1C* genes responsible for metal detoxification (Helaoui *et al.*, 2020). ABC transporter and metallothionein transcripts displayed differential expressions in Ni-stressed *Festuca arundinacea*, which corresponded to Ni uptake and metal chelation, respectively (Shabani *et al.*, 2016). In *Solanum nigrum*, Ni toxicity led to significant upregulation in the transcripts that encode the metallothioneins *MT2a*, *MT2c*, *MT2d* and *MT3*, indicating the involvement of these genes in metal homeostasis and detoxification (Ferraz *et al.*, 2012). Nkongolo *et al.* (2018a) reported upregulation of genes involved in the transport of Ni in the cytosol of resistant genotypes, while transcripts encoding translation in the ribosome were more abundant in the susceptible genotypes of *Acer rubrum*. Transcriptome analysis performed by Theriault *et al.* (2016) identified five candidate genes, i.e., a thioredoxin-family protein, two Nramp transporters, a putative transmembrane protein and a glutathione-S-transferase, in Ni-stressed *Betula papyrifera* involved in metal resistance. Furthermore, a qPCR study revealed that Ni toxicity led to the upregulation of miR319 and miR159 in leaves and roots of castor beans, along with a significant change in the expression levels of miR395 and its putative genes in leaves, which are responsible for sulphur assimilation and metal stress tolerance (Çelik and Akdaş, 2019).

4.3. *Osmolytes accumulation*

In addition, the synthesis of osmolytes, such as total soluble sugars (TSSs), free amino acids (FAAs) and proline (Pro), is a common defensive response against Ni toxicity (Rizwan *et al.*, 2018; Valivand and Amooaghaie, 2020). These osmotically active compounds stabilise proteins and membranes as well as reduce osmotic potential, with the aim of preventing cellular dehydration (Yoon *et al.*, 2017). In a recent study, the contents of soluble sugars significantly increased in Zucchini leaf and root under Ni stress, where the soluble sugars acted as ROS scavengers and stabilised the cellular membranes (Valivand and Amooaghaie, 2020). Sugars, such as disaccharides, sugar alcohols, glucose and fructans, can scavenge cell membranes as well as protoplast, ·OH radicals and can restrict LPO under adverse environmental conditions (Keunen *et al.*, 2013; Guo *et al.*, 2015). In addition, soluble sugars are major photosynthetic products in higher plants and play important roles in carbohydrate metabolism and storage (Nie *et al.*, 2015). Moreover, the accumulation of proline in plant cells due to HM stress has been demonstrated as a potent indicator of stress tolerance (Ahmad *et al.*, 2015; Rizwan *et al.*, 2017). It protects the cells from oxidative damage by scavenging ROS and inhibiting LPO, thereby ameliorating the oxidative burden (Zaid *et al.*, 2019). In plants, Pro is synthesised from glutamate via successive reductions with two enzymes, pyrroline-5-carboxylate synthase (P5CS) and pyrroline-5-carboxylate reductase (P5CR) (Forlani *et al.*, 2019), and from ornithine (Orn) by Orn-δ-aminotransferase (OAT) (Trovato *et al.*, 2019). Pro degradation is the reverse process and is catalysed back to glutamate by Pro dehydrogenase (ProDH) (Kishor *et al.*, 2005; Siddique *et al.*, 2018). A high level of Ni triggered the production of proline in *Brassica oleracea* (Pandey and Sharma, 2002), *Pisum sativum* (Gajewska and Skłodowska, 2005), *Triticum aestivum* (Gajewska and Skłodowska, 2009) and rice (Rizwan *et al.*, 2017), which have been reported to ameliorate stress-mediated photoinhibition while also leading to cellular homeostasis. Moreover, under Cd and Zn toxicity, enhanced Pro accumulation along with increased P5CS and P5CR have been reported in two pigeon pea genotypes (Garg and Singh, 2018).

5. Conclusion

Ni is an essential nutrient for the proper growth of plants; however, its elevated concentrations in soil have devastating effects on various physiological and metabolic processes, such as photosynthesis, transpiration, root growth, enzymatic activities and mineral nutrition. Moreover, Ni toxicity alters the permeability of membranes, plant water status and the osmotic balance of plants by inducing the generation of ROS, which ultimately leads to inhibition of growth and productivity. Although plants adopt certain inherent tolerance strategies, such as upregulation of antioxidant defence machinery, thiol derivatives, metal chelators, etc., in order to ameliorate Ni toxicity, the extent of activation of these mechanisms varies from species to species and thus needs further investigations at proteomic and transcript levels.

References

Abd_Allah E.F., Hashem A., Alam P., Ahmad P. (2019). Silicon alleviates nickel-induced oxidative stress by regulating antioxidant defense and glyoxalase systems in mustard plants. *J. Plant Growth Regul.* 38:1260–1273.

Afzal F., Khan T., Khan A., Khan S., Raza H., Ihsan A., Ahanger M.A., Kazi A.G. (2015). Nutrient deficiencies under stress in legumes: An overview. In: Ahmad P (ed.) *Legumes Under Environmental Stress: Yield, Improvement and Adaptations*, John Wiley & Sons, Ltd., US, pp. 53–65.

Agrawal B., Czymmek K.J., Sparks D.L., Bais H.P. (2013). Transient influx of nickel in root mitochondria modulates organic acid and reactive oxygen species production in nickel hyperaccumulator *Alyssum murale*. *J. Biol. Chem.* 288:7351–7362.

Ahmad A., Hadi F., Ali N. (2015). Effective phytoextraction of cadmium (Cd) with increasing concentration of total phenolics and free proline in *Cannabis sativa* (L.) plant under various treatments of fertilizers, plant growth regulators and sodium salt. *Int. J. Phytorem.* 17:56–65.

Ahmad M.S., Ashraf M. (2011). Essential roles and hazardous effects of nickel in plants. *Rev. Environ. Contam. Toxicol.* 214:125–167.

Ahmad M.S.A., Riffat A., Hussain M., Hameed M., Alvi A.K. (2023). Toxicity and tolerance of nickel in sunflower (*Helianthus annuus* L.). *Environ. Sci. Pollut. Res.* 30:50346–50363.

Ahmad P., Jaleel C.A., Azooz M.M., Nabi G. (2009). Generation of ROS and non-enzymatic antioxidants during abiotic stress in plants. *Bot. Res. Int.* 2:11–20.

Akram N.A., Shafiq F., Ashraf M. (2017). Ascorbic acid-a potential oxidant scavenger and its role in plant development and abiotic stress tolerance. *Front Plant Sci.* 8:1–17.

Alam M.M., Hayat S., Ali B., Ahmad A. (2007). Effect of 28-homobrassinolide treatment on nickel toxicity in *Brassica juncea. Photosynthetica* 45:139–142.

Alamillo J.M., Díaz-Leal J.L., Sánchez-Moran M.V., Pineda M. (2010). Molecular analysis of ureide accumulation under drought stress in *Phaseolus vulgaris* L. *Plant Cell Environ.* 33:1828–1837.

Alfano M., Cavazza C. (2020). Structure, function and biosynthesis of nickel-dependent enzymes. *Protein Sci.* 29:1071–1089.

Amari T., Souid A., Ghabriche R., Porrini M., Lutts S., Sacchi G.A., Abdelly C., Ghnaya T. (2020). Why does the halophyte Mesembryanthemum crystallinum better tolerate Ni toxicity than *Brassica juncea*: Implication of antioxidant defense systems. *Plants* 9:1–17.

Amjad M., Raza H., Murtaza B., Abbas G., Imran M., Shahid M., Naeem M.A., Zakir A., Iqbal M.M. (2020). Nickel toxicity induced changes in nutrient dynamics and antioxidant profiling in two maize (*Zea mays* L.) hybrids. *Plants* 9:1–15.

Anjum S.A., Ashraf U., Tanveer M., Khan I., Hussain S., Shahzad B., Zohaib A., Abbas F., Saleem M.F., Ali I., Wang L.C. (2017). Drought induced changes in growth, osmolyte accumulation and antioxidant metabolism of three maize hybrids. *Front. Plant Sci.* 8:1–12.

Ansari M.K.A., Ahmad A., Umar S., Zia M.H., Iqbal M., Owens G. (2015). Genotypic variation in phytoremediation potential of Indian mustard exposed to nickel stress: A hydroponic study. *Int J. Phytorem.* 17:135–144.

Appenroth K.J., Krech K., Keresztes A., Fischer W., Koloczek H. (2010). Effects of nickel on the chloroplasts of the duckweeds *Spirodela polyrhiza* and *Lemna minor* and their possible use in biomonitoring and phytoremediation. *Chemosphere* 78:216–223.

Aqeel M., Khalid N., Tufail A., Ahmad R.Z., Akhter M.S., Luqman M., Javed M.T., Irshad M.K., Alamri S., Hashem M., Noman A. (2021). Elucidating the distinct interactive impact of cadmium and nickel on growth, photosynthesis, metal-homeostasis, and yield responses of mung bean (*Vigna radiata* L.) varieties. *Environ. Sci. Pollut. Res.* 28:27376–27390.

ATSDR (2022). https://www.atsdr.cdc.gov/spl/index.html#2022spl.

Badawy I.H., Hmed A.A., Sofy M.R., Al-Mokadem A.Z. (2022) Alleviation of cadmium and nickel toxicity and phyto-stimulation of tomato plant L. by endophytic *Micrococcus luteus* and *Enterobacter cloacae. Plants* 11:2018.

Barcelos J.P.Q., Reis H.P.G., Godoy C.V., Gratão P.L., Furlani J.E., Putti F.F., Reis A.R. (2018). Impact of foliar nickel application on urease activity, antioxidant metabolism and control of powdery mildew (*Microsphaera diffusa*) in soybean plants. *Plant Pathol.* 67:1502–1513.

Beri A., Sharma R. (2016). Nickel toxicity to photosynthetic attributes in the leaves of lentil (*Lens culnaris* Medic. Masar). *Int. J. Appl. Res.* 2:239–242.

Bhalerao S.A., Sharma A.S., Poojari A.C. (2015). Toxicity of nickel in plants. *Int. J. Pure Appl. Biosci.* 3:345–355.

Bhalla S., Bisht A., Garg N. (2023). Silicon and arbuscular mycorrhizal species complement in improving soil characteristics, sulfur metabolism and antioxidant defense responses in arsenic stressed *Cajanus cajan* (L.) Millsp. *Archiv. Agron. Soil Sci.* 19:1–9.

Bielen A., Remans T., Vangronsveld J., Cuypers A. (2013). The influence of metal stress on the availability and redox state of ascorbate, and possible interference with its cellular functions. *Int. J. Mol. Sci.* 14:6382–6413.

Bisht A., Bhalla S., Kumar A., Kaur J., Garg N. (2022). Arbuscular mycorrhizae impart Cd tolerance in *Cajanus Cajan* (L.) Millsp. by upregulating the expression of metallothionein (*CcMT1*) and phytochelatin synthase (*CcPCS1*) genes. *J. Plant Grow. Regul.* 9:1–20.

Bisht A., Garg N. (2022). AMF modulated rhizospheric microbial enzyme activities and their impact on sulphur assimilation along with thiol metabolism in pigeon pea under Cd stress. *Rhizosphere* 21:100478.

Biswas K., Adhikari S., Tarafdar A., Kumar R., Saha S., Ghosh P. (2020). Reactive oxygen species and antioxidant defence systems in plants: Role and crosstalk under biotic stress. In: Roychowdhury R., Choudhury S., Hasanuzzaman M., Srivastava S. (eds.) *Sustainable Agriculture in the Era of Climate Change*. Springer, Cham, pp. 265–292.

Boyd M., Nkongolo K.K. (2021). Nickel induces changes in expression of genes encoding 1-aminocyclopropane-1-carboxylic acid deaminase and glutathione reductase in *Picea glauca. Chem Ecol.* 37:589–602.

Callahan D.L., Kolev S.D., O'Hair R.A., Salt D.E., Baker A.J. (2007). Relationships of nicotianamine and other amino acids with nickel, zinc and iron in *Thlaspi* hyperaccumulators. *New Phytol.* 176:836–848.

Carrier P., Baryla A., Havaux M. (2003). Cadmium distribution and microlocalization in oilseed rape (*Brassica napus*) after long-term growth on cadmium-contaminated soil. *Planta* 216:939–950.

Çelik Ö., Akdaş E.Y. (2019). Tissue-specific transcriptional regulation of seven heavy metal stress-responsive miRNAs and their putative targets in nickel indicator castor bean (*R. communis* L.) plants. *Ecotoxicol. Environl. Saf.* 170:682–690.

Dalir N., Khoshgoftarmanesh A.H. (2014). Symplastic and apoplastic uptake and root to shoot translocation of nickel in wheat as affected by exogenous amino acids. *J. Plant Physiol.* 171:531–536.

Dalir N., Khoshgoftarmanesh A.H. (2015). Root uptake and translocation of nickel in wheat as affected by histidine. *J. Plant Physiol.* 184:8–14.

de Macedo F.G., Bresolin J.D., Santos E.F., Furlan F., da Silva L., Wilson T., Polacco J.C. Lavres J. (2016). Nickel availability in soil as influenced by liming and its role in soybean nitrogen metabolism. *Front. Plant Sci.* 7:1–12.

de Queiroz Barcelos J.P., de Souza Osório C.R.W., Leal A.J.F., Alves C.Z., Santos E.F., Reis H.P.G., dos Reis A.R. (2017). Effects of foliar nickel (Ni) application on mineral nutrition status, urease activity and physiological quality of soybean seeds. *Aus J. Crop Sci.* 11:184–192.

Demidchik V., Maathuis F.J. (2007). Physiological roles of nonselective cation channels in plants: From salt stress to signalling and development. *New Phytol.* 175:387–404.

Deng T.H.B., Cloquet C., Tang Y.T., Sterckeman T., Echevarria G., Estrade N., Morel J.L., Qiu R.L. (2014). Nickel and zinc isotope fractionation in hyperaccumulating and nonaccumulating plants. *Environ. Sci. Technol.* 48:11926–11933.

Dhaliwal S.S., Naresh R.K., Mandal A., Singh R., Dhaliwal M.K. (2019). Dynamics and transformations of micronutrients in agricultural soils as influenced by organic matter build-up: A review. *Environ. Sustainab. Indicat.* 1:1–14.

Drążkiewicz M., Baszyński T. (2010). Interference of nickel with the photosynthetic apparatus of *Zea mays*. *Ecotoxicol. Environ. Saf.* 73:982–986.

Duan Y., Sangani C.B., Muddassir M., Soni K.V. (2020). Copper, chromium and nickel heavy metal effects on total sugar and protein content in *Glycine max*. *Res. Squar.* 1–20. (pre-print server).

Einhardt A.M., Ferreira S., Rodrigues F.Á. (2021). Biochemical and physiological responses of soybean [*Glycine max* (L.) Merrill] to nickel toxicity. *Bragantia* 80:1–10.

Emamverdian A., Ding Y., Mokhberdoran F., Xie Y. (2015). Heavy metal stress and some mechanisms of plant defense response. *Sci. World J.* 2015:1–19.

Fabiano C., Tezotto T., Favarin J.L., Polacco J.C., Mazzafera P. (2015). Essentiality of nickel in plants: A role in plant stresses. *Front. Plant Sci.* 6:1–4.

Fashola M.O., Ngole-Jeme V.M., Babalola O.O. (2016). Heavy metal pollution from gold mines: Environmental effects and bacterial strategies for resistance. *Int. J. Environ. Res Public Health* 13:1–20.

Ferraz P., Fidalgo F., Almeida A., Teixeira J. (2012). Phytostabilization of nickel by the zinc and cadmium hyperaccumulator *Solanum nigrum* L. Are metallothioneins involved? *Plant Physiol. Biochem.* 57:254–260.

Fisher N., Bricker T.M., Kramer D.M. (2019). Regulation of photosynthetic cyclic electron flow pathways by adenylate status in higher plant chloroplasts. *Biochim. Biophys. Acta Bioenerg.* 1860:1–8.

Forlani G., Trovato M., Funck D., Signorelli S. (2019) Regulation of proline accumulation and its molecular and physiological functions in stress defence. In: Hossain M., Kumar V., Burritt D., Fujita M., Mäkelä P. (eds.) *Osmoprotectant-Mediated Abiotic Stress Tolerance in Plants*. Springer, Cham, pp. 73–97.

Franchi N., Piccinni E., Ferro D., Basso G., Spolaore B., Santovito G., Ballarin L. (2014). Characterization and transcription studies of a phytochelatin synthase gene from the solitary tunicate *Ciona intestinalis* exposed to cadmium. *Aqua. Toxicol.* 152:47–56.

Freeman J.L., Garcia D., Kim D., Hopf A., Salt D.E. (2005). Constitutively elevated salicylic acid signals glutathione-mediated nickel tolerance in *Thlaspi* nickel hyperaccumulators. *Plant Physiol.* 137:1082–1091.

Freitas D.S., Rodak B.W., dos Reis A.R., de Barros Reis F., de Carvalho T.S., Schulze J., Carbone Carneiro M.A., Guilherme L.R.G. (2018). Hidden nickel deficiency? Nickel fertilization via soil improves nitrogen metabolism and grain yield in soybean genotypes. *Front. Plant Sci.* 9:614.

Gajewska E., Skłodowska M. (2005). Antioxidative responses and proline level in leaves and roots of pea plants subjected to nickel stress. *Acta Physiol. Plant* 27:329–340.

Gajewska E., Skłodowska M. (2007). Effect of nickel on ROS content and antioxidative enzyme activities in wheat leaves. *Biometals* 20:27–36.

Gajewska E., Skłodowska M. (2009). Nickel-induced changes in nitrogen metabolism in wheat shoots. *J. Plant Physiol.* 166:1034–1044.

García-Caparrós P., De Filippis L., Gul A., Hasanuzzaman M., Ozturk M., Altay V., Lao M.T. (2020). Oxidative stress and antioxidant metabolism under adverse environmental conditions: A review. *Bot Rev.* 87:1–46.

Garg N., Bhandari P. (2016). Interactive effects of silicon and arbuscular mycorrhiza in modulating ascorbate-glutathione cycle and antioxidant scavenging capacity in differentially salt-tolerant *Cicer arietinum* L. genotypes subjected to long-term salinity. *Protoplasma* 253:1325–1345.

Garg N., Kashyap L. (2019). Joint effects of Si and mycorrhiza on the antioxidant metabolism of two pigeonpea genotypes under As (III) and (V) stress. *Environ. Sci. Pollut. Res.* 26:7821–7839.

Garg N., Manchanda G. (2009). ROS generation in plants: Boon or bane? *Plant Biosyst.* 143:81–96.

Garg N., Saroy K. (2019). Interactive effects of polyamines and arbuscular mycorrhiza in modulating plant biomass, N_2 fixation, ureide, and trehalose metabolism in *Cajanus cajan* (L.) Millsp. genotypes under nickel stress. *Environ. Sci. Pollut. Res.* 27:3043–3064.

Garg N., Singh S. (2018). Arbuscular mycorrhiza Rhizophagus irregularis and silicon modulate growth, proline biosynthesis and yield in *Cajanus cajan* L. Millsp. (pigeonpea) genotypes under cadmium and zinc stress. *J. Plant Growth Regul.* 37:46–63.

González E., Gálvez L., Royuela M., Aparicio-Tejo P., Arrese-Igor C. (2001). Insights into the regulation of nitrogen fixation in pea nodules: Lessons from drought, abscisic acid and increased photoassimilate availability. *Agronomie* 21:607–613.

González-Guerrero M., Matthiadis A., Sáez Á., Long T.A. (2014). Fixating on metals: New insights into the role of metals in nodulation and symbiotic nitrogen fixation. *Front. Plant Sci.* 5:1–6.

Gonzalez-Mendoza D., Moreno A.Q., Zapata-Perez O. (2007). Coordinated responses of phytochelatin synthase and metallothionein genes in black mangrove, *Avicennia germinans*, exposed to cadmium and copper. *Aqua. Toxicol.* 83:306–314.

Guo S., Duan J.A., Qian D., Tang Y., Wu D., Su S., Zhao Y. (2015). Content variations of triterpenic acid, nucleoside, nucleobase, and sugar in jujube (*Ziziphus jujuba*) fruit during ripening. *Food Chem.* 167:468–474.

Gupta D.K., Chatterjee S., Datta S., Veer V., Walther C. (2014). Role of phosphate fertilizers in heavy metal uptake and detoxification of toxic metals. *Chemosphere* 108:134–144.

Haddad S.A., Tabatabai M.A., Abdel-Moneim A.M.A., Loynachan T.E. (2015). Inhibition of nodulation and nitrogen nutrition of leguminous crops by selected heavy metals. *Air Soil Water Res.* 8:1–7.

Hao C.Y., Zhang X.Y., Wang L.F., Dong Y.S., Shang X.W., Jia J.Z. (2006). Genetic diversity and core collection evaluations in common wheat germplasm from the Northwestern Spring Wheat Region in China. *Mol. Breed.* 17:69–77.

Harasim P., Filipek T. (2015). Nickel in the environment. *J. Elementol.* 20:525–534.

Harshavardhan V.T., Wu T.M., Hong C.Y. (2017). Glutathione reductase and abiotic stress tolerance in plants. In: Hossain M.A., Mostofa M.G., Diaz-Vivancos P., Burritt D.J., Fujita M., Tran L.P. (eds.) *Glutathione in Plant Growth, Development, and Stress Tolerance*. Springer, Cham, pp. 265–286.

Hasan M., Ahammed G.J., Yin L., Shi K., Xia X., Zhou Y., Zhou J. (2015). Melatonin mitigates cadmium phytotoxicity through modulation of phytochelatins biosynthesis, vacuolar sequestration, and antioxidant potential in *Solanum lycopersicum* L. *Front. Plant Sci.* 6:601.

Hasanuzzaman M., Nahar K., Hossain M.S., Anee T.I., Parvin K., Fujita M. (2017). Nitric oxide pretreatment enhances antioxidant defense and glyoxalase systems to confer PEG-induced oxidative stress in rapeseed. *J. Plant Interact.* 12:323–331.

Haseeb M., Iqbal S., Hafeez M.B., Saddiq M.S., Zahra N., Raza A., Siddiqui M.H. (2022). Phytoremediation of nickel by quinoa: Morphological and physiological response. *Plos One* 17:e0262309.

Hassan M.U., Chattha M.U., Khan I., Chattha M.B., Aamer M., Nawaz M., Ali A., Khan M.A.U., Khan T.A. (2019). Nickel toxicity in plants: Reasons, toxic effects, tolerance mechanisms, and remediation possibilities — A review. *Environ. Sci. Pollut. Res.* 26:12673–12688.

Helaoui S., Boughattas I., Hattab S., Mkhinini M., Alphonse V., Livet A., Bousserrhine N., Banni M. (2020). Physiological, biochemical and transcriptomic responses of *Medicago sativa* to nickel exposure. *Chemosphere* 249:126121.

Iyaka Y.A. (2011). Nickel in soils: A review of its distribution and impacts. *Sci. Res. Essays* 6:6774–6777.

Jahan M.S., Guo S., Baloch A.R., Sun J., Shu S., Wang Y., Ahammed G.J., Kabir K., Roy R. (2020) Melatonin alleviates nickel phytotoxicity by improving photosynthesis, secondary metabolism and oxidative stress tolerance in tomato seedlings. *Ecotoxicol. Environ. Saf.* 197:1–12.

Jan R., Khan M.A., Asaf S., Lee I.J., Kim K.M. (2019). Metal resistant endophytic bacteria reduces cadmium, nickel toxicity, and enhances expression of metal stress related genes with improved growth of *Oryza Sativa*, via regulating its antioxidant machinery and endogenous hormones. *Plants* 8:1–23.

Jan S., Parray J.A. (2016). Metal tolerance strategy in plants. In: *Approaches to Heavy Metal Tolerance in Plants*. Springer, Singapore, pp. 19–32.

Janků M., Luhová L., Petřivalský M. (2019). On the origin and fate of reactive oxygen species in plant cell compartments. Antioxidants 8:1–15.

Kamran M.A., Eqani S.A.M.A.S., Bibi S., Xu R.K., Monis M.F.H., Katsoyiannis A., Chaudhary H.J. (2016). Bioaccumulation of nickel by E sativa and role of plant growth promoting rhizobacteria (PGPRs) under nickel stress. *Ecotoxicol. Environ. Saf.* 126:256–263.

Kashyap L., Garg N. (2018). Arsenic toxicity in crop plants: Responses and remediation strategies. In: Hasanuzzaman M., Nahar K., Fujita M. (eds.) *Mechanisms of Arsenic Toxicity and Tolerance in Plants*. Springer, Singapore, pp. 129–169.

Kasprzak K.S. (1987). Nickel. *Adv. Mod. Environ. Toxicol.* 11:145–183.

Keunen E.L.S., Peshev D., Vangronsveld J., Van Den Ende W.I.M., Cuypers A.N.N. (2013) Plant sugars are crucial players in the oxidative challenge during abiotic stress: Extending the traditional concept. *Plant Cell Environ.* 36:1242–1255.

Khaliq A., Ali S., Hameed A., Farooq M.A., Farid M., Shakoor M.B., Mahmood K., Ishaque W., Rizwan M. (2016). Silicon alleviates nickel toxicity in cotton seedlings through enhancing growth, photosynthesis, and suppressing Ni uptake and oxidative stress. *Arch. Agron. Soil Sci.* 62:633–647.

Khoshgoftarmanesh A.H., Bahmanziari H., Sanaeiostovar A. (2014). Responses of cucumber to deficient and toxic amounts of nickel in nutrient solution containing urea as nitrogen source. *Biol. Plant* 58:524–530.

Kim E.J., Kim H.P., Hah Y.C., Roe J.H. (1996). Differential expression of superoxide dismutases containing Ni and Fe/Zn in *Streptomyces coelicolor*. *Eur. J. Biochem.* 241:178–185.

King C.A., Purcell L.C. (2005). Inhibition of N_2 fixation in soybean is associated with elevated ureides and amino acids. *Plant Physiol.* 137:1389–1396.

Kishor P.K., Sangam S., Amrutha R.N., Laxmi P.S., Naidu K.R., Rao K.S., Sreenivasulu N. (2005). Regulation of proline biosynthesis, degradation, uptake and transport in higher plants: Its implications in plant growth and abiotic stress tolerance. *Curr Sci.* 88:424–438.

Klein C., Costa M.A.X. (2015). *Nickel*. In: Nordberg G.F., Fowler B.A., Nordberg (eds.) *Handbook on the Toxicology of Metals*. Academic Press, USA, pp. 1091–1111.

Koch H.G., Schneider D. (2016). Assembly of transmembrane *b*-type cytochromes and cytochrome complexes. In: Cramer W., Kallas T. (eds.) *Cytochrome Complexes: Evolution, Structures, Energy Transduction, and Signaling. Advances in Photosynthesis and Respiration (Including Bioenergy and Related Processes)*, Springer, pp. 555–584.

Krämer U., Cotter-Howells J.D., Charnock J.M., Baker A.J., Smith J.A.C. (1996). Free histidine as a metal chelator in plants that accumulate nickel. *Nature* 379:635–638.

Kucharski J., Boros E., Wyszkowska J. (2009). Biochemical activity of nickel-contaminated soil. *Pol. J. Environ. Stud.* 18:1039–1044.

Küpper H., Kroneck P.M. (2007). Nickel in the environment and its role in the metabolism of plants and cyanobacteria. *Met. Ions Life Sci.* 2:31–62.

Laghlimi M., Baghdad B., El Hadi H., Bouabdli A. (2015). Phytoremediation mechanisms of heavy metal contaminated soils: A review. *Open J. Ecol.* 5:375–388.

Lee B.D., Hwang S. (2015). Tobacco phytochelatin synthase (NtPCS1) plays important roles in cadmium and arsenic tolerance and in early plant development in tobacco. *Plant Biotechnol. Rep.* 9:107–114.

Lee S., Moon J.S., Ko T.S., Petros D., Goldsbrough P.B., Korban S.S. (2003). Overexpression of *Arabidopsis* phytochelatin synthase paradoxically leads to hypersensitivity to cadmium stress. *Plant Physiol.* 131:656–663.

Liao L., Selim H.M. (2018). Nickel adsorption–desorption and mobility in soils: Evidence of kinetics. In: Tsadilas C.D., Rinklebe J., Selim H.M. (eds.) *Nickel in Soils and Plants*. CRC Press, pp. 243–264.

Matraszek R., Hawrylak-Nowak B., Chwil S., Chwil M. (2016). Macronutrient composition of nickel-treated wheat under different sulfur concentrations in the nutrient solution. *Environ. Sci. Pollut. Res.* 23:5902–5914.

Mehrandish R., Rahimian A., Shahriary A. (2019). Heavy metals detoxification: A review of herbal compounds for chelation therapy in heavy metals toxicity. *J. Herbmed Pharmacol.* 8:69–77.

Mir M.A., Sirhindi G., Alyemeni M.N., Alam P., Ahmad P. (2018). Jasmonic acid improves growth performance of soybean under nickel toxicity by regulating nickel uptake, redox balance, and oxidative stress metabolism. *J. Plant Growth Regul.* 37:1195–1209.

Mishra P., Sharma P. (2019). Superoxide Dismutases (SODs) and their role in regulating abiotic stress induced oxidative stress in plants. In: Hasanuzzaman M., Fotopoulos V., Nahar K., Fujita M. (eds.) *Reactive Oxygen, Nitrogen and Sulfur Species in Plants: Production, Metabolism, Signaling and Defense Mechanisms*, John Wiley & Sons Ltd, US, pp. 53–88.

Mishra S, Srivastava S., Tripathi R.D., Govindarajan R., Kuriakose S.V., Prasad M.N.V. (2006). Phytochelatin synthesis and response of antioxidants during cadmium stress in *Bacopa monnieri* L. *Plant Physiol. Biochem.* 44:25–37.

Mitra G. (2018). Molecular approaches to nutrient uptake and cellular homeostasis in plants under abiotic stress. In: Hasanuzzaman M., Fujita M., Oku H., Nahar K., Hawrylak-Nowak B. (eds.) *Plant Nutrients and Abiotic Stress Tolerance*. Springer, pp. 525–590.

Mizuno N., Nosaka S., Mizuno T., Horie K., Obata H. (2003). Distribution of Ni and Zn in the leaves of *Thlaspi japonicum* growing on ultramafic soil. *Soil Sci. Plant Nutr.* 49:93–97.

Moarefi N., Nkongolo K.K. (2022). Contrasting tolerance and gene expression between white pine (*Pinus strobus*) and jack pine (*P. banksiana*) exposed to an increasing nickel concentration. *Ecol. Genet. Genom.* 24:100124.

Molas J. (1998). Identification and localization of nickel in callus tissue of *Brassica oleracea* L. *Acta Physiol. Planta* 20.

Molas J., Baran S. (2004). Relationship between the chemical form of nickel applied to the soil and its uptake and toxicity to barley plants (*Hordeum vulgare* L.). *Geoderma* 122:247–255.

Moreno J.L., Garcia C., Hernandez T. (2003). Toxic effect of cadmium and nickel on soil enzymes and the influence of adding sewage sludge. *Eur. J. Soil Sci.* 54:377–386.

Mosa A., El-Banna M.F., Gao B. (2016). Biochar filters reduced the toxic effects of nickel on tomato (*Lycopersicon esculentum* L.) grown in nutrient film technique hydroponic system. *Chemosphere* 149:254–262.

Mujeeb A., Iqbal M.Z., Shafiq M., Kabir M., Farooqi Z.U.R. (2019). Effects of nickel toxicity on seedling growth, photosynthetic pigments, carotenoids and phenols contents of cowpea *Vigna unguiculata* (L.) *Int. J. Environ. Agric. Biotech.* 4:2456–1878.

Nabi A., Naeem M., Aftab T., Khan M.M.A. (2020). Alterations in photosynthetic pigments, antioxidant machinery, essential oil constituents and growth of menthol mint (*Mentha arvensis* L.) upon nickel exposure. *Bras. J. Bot.* 43:721–731.

Nahar K., Hasanuzzaman M., Alam M.M., Fujita M. (2015). Exogenous glutathione confers high temperature stress tolerance in mung bean (*Vigna radiata* L.) by modulating antioxidant defense and methylglyoxal detoxification system. *Environ. Exp. Bot.* 1124:4–54.

Nakazawa R., Ozawa T., Naito T., Kameda Y., Takenaga H. (2001). Interactions between cadmium and nickel in phytochelatin biosynthesis and the detoxification of the two metals in suspension-cultured tobacco cells. *Biol. Plant* 44:627–630.

Nasibi F., Heidari T., Asrar Z., Mansoori H. (2013). Effect of arginine pretreatment on nickel accumulation and alleviation of the oxidative stress in *Hyoscyamus niger*. *J. Soil Sci. Plant Nutr.* 13:680–689.

Naz M., Ghani M.I., Sarraf M., Liu M., Fan X. (2022). Ecotoxicity of nickel and its possible remediation. In: Bhat R.F., Tonelli F.M.P.T., Hakeem K., (eds.) *Phytoremediation*. Academic Press, US, pp. 297–322.

Nie J., Pan Y., Shi J., Guo Y., Yan Z., Duan X., Xu M. (2015). A comparative study on the uptake and toxicity of nickel added in the form of different salts to maize seedlings. *Int. J. Environ. Res. Public Health* 12:15075–15087.

Nkongolo K., Theriault G., Michael P. (2018a). Differential levels of gene expression and molecular mechanisms between red maple (*Acer rubrum*) genotypes resistant and susceptible to nickel toxicity revealed by transcriptome analysis. *Ecol. Evol.* 8:4876–4890.

Nkongolo K., Theriault G., Michael P. (2018b). Nickel-induced global gene expressions in red maple (*Acer rubrum*): Effect of nickel concentrations. *Plant Gene.* 14:29–36.

Ojeda-Barrios D.L., Sánchez-Chávez E., Sida-Arreola J.P., Valdez-Cepeda R., Balandran-Valladares M. (2016). The impact of foliar nickel fertilization on urease activity in pecan trees. *J. Soil Sci. Plant Nutr.* 16:237–247.

Ozdener Y., Kutbay H.G. (2009). Toxicity of copper, cadmium, nickel, lead and zinc on seed germination and seedling growth in *Eruca sativa*. *Fresenius Environ. Bull.* 18:26–31.

Page V., Feller U. (2015). Heavy metals in crop plants: Transport and redistribution processes on the whole plant level. *Agronomy* 5:447–463.

Panda G.C., Das S.K., Bandopadhyay T.S., Guha A.K. (2007). Adsorption of nickel on husk of *Lathyrus sativus*: Behavior and binding mechanism. *Colloids Surf. B Biointerfaces* 57:135–142.

Pandey N., Sharma C.P. (2002). Effect of heavy metals Co^{2+}, Ni^{2+} and Cd^{2+} on growth and metabolism of cabbage. *Plant Sci.* 163:753–758.

Pandey V.K., Rajeev G. (2010). Nickel toxicity effects on growth and metabolism of eggplant. *Int. J. Veg. Sci.* 16:351–360.

Patra A., Choudhary S., Dutta A., Jatav S., Chattopadhyay A. (2019). Horizon of nickel as essential to toxic element. *Int. J. Comp. Sociol.* 7:1185–1191.

Persans M.W., Nieman K., Salt D.E. (2001). Functional activity and role of cation-efflux family members in Ni hyperaccumulation in *Thlaspi goesingense*. *Proc. Natl. Acad. Sci.* 98:9995–10000.

Piotrowska-Niczyporuk A., Bajguz A., Kotowska U., Zambrzycka-Szelewa E., Sienkiewicz A. (2020). Auxins and cytokinins regulate phytohormone homeostasis and thiol-mediated detoxification in the green alga *Acutodesmus obliquus* exposed to lead stress. *Sci. Rep.* 10:1–14.

Poonkothai M.V.B.S., Vijayavathi B.S. (2012). Nickel as an essential element and a toxicant. *Int. J. Environ. Sci.* 1:285–288.

Prajapati R.C., Gautam S., Surve S., Shukla V.N. (2019). Structural, magnetic, electrical and optical studies of Cr doped nanostructured ZnO thin films for spintronics application. *Mater. Res. Express* 6:1–11.

Rehman A.U., Nazir S., Irshad R., Tahir K., ur Rehman K., Islam R.U., Wahab Z. (2020). Toxicity of heavy metals in plants and animals and their uptake by magnetic iron oxide nanoparticles. *J. Mole. Liq.* 321:1–33.

Rehman M.Z.U., Rizwan M., Ali S., Fatima N., Yousaf B., Naeem A., Sabir M., Ahmad H.R., Ok Y.S. (2016). Contrasting effects of biochar, compost and farm manure on alleviation of nickel toxicity in maize (*Zea mays* L.) in relation to plant growth, photosynthesis and metal uptake. *Ecotoxicol. Environ. Saf.* 133:218–225.

Rizwan M., Imtiaz M., Dai Z., Mehmood S., Adeel M., Liu J., Tu S. (2017). Nickel stressed responses of rice in Ni subcellular distribution, antioxidant production, and osmolyte accumulation. *Environ. Sci. Pollut. Res.* 24: 20587–20598.

Rizwan M., Mostofa M.G., Ahmad M.Z., Imtiaz M., Mehmood S., Adeel M., Tu S. (2018). Nitric oxide induces rice tolerance to excessive nickel by regulating nickel uptake, reactive oxygen species detoxification and defense-related gene expression. *Chemosphere* 191:23–35.

Rizwan, M., Mostofa, M.G., Ahmad, M. Z., Zhou, Y., Adeel, M., Mehmood, S., Ahmad M.A., Javed R., Imtiaz M., Aziz O., Ikram M., Liu, Y. (2019). Hydrogen sulfide enhances rice tolerance to nickel through the prevention of chloroplast damage and the improvement of nitrogen metabolism under excessive nickel. *Plant Physiol. Biochem.*, 138:100–111.

Rokka A., Suorsa M., Saleem A., Battchikova N., Aro E.M. (2005). Synthesis and assembly of thylakoid protein complexes: Multiple assembly steps of photosystem II. *Biochem J.* 388:159–168.

Saad R., Kobaissi A., Robin C., Echevarria G., Benizri E. (2016). Nitrogen fixation and growth of Lens culinaris as affected by nickel availability: A prerequisite for optimization of agromining. *Environ. Exp. Bot.* 131:1–9.

Sachan P., Lal N. (2017). An overview of nickel (Ni^{2+}) essentiality, toxicity and tolerance strategies in plants. *Asian J. Biol.* 2:1–15.

Sanz-Cobena A., Misselbrook T.H., Arce A., Mingot J.I., Diez J.A., Vallejo A. (2008). An inhibitor of urease activity effectively reduces ammonia emissions from soil treated with urea under Mediterranean conditions. *Agric. Ecosyst. Environ.* 126:243–249.

Saroy K., Garg N. (2021). Relative effectiveness of arbuscular mycorrhiza and polyamines in modulating ROS generation and ascorbate-glutathione cycle in *Cajanus cajan* under nickel stress. *Environ. Sci. Pollut. Res.* 28:1–18.

Seregin I., Kozhevnikova, A.D. (2006). Physiological role of nickel and its toxic effects on higher plants. *Russ. J. Plant Physiol.* 53:257–277.

Shabani L., Sabzalian M.R., Mostafavi pour S. (2016). Arbuscular mycorrhiza affects nickel translocation and expression of ABC transporter and metallothionein genes in *Festuca arundinacea. Mycorrhiza* 26:67–76.

Shafeeq A., Butt Z.A., Muhammad S. (2012). Response of nickel pollution on physiological and biochemical attributes of wheat (*Triticum aestivum* L.) var. Bhakar-02. *Pak. J. Bot.* 44:111–116.

Shahzad B., Tanveer M., Rehman A., Cheema S.A., Fahad S., Rehman S., Sharma A. (2018). Nickel; whether toxic or essential for plants and environment-A review. *Plant Physiol. Biochem.* 132:641–651.

Sharma A., Kumar V., Shahzad B., Ramakrishnan M., Sidhu G.P.S., Bali A.S., Zheng B. (2019). Photosynthetic response of plants under different abiotic stresses: A review. *J. Plant Growth Regul.* 39:1–23.

Sharma P., Bhardwaj R., Arora N., Arora H.K., Kumar A. (2008). Effects of 28-homobrassinolide on nickel uptake, protein content and antioxidative defence system in *Brassica juncea. Biol. Planta* 52:767–770.

Sharma P., Jha A.B., Dubey R.S., Pessarakli M. (2012). Reactive oxygen species, oxidative damage, and antioxidative defense mechanism in plants under stressful conditions. *J Bot.* 2012:1–27.

Sharma S.S., Dietz K.J., Mimura T. (2016). Vacuolar compartmentalization as indispensable component of heavy metal detoxification in plants. *Plant Cell Environ.* 39:1112–1126.

Shaw W.H., Raval D.N. (1961). The inhibition of urease by metal ions at pH 8.9. *J. Am. Chem. Soc.* 83:3184–3187.

Sheoran I.S., Aggarwal N., Singh R. (1990). Effects of cadmium and nickel on *in vivo* carbon dioxide exchange rate of pigeon pea (*Cajanus cajan* L.). *Plant Soil* 129:243–249.

Siddique A., Kandpal G., Kumar P. (2018). Proline accumulation and its defensive role under diverse stress condition in plants: An overview. *J. Pure Appl. Microbiol.* 12:1655–1659.

Singh G., Agnihotri R.K., Reshma R.S., Ahmad M. (2012). Effect of lead and nickel toxicity on chlorophyll and proline content of Urd (*Vigna mungo* L.) seedlings. *Int. J. Plant Physiol. Biochem.* 4:136–141.

Sirhindi G., Mir M.A., Abd-Allah E.F., Ahmad P., Gucel S. (2016). Jasmonic acid modulates the physio-biochemical attributes, antioxidant enzyme activity, and gene expression in *Glycine max* under nickel toxicity. *Front. Plant Sci.* 7:1–12.

Snezhkina A.V., Kudryavtseva A.V., Kardymon O.L., Savvateeva M.V., Melnikova N.V., Krasnov G.S., Dmitriev A.A. (2019). ROS generation and antioxidant defense systems in normal and malignant cells. *Oxid. Med. Cell Longev.* 2019:1–17.

Soares C., de Sousa A., Pinto A., Azenha M., Teixeira J., Azevedo R.A., Fidalgo F. (2016). Effect of 24-epibrassinolide on ROS content, antioxidant system, lipid peroxidation and Ni uptake in *Solanum nigrum* L. under Ni stress. *Environ. Exp. Bot.* 122:115–125.

Soliman M., Alhaithloul H.A., Hakeem K.R., Alharbi B.M., El-Esawi M., Elkelish A. (2019). Exogenous nitric oxide mitigates nickel-induced oxidative damage in eggplant by upregulating antioxidants, osmolyte metabolism, and glyoxalase systems. *Plants* 8:1–17.

Sreekanth T.V.M., Nagajyothi P.C., Lee K.D., Prasad T.N.V.K.V. (2013). Occurrence, physiological responses and toxicity of nickel in plants. *Int. J. Environ. Sci. Technol.* 10:1129–1140.

Sun Q., Wang X.R., Ding S.M., Yuan X.F. (2005). Effects of interactions between cadmium and zinc on phytochelatin and glutathione production in wheat (*Triticum aestivum* L.). *Environ. Toxicol. Int. J.* 20:195–201.

Takagi H., Watanabe S., Tanaka S., Matsuura T., Mori I.C., Hirayama T., Shimada H., Sakamoto A. (2018). Disruption of ureide degradation affects plant growth and development during and after transition from vegetative to reproductive stages. *BMC Plant Biol.* 18:1–16.

Tanveer M. (2020). An overview of nickel toxicity in plants. Metal toxicology handbook. In: Bagchi D., Bagchi M. (eds.) CRC Press, Boca Raton, pp. 465–474.

Theriault G., Michael P., Nkongolo K. (2016). Comprehensive transcriptome analysis of response to nickel stress in white birch (*Betula papyrifera*). *PLoS One* 11:e0153762.

Todd C.D., Tipton P.A., Blevins D.G., Piedras P., Pineda M., Polacco J.C. (2006). Update on ureide degradation in legumes. *J. Exp. Bot.* 57:5–12.

Tripathy B.C., Bhatia B., Mohanty P. (1981). Inactivation of chloroplast photosynthetic electron-transport activity by Ni^{2+}. *Biochim. Biophys. Acta Bioenerg.* 638:217–224.

Tripathy S., Mohanty P.K. (2017). Reactive oxygen species (ROS) are boon or bane. *Int. J. Pharm. Sci. Res.* 8:1–17.

Trovato M., Forlani G., Signorelli S., Funck D. (2019). Proline metabolism and its functions in development and stress tolerance. In: Hossain M., Kumar V., Burritt D., Fujita M., Mäkelä P. (eds.) *Osmoprotectant-Mediated Abiotic Stress Tolerance in Plants*. Springer, pp. 41–72.

Ulrich K., Jakob U. (2019). The role of thiols in antioxidant systems. *Free Radic. Biol. Med.* 140:14–27.

Vacchina V., Mari S., Czernic P., Marquès L., Pianelli K., Schaumlöffel D., Lebrun M., Łobiński R. (2003). Speciation of nickel in a hyperaccumulating plant by high-performance liquid chromatography — Inductively coupled plasma mass spectrometry and electrospray MS/MS assisted by cloning using yeast complementation. *Anal. Chem.* 75:2740–2745.

Valivand M., Amooaghaie R. (2020). Sodium hydrosulfide modulates membrane integrity, cation homeostasis, and accumulation of phenolics and osmolytes in zucchini under nickel stress. *J. Plant Growth Regul.* 40:313–328.

Van der Pas L., Ingle R.A. (2019). Towards an understanding of the molecular basis of nickel hyperaccumulation in plants. *Plants* 8:11.

Vatansever R., Ozyigit I.I., Filiz E. (2017). Essential and beneficial trace elements in plants, and their transport in roots: A review. *App. Biochem. Biotechnol.* 181:464–482.

Vijayakumar V. (2018). Symbiotic tripartism in the model plant family of legumes and soil sustainability. In: Meena R., Das A., Yadav G., Lal R. (eds.) *Legumes for Soil Health and Sustainable Management*. Springer, Singapore, pp. 173–203.

Vischetti C., Marini E., Casucci C., De Bernardi A. (2022). Nickel in the environment: Bioremediation techniques for soils with low or moderate contamination in European Union. *Environments* 9:133.

Wang H., Feng T., Peng X., Yan M., Tang X. (2009). Up-regulation of chloroplastic antioxidant capacity is involved in alleviation of nickel toxicity of *Zea mays* L. by exogenous salicylic acid. *Ecotoxicol Environ. Saf.* 72:1354–1362.

Wheeler C.T., Hughes L.T., Oldroyd J., Pulford I.D. (2001). Effects of nickel on *Frankia* and its symbiosis with *Alnus glutinosa* (L.) Gaertn. *Plant Soil* 231:81–90.

Wu Z., Liu S., Zhao J., Wang F., Du Y., Zou S., Huang Y. (2017). Comparative responses to silicon and selenium in relation to antioxidant enzyme system and the glutathione-ascorbate cycle in flowering Chinese cabbage (*Brassica campestris* L. ssp. chinensis var. utilis) under cadmium stress. *Environ. Exp Bot.* 133:1–11.

Xiao Q., Wang Y., Lü Q., Wen H., Han B., Chen S., Lin R. (2020). Responses of glutathione and phytochelatins biosysthesis in a cadmium accumulator of Perilla frutescens (L.) Britt. under cadmium contaminated conditions. *Ecotoxicol. Environ. Saf.* 201:1–10.

Yasin U.A., Horo J.T., Gebre B.A. (2019). Physicochemical and sensory properties of tofu prepared from eight popular soybean [*Glycine max* (L.) Merrill] varieties in Ethiopia. *Sci. Afr.* 6:e00179.

Yoon Y.E., Kuppusamy S., Cho K.M., Kim P.J., Kwack Y.B., Lee Y.B. (2017). Influence of cold stress on contents of soluble sugars, vitamin C and free amino acids including gamma-aminobutyric acid (GABA) in spinach (*Spinacia oleracea*). *Food Chem.* 215:185–192.

Zaid A., Mohammad F., Wani S.H., Siddique K.M. (2019). Salicylic acid enhances nickel stress tolerance by up-regulating antioxidant defense and glyoxalase systems in mustard plants. *Ecotoxicol. Environ. Saf.* 180:575–587.

Zhao J., Shi G., Yuan Q. (2008). Polyamines content and physiological and biochemical responses to ladder concentration of nickel stress in *Hydrocharis dubia* (Bl.) Backer leaves. *Biometals* 21:665–674.

Zheng M., Zhou Z., Luo Y., Zhao P., Mo J. (2019). Global pattern and controls of biological nitrogen fixation under nutrient enrichment: A meta-analysis. *Global. Change Biol.* 25:3018–3030.

Zobiole L.H.S., Oliveira Jr R.S., Kremer R.J., Constantin J., Yamada T., Castro C., Oliveira F.A., Oliveira Jr A. (2010). Effect of glyphosate on symbiotic N_2 fixation and nickel concentration in glyphosate-resistant soybeans. *Appl. Soil Ecol.* 44:176–180.

https://doi.org/10.1142/9789811283123_0005

Chapter 5

Approaches to Antioxidant Defence System: An Overview of Coping Mechanism Against Lithium/Nickel Exposure in Plants

Rania M. Mahmoud and Mai Sayed Fouad*

Botany Department, Faculty of Science,
Fayoum University, Fayoum 63514, Egypt
*msf00@fayoum.edu.eg

Abstract

Lithium (Li) and nickel (Ni) are two of the most widely used metals in various industrial applications. Since they are available for plants' uptake from the soil, excessive plant exposure to high doses of both metals may be tolerable or not for accumulating species and sensitive species, respectively. Many plants adopt a number of powerful detoxification techniques in their fight for survival. Among them, the antioxidant defence system is a crucial mechanism that helps plants cope with toxic metals, including Li and Ni. Understanding the different approaches that plants use to activate this system can provide insights into how we can improve plant resilience and protect them from environmental stressors. Besides the natural endogenously reacting antioxidant system in the plant's own body, exogenously applied antioxidants have proven to be effective in mitigating the negative effects of the aforementioned metals.

Herein, we review the traditional as well as recent advances used in overcoming the toxicity of Li and Ni to plants.

Keywords: Lithium, nickel, reactive species, free radicals, oxidative stress, antioxidants, phytohormones

1. Introduction

Lithium (Li) and nickel (Ni) are naturally occurring trace elements, constituting 0.006% and 0.008% of the Earth's crust, respectively (Sachan and Lal, 2017; Wietelmann and Bauer, 2000). Both metals are widely distributed terrestrially and aquatically (Habashi, 2013; Hertel *et al.*, 1991). Anthropogenic activities, particularly certain industries, are the major emission sources of Li and Ni in the ecosystem, thus bringing the contamination problem to the surface (Al-Thyabat *et al.*, 2013; Hull *et al.*, 2014; Maleva *et al.*, 2009; Rathor *et al.*, 2014; Shacklette and Boerngen, 1984). It was a great debate concerning not only the necessity of Li and Ni for plant growth but also the mechanism of their uptake from the soil. Lately, many researchers have come to an agreement with regard to Ni; however, with Li, the processes are yet to be uncovered holistically. Li and Ni are important for plant nutrition and development, but in precise amounts (Aral and Vecchio-Sadus, 2008; Cannon *et al.*, 1975; Kljakovic, 1984; Nishida *et al.*, 2011).

Increasing concentrations of Li and Ni may lead to the appearance of toxicity symptoms in exposed plants (Cannon *et al.*, 1975). Intuitively, a great majority of these symptoms are attributed to the production of reactive oxygen species (ROS) (Kiełczykowska *et al.*, 2004; Oktem *et al.*, 2005; Yadav and Sharma, 2016). This study is concerned with reactive species, which are the main products of oxidative stress caused by exposure to metals, whether heavy, like Ni, or not, like Li (Shahzad *et al.*, 2016; Yadav and Sharma, 2016; Sachan and Lal, 2017). Notably, several metabolic activities were deteriorated, which is clearly visible as undesirable changes in photosynthesis, respiration, the integrity of the cell membrane and ion homeostasis (Sachan and Lal, 2017). Chloroplast damage and chlorophyll breakdown are mainly responsible for leaf necrosis (Molas, 2002). Malondialdehyde (MDA) production is a result of ROS attacking membrane fatty acids (Hawrylak-Nowak *et al.*, 2012; Yilmaz and Parlak, 2011). At the molecular level, both DNA and proteins are subjected to deformation. Plants possess a non-enzymatic defence, along with an endogenous defence of an enzymatic nature for scavenging free radicals during their defensive actions (Bach and Gallicchio, 1990). Antioxidative enzymes, such as ascorbate peroxidase (APX), peroxidase (POD), superoxide dismutase (SOD) and catalase (CAT), together with non-enzymatic antioxidants, such as glutathione (GSH), carotenoids, ascorbic acid (AsA) and osmolytes, including proline, act as protectors (Bakhat *et al.*, 2020; Gupta *et al.*, 2017; Kuloğlu *et al.*, 2022). Other protection stimulators, such as chelators, mutagens and phytohormones, may be supplied exogenously to augment plant resistance (Kavanagh *et al.*, 2018; Khan and Khan, 2014).

This chapter presents a collective review of Li/Ni uptake in plants and discusses the degree of their toxicities. Moreover, the discussion is extended to frame out the combat mechanisms of plants for overcoming oxidative stress.

2. Lithium/Nickel Occurrence

Li is known as 'white gold' and 'green energy metal' (Graham *et al.*, 2021; Lu *et al.*, 2010). Li is not considered an inert element; it is always found in conjugation with other metals or salts, such as lithium chloride (LiCl), lithium carbonate (Li_2CO_3) and lithium hydroxide (LiOH) (Dessemond *et al.*, 2019; Speirs *et al.*, 2014). Li is terrestrially as well as aquatically widespread. It occupies the 27th rank among the most abundant elements (Habashi, 2013; Kszos and Stewart, 2003).

Moreover, Li amounts to 0.006% (by weight) of the Earth's crust; it reaches 0.2 ppm in marine water and 0.03 ppm in fresh water (Wietelmann and Bauer, 2000; Habashi, 2013). Li batteries, Li brines and newly appearing Li industrial applications are leading to its emission into the ecosystem and may be a possible cause of environmental contamination with Li (Al-Thyabat *et al.*, 2013; Hull *et al.*, 2014; Shacklette and Boerngen, 1984).

Ni is a trace metal that constitutes almost 0.008% of the Earth's crust and is present everywhere in the biosphere, including the soil, water and air. The levels of Ni in natural waters have been found to range from 2 to 10 $\mu g\ dm^{-3}$ (fresh water) and from 0.2 to 0.7 $\mu g\ dm^{-3}$ (marine) (Hertel *et al.*, 1991; Sachan and Lal, 2017). It is one of the heavy metals (HMs) that are regularly found in industrial effluents. The concentration of Ni in the environment has reportedly increased as a result of anthropogenic activities, such as mining, the production of steel, storage batteries, spark plugs, magnets, electroplating, motor vehicles, aircraft paint, textiles, machine tools and the burning of coal and oil (Maleva *et al.*, 2009; Rathor *et al.*, 2014). As a result, elevated Ni concentrations have been found in several regions around the world, suggesting the global nature of Ni contamination as a safety concern for the environment (Chen *et al.*, 2019).

3. Lithium/Nickel Uptake in Plants

Li is found in all plants and may be necessary for the growth and development of some plants (Aral and Vecchio-Sadus, 2008; Cannon *et al.*, 1975; Kljakovic, 1984). According to (Vlasiuk and Okhrimenko, 1969), Li increases yield by promoting the photochemical activity of chloroplasts and the concentration of chlorophyll in tomato and pepper leaves. The protein part of the leaf contains Li, and according to (Borovik-Romanova, 1965), Li ions significantly boost protein, carbohydrates and nitrogen metabolism, as well as respiration, nucleic and organic acid content and biomass accumulation in the plant. The levels of Li that promote growth and those that cause poisoning symptoms depend on the tolerance of the plant species and the kind of salt applied (Cannons *et al.*, 1975). Li's effects on plants are framed as being dependent on the extent of dose, exposure in terms of time or application and species.

The relationship between Li, which can be extracted from the soil, and the amount of Li absorbed by plants is not well understood. White gold is an alkali with a small radius, similar to Na and K, and thus it may follow a similar route for its uptake (Shahzad *et al.*, 2016). Li is only stable in its ionic state in biological systems. Despite the competitiveness of Li with Ca, Mg, K and Na, it is not listed as an essential element for plants. Mg almost precedes Li in complex formation, which is why it is difficult to be substituted by Li. Contrarily, Li has strong replacement ability over K and Na in occupying binding sites (Birch and Nadelmann, 1974). Epstein (1960) came to the conclusion that Li and Ca shared binding sites. When Li was present, Ca tended to boost K absorption while decreasing Li absorption contradicted (Epstein, 1960) claim that Li in the soil can cause plants to absorb calcium in a backwards manner.

Several studies showed that Ca supplementation could mitigate Li harm. Experiments in Germany found that Li could not adequately replace K and reduced fresh weight biomass at a Li concentration of 69 ppm in solution with oats (*Avena sativa*), wheat (*T. aestivum*), rye (*Secale cereale*) and barley (*H. vulgare*) (Bach and Galliccchio, 1990).

Plants usually display different capabilities towards Li absorption, ranging from 0.2 to 30 ppm and somewhere between 0.2 and 6,000 mg kg^{-1} (Kavanagh *et al.*, 2018). *Cirsium arvense* and *Solanum dulcamera* exhibit unprecedented Li accumulation that amounts to six times that found across other plants. Nightshade species can accumulate up to 1 mg g^{-1} of Li (Shacklette and Boerngen, 1984). Halophils rely on Li for metabolism. Evans and Sorger (1966) stated that both Na and Li stimulate at least one halophile enzyme. Halophytes are NaCl accumulators, which may explain why a number of members of the family Chenopodiaceae contain high quantities of Li (Cannon *et al*., 1975; Shacklette and Boerngen, 1984; Kabata-pendias and Mukherjee, 2007).

Li concentrations of 99.6–226.4 mg g^{-1} have been recorded in salt-tolerant plants, including *Carduus arvense* and *Holoschoenus vulgaris* (Shacklette and Boerngen, 1984). In addition to the Solanaceae and Asteraceae families, it is also believed that the Ranunculaceae and Rosaceae families contain species of plants that are tolerant to Li (Kavanagh *et al.*, 2018). *Apocynum venetum,* a novel Li accumulator plant, was uncovered in China in 2013. Compared to *Cirsium arvense* and *Solanum dulcamera,* whose Li accumulation amounts to 1,000 mg kg^{-1}, *Apocynum venetum* fared better but was on par with *Apocynum pictum.* The bioconcentration factor (BCF) and translocation factor (TF) were found to be higher than 1.0 for *Apocynum pictum* (Jiang *et al.*, 2018).

Li concentrations of greater than 300 mg kg^{-1} have been found in *Brassica juncea* following an investigation into its capacity to accumulate Li, V and Cr from Li mine tailings (Kavanagh *et al.*, 2018). It has been proposed that the amount of Li in plants can serve as a reliable indicator of its amount in the soil. (Kavanagh *et al.*, 2018). For Na-accumulating species, the ratio between Li concentration in plants and that in soil is high, amounting to 10 in pickle weed and desert holly. Lycium could not develop in NaCl-rich soils, though it is definitely an accumulator as well as an indicator of Li (Cannon *et al.*, 1975).

It was found that Li accumulation is positively affected by clayey soil (McStay *et al.*, 1980). Acidic soils trigger plants' Li uptake by increasing the solubilities of some HMs (Fe, Ni, Co, Mn, Cu, Pb, Al and Cd), as Li is preferentially impacted by the presence of these metals (Schrauzer, 2002). The study of arid basins in the Western United States reveals greater hormetic Li uptake by native plants than in wet environments (Bradley *et al.*, 2017). (Borovik-Romanova, 1965) examined 138 plant species from eight different classes of soil and concluded that the majority of plants contain 0.15–0.3 ppm Li, while those growing in salty soils have two- to tenfold that level.

Generally, Li sequestration starts in the root; afterwards, it will be translocated into older leaves, which is the fate of Li and where it becomes immobilised (Kent, 1941). Li is easily transported in leaves because it passes through the cell membrane via the same transport carrier as K does (Kabata-pendias and Mukherjee, 2007). It is well known that leaf machinery is better able to accumulate Li than roots (Bradford, 1966). A study (Birch-Hirschfeld, 1920) has shown a low rate of translocation of Li in the phloem. Hinz and Fischer (1976) found that Li behaves similarly to the alkaline earth metals Ca, Sr and Ba in its very slight mobility in the phloem (Bach and Galliccchio, 1990).

Because it is both contentious and essential, Ni has emerged as an interesting metal. It is beneficial in plant nutrition and crucial to certain metalloenzymes composition. Ni uptake from acidic soils is greater than from alkaline soils (Yusuf *et al.*, 2011). Although Ni ions can enter plant roots passively, (Nishida *et al.*, 2011) found that active absorption processes using weakly selective ZRT/IRT-like (ZIP) cation transporters were more important for plant Ni uptake. Ni is absorbed via the Mg ion transport system (Ravenna and Maugini, 1912). Ca^{2+} and Mg^{2+} have been reported to decrease *Berkheya coddii*'s uptake of Ni^{2+} (Rankin, 1917). However, it has been observed that non-competitive inhibitors of Ni^{2+} influx exist in the Ca^{2+} and Mg^{2+} roots of cut barley (Chen *et al.*, 2009). Ni is relatively mobile within the plant body, as evidenced by the fact that, in non-hyperaccumulator plant species, Ni taken up by the roots is transported quickly in the xylem through the transpiration stream and is then followed by its re-translocation in the phloem (Guo *et al.*, 1996; Neumann, 1986).

According to Krämer *et al.* (2000), the range of Ni concentrations in plants is greater than 10 mg kg^{-1} dry mass (DM) in sensitive species, 50 mg kg^{-1} DM in moderately tolerant species and 1,000 mg kg^{-1} DM in Ni hyperaccumulator plants, such as *Alyssum* and *Thlaspi* species. Waterweed's roots can absorb trace metals from sediments and interstitial fluids. Particularly, short-term (24 h) exposure to Ni in concentrations of 3.5–30 μM in a culture medium of *E. canadensis* resulted in an efficient uptake of Ni. Additionally, it was demonstrated that Ni buildup influenced the uptake of nitrogen and phosphorus after continuous exposure, thus hindering the growth of *E. canadensis* (Maleva *et al.*, 2009).

4. Lithium/Nickel Toxicity in Plants

Reactive species (RS) are a collective group that comprises oxygen (ROS), nitrogen (reactive nitrogen species, RNS), sulphur (reactive sulphur species, RSS) and other species, many of which are free radicals. Although free radicals have a limited life span, measured in milli, micro or nanoseconds, all living things have built-in antioxidants, which are chemicals that protect cells from oxidative stress (Kapoor *et al.*, 2019). However, due to their buildup within the cell to a level that exceeds the capacity to eliminate them, each of these species has the potential to cause oxidative stress (Choi *et al.*, 2017). Even the tolerance mechanism is ambiguous; it could be proposed that adverse symptoms arising from Li exposure are due to ROS formation, followed by oxidative stress (Oktem *et al.*, 2005). ROS, such as singlet oxygen ($^1O_2$), superoxide (O^{2-}), hydroxyl radical ($HO^•$) and hydrogen peroxide (H_2O_2), are produced in chloroplasts, mitochondria and peroxisomes in various plant species (Hattab *et al.*, 2016; Liu *et al.*, 2008). Despite the fact that ROS are the most commonly occurring RS, Ni toxicity is supposed to create ROS, RNS and RSS (Kolbert *et al.*, 2020).

Both Li and Ni are reported to produce RS through a Fenton-type reaction. The intracellular levels of (O^{2-}) and ($HO^•$) radicals are increased with Li and Ni toxicities, which promote lipid peroxidation. It is possible that Li and Ni can impede the activation of antioxidant enzymes or the transcription of genes that produce antioxidants, which would lead to increased oxidation (Shahzad *et al.*, 2016; Yadav and Sharma, 2016.)

Ni disturbs the homeostasis of ROS, RNS and RSS in Ni-sensitive *Arabidopsis thaliana*. The amounts of significant ROS (superoxide anion and hydrogen peroxide), RNS (nitric oxide and peroxynitrite) and RSS (hydrogen sulphide) are determined in the root system to evaluate the redox signalling in Ni-exposed plants. In *Arabidopsis thaliana* roots grown in the presence of Ni, the levels of all major ROS, RNS and RSS dramatically increased, indicating amplified redox signalling in this species. *Brassica juncea*'s root system, in contrast, did not modify its ROS, RNS or RSS levels in response to high levels of Ni. Peroxynitrite is an exception, as Ni increased its levels in both species to a comparable extent (Kolbert *et al.*, 2020).

Ni enhanced the abundance of the GSNOR protein, which controls GSNO metabolism and, as a result, protein nitration, providing additional support for Ni-induced RNS signalling in *Arabidopsis thaliana* (Kolbert *et al.*, 2020). An increase in protein nitration in the root has been associated with Ni stresses in *Brassica napus* (Corpas *et al.*, 2021; Feigl *et al.*, 2020). This nitration process is regarded as a reliable marker for nitro-oxidative stress and is one of the post-translational changes that results from the amplification of RNS signalling caused by a variety of stress situations.

In the presence of oxidative stress, macromolecules such as DNA, lipids and proteins are damaged. Phosphate groups in nucleic acids have a significant affinity for cation binding. The Na/K ratio is known to control the condition for DNA condensation, and in this regard, the Li^+ ions can bind more easily and cause abnormalities in DNA conformation (Bach and Galliccchio, 1990). At low Li concentrations, macromolecules are built with increasing amounts of adsorbed water, whereas at high concentrations, Li tends to compete with macromolecules for water, thus acting as a dehydrating agent (Vlasyuk *et al.*, 1979).

Li, similar to Ni, may obstruct several metabolic processes, including respiration, photosynthesis and many other metabolic pathways (Sachan and Lal, 2017). The appearance of necrotic spots and decreased chlorophyll content in lettuce leaves, red kidney bean, sour orange and tomato could be ascribed to the breakdown of chlorophyll contents in leaves in the presence of Li (Bingham *et al.*, 1964; Kalinowska *et al.*, 2013; Shahzad *et al.*, 2016). According to (Hawrylak-Nowak *et al.*, 2012), a 25 mg dm^{-3} Li exposure to sunflower resulted in a 16% drop in carotenoid content and the development of necrotic patches on maize leaf surfaces. In chicory (*Cichorium intybus*) root slices, it slowed down the typical increase in respiration linked to 'wound-response' (Chares, 1990).

At high Ni administration, water spinach leaves exhibited chlorosis and necrotic areas (Sun and Wu, 1998). Damage to the chloroplastic ultrastructure, the inhibition of electron chain reactions, the replacement of Mg in chlorophyll, reduced stomatal conductance, the inhibition of the Calvin cycle or the breakdown of plastoquinone and carotenoids may all be reasons for reduced photosynthetic activity in the presence of Ni (Molas, 2002).

Li first affects the membrane transporting system, which controls ion gateways, and the ultimate impact, either directly or indirectly, is focused on the metabolic system (Clemens, 2006; Singh *et al.*, 2010). Many authors correlated Li risk with the applied dose during exposure. MDA formation in the leaves and roots of sunflower and maize plants cultivated in the presence of 50 mg dm^{-3} Li revealed elevated levels of lipid peroxidation, as documented by (Hawrylak-Nowak *et al.*, 2012).

At increased Ni concentrations, it was accepted as plausible that cell membrane fatty acids are attacked by free radicals, terminating in MDA production and cell membrane damage in *Lemna gibba* plant (Yilmaz and Parlak, 2011). MDA production, in turn, increases plasma membrane permeability and results in K^+ ion leakage, amino acid oxidation and, ultimately, cell death (Gratão *et al.*, 2005; Yilmaz and Parlak, 2011).

Gayathri *et al.* (2022) stated that the germination rate of *A. cepa* dropped with increasing doses of Li, with germination rates of 73% at 50 ppm, 57% at 75 ppm and 41% at 100 ppm. One can infer that extreme and prolonged Li exposure results in the creation of free radicals and oxidative stress. Some signals in a physiological response alter cellular pumping or the leakage of ions, such as those of sodium or hydrogen, obeying a 'relay and amplification' mechanism. The K level in cells is subsequently regulated by this. Only above a certain K level in cells is it possible to biosynthesise a small number of ion-specific proteins. The K concentration of these cells would drop off as a result of Li accumulation brought on by an effective pump with minimal leakage. Li might stop the biosynthesis of particular proteins in this way (Jahan *et al.*, 2020).

An investigation found that Li did not directly cause the solute uptake machinery to be turned off during transport in the case of the sycamore (*Acer pseudoplatanus*). Instead, Li interfered with transcription and translation in some way to block the development of the solute uptake mechanism. Li was therefore suggested to have a major impact on protein production (Chares, 1990). Possible plant strategies are configured in Figure 1 and will be discussed in the following section.

5. Plant Antioxidant Defence Against Lithium/Nickel Stress

5.1. *Nitrogen-containing compounds*

5.1.1. *Polyamines*

When plants are exposed to high Ni concentrations, non-enzymatic low-molecular-weight antioxidative compounds essential for the healthy

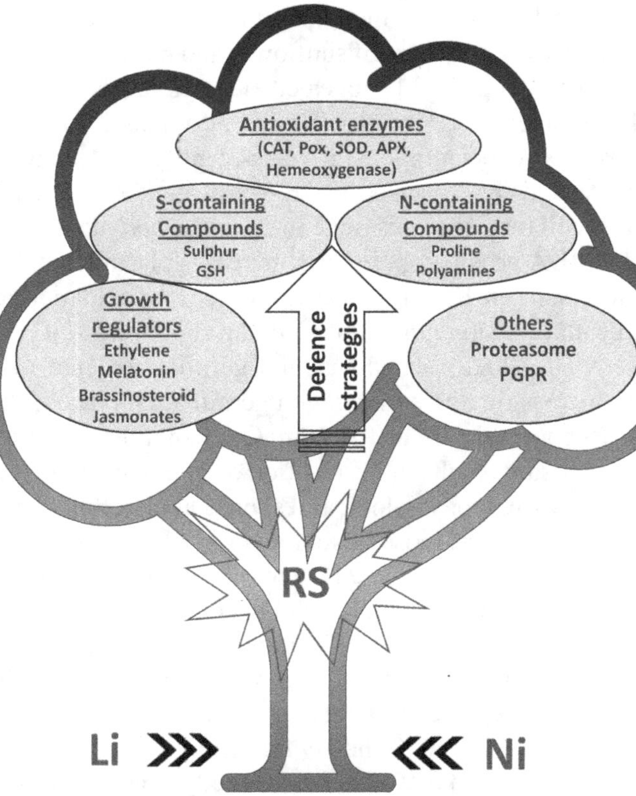

Figure 1. Plant defence strategies against Li and Ni stressors.

growth of living things, called polyamines (PAs), are altered (Handa and Mattoo, 2010). Diamino oxidase (DAO, EC 1.4.3.6), and polyamine oxidase (PAO, EC 1.5.3.3) catabolise PAs in plants, releasing hydrogen peroxide, which stimulates the activity of antioxidative enzymes (Pietrini *et al.*, 2015).

According to (Galardi *et al.*, 2007), Ni hyperaccumulator plant species from the genera *Alyssum* and *Thlaspi* can collect more than 1,000 mg kg^{-1} dry weight (DW) in their above-ground organs without displaying hazardous effects. Depending on the PA molecule, the plant species, the type and the intensity of the stress, PA content alterations were frequently seen as and linked to being both a toxicity symptom and a defence response. (Shevyakova *et al.* 2011) reported that Pas play a protective role in *Amaranthus* plants by chelating Ni. *Amaranthus* plants can continue to function normally up to a concentration of 50 $MniCl_2$, demonstrating Ni tolerance even at high metal concentrations in their organs (Pietrini *et al.*, 2015).

5.1.2. *Osmolytes* (*Proline*)

Proline concentrations considerably increased under HMs compared to those in non-stressed plants. The highest levels of proline were found at 50 M Ni. (Badawy *et al.*, 2022). To combat the effects of Ni stress, osmolyte synthesis was boosted (Kumar *et al.*, 2022). In response to HM stress, plants frequently accumulate free proline (Bassi and Sharma, 1993; Schat *et al.*, 1997). According to (Kaul *et al.*, 2008), proline may serve as a free radical scavenger. Intriguingly, the majority of Ni hyperaccumulators are members of the *Brassicaceae* family. According to the well-known HM-dependent proline buildup in many plant species, the Ni-stressed seedlings showed increased amounts of free proline. As HM stress is linked to worsening plant–water relationships, proline is likely involved in at least one of the following: (i) osmoregulation, as HM stress is associated with deterioration of plant–water relations, (ii) cellular anti-oxidative defence, as shown in proline overproducing transgenic *Chlamydomonas reinhardtii* and in an *in vitro* assay system and (iii) the protection of enzymes against HM poisoning, as demonstrated in the case of Cd and lead (Thakur and Sharma, 2016).

In comparison to control, the proline content in rice plants increased by 12.57%, 12.22%, 36.77% and 114.69% in the shoot at 10, 50, 100 and 200, respectively. While 12.45%, 35.11% and 57.44% in the root at 10, 100 and 200 M Ni, respectively. Proline content was found to significantly decrease in Ni-treated roots at 50 M. Following Ni treatment, the concentration of soluble sugar in the rice seedling's shoots and roots decreased. According to Rizwan *et al.* (2017), proline reduces lipid peroxidation and shields cells from oxidative damage caused by scavenging ROS. When Ni treatments were applied, the generation of proline increased significantly in sweet potato, reaching a maximum increase of 1,013.6% under a 60 mg L^{-1} Ni treatment in contrast to the control (Kumar *et al.*, 2022). The proline content of *L. gibba* and MDA concentration had a considerably positive association, whereas APX had a negative correlation (Yilmaz and Parlak, 2011).

Maximum proline accumulation occurred in mustard plants treated with the combination of Ni and ethephon therapy at 60 days after sowing (DAS) in comparison to the control, with ethephon further boosting proline accumulation under Ni stress (Khan *et al.*, 2020b). Proline and soluble sugars alleviate Li toxicity during the germination stage of *Apocynumpictum* (Salieri *et al.*, 1995; Jiang *et al.*, 2018). Significant stability of growth and pigments even at 75 mg kg^{-1} Li may explain why *Apocynumpictum* is considered a tolerant species (Bingham *et al.*, 1964).

5.2. *Antioxidant enzymes*

5.2.1. *SOD activity*

Ni-treated seedlings had increased levels of SOD activity. In Kranti, Varuna and Pusa Jai Kisan, the activity rose by 32%, 20% and 14%, respectively, at 200 M Ni. Nevertheless, Varuna had the highest and Kranti had the lowest SOD activity (Thakur and Sharma, 2016). With a progressive increase in Ni concentration in the nutritional solution, SOD activity showed a persistent decrease in both the shoot and root of rice (Rizwan *et al.*, 2017). When SOD activity was compared to control at 10, 50, 100 and 200 M Ni, it fell by 11.67%, 39.27%, 41.22% and 56.35% in the shoot and 5.68, 45.86, 54.70, and 77.87% in the root, respectively (Rizwan *et al.*, 2017). It was noticed that SOD and CAT production was enhanced at 25 and 50 mg L^{-1} Li in Allium cepa and then decreased at 100 mg L^{-1} but still higher than the control. (Bakhat *et al.*, 2020) found that applying Li to spinach shoots boosted the activity of the enzymes SOD, CAT and APX.

5.2.2. *POD activity*

After Ni treatment, there was a noticeable increase in POD activity in the wheat shoots, where, after being exposed to Ni, POD activity in the shoots increased by over 11-fold (Gajewska and Skłodowska, 2010). The Ni concentration in the growth culture increased, although POD and CAT activities in the Ni-stressed rice plants showed an increasing trend (Rizwan *et al.*, 2017). Ni stress significantly changed the POD activity in rice seedlings' roots and shoots. Ni concentration increased along with POD activity, which peaked at 200 M in the shoots at 104.81%.

This revealed a positive correlation between POD activity and Ni concentration during Ni stress. At first, a comparable rise in POD activity was seen with increasing Ni concentrations up to 100 M (46.54% more than control). The POD activity in roots remained unchanged even when the Ni concentration was increased to 200 M (Rizwan *et al.*, 2017).

5.2.3. *CAT activity*

With Ni stress, CAT activity was increased in wheat plants by roughly 1.2 times (Gajewska and Skłodowska, 2010). While the activity of SOD was highest in *Elodea canadensis* plants treated with 10 μM of Ni, the CAT activity increased by approximately 20% in leaves treated with 1 μM of Ni but dropped by 30% and 43% in plant material treated with 10 and 50 μM of Ni, respectively (Maleva *et al.*, 2009). However, after exposure to Ni, the activity of the H_2O_2-scavenging enzymes CAT and APX significantly decreased. Thus, in 200 M Ni-treated seedlings of Kranti, Varuna and Pusa Jai Kisan, CAT activity dropped by 14%, 44% and 34%, respectively (Thakur and Sharma, 2016). With the exception of 10 M Ni, a similar upward pattern in CAT activity was seen in shoots and roots, and the CAT activity in the 200 M Ni treatment increased by 25.19% and 281.40% when compared to the control, respectively. The rice shoot's CAT activity was 25% lower at 10 M Ni than that of the control (Rizwan *et al.*, 2017). In comparison to the control (0.0179 U mg^{-1} protein), the addition of 20 mg L^{-1} Ni (0.0177 U mg^{-1} protein) did not significantly alter the CAT activity. Even at the greatest Ni concentration (20 mg L^{-1}), CAT activity was significantly reduced. *Lemna gibba* plants exposed to Ni showed statistically significant differences between treatments. Additionally, a highly significant positive association between CAT and APX, as well as between CAT and glutathione reductase (GR) activity, was discovered in this plant (Yilmaz and Parlak, 2011).

5.2.4. *Heme oxygenase*

Recent discoveries have helped in better understanding how heme oxygenase (HO), a novel antioxidant enzyme, protects plant cells from oxidative stress. HOs are a universal and highly active family of enzymes which catalyse the oxidative degradation of Fe(III) protoporhyrin IX (heme) to biliverdin, Fe(II) and CO in the presence of reducing equivalents

(Shekhawat and Verma, 2010). A wide range of biological systems, including mammals, algae, cyanobacteria, cryptophytes, higher plants and pathogenic bacteria, have already yielded the genes responsible for producing HOs (Mahawar *et al.*, 2018). (Mahawar *et al.*, 2018) looked at how the heme oxygenase (HO1) affected the uptake of a HM, Ni, and the impacts of Ni toxicity in *Vigna radiata* var. PDM 54 seedlings grown hydroponically. The highest levels of (HO1) activity were discovered in leaves at 60 mM $NiSO_4$. At this metal concentration, the largest (HO1) activity was evident from the decrease in H_2O_2 content, which is coupled with the rising activity of other antioxidant enzymes (CAT and APX). *V. radiata* seedlings were more resistant to metal-induced oxidative stress, in which (HO1) is localised in its residence (plastids) (Mahawar *et al.*, 2018).

5.3. *Sulphur and sulphur-containing compounds*

5.3.1. *Sulphur*

An optimal level of S can improve the growth of plants under trace element stress because S has an important protective function against trace element stress. The sulphydryl functional groups (–SH) of ligands (GSH and phytochelatins), which can form complexes with trace metals, are related to the protective role of S compounds against excessive levels of trace metals (Zagorchev *et al.*, 2013).

The shoot and root biomass of Ni-treated *Brassica juncea* plants under a S deficit was considerably lower than it was with an adequate S supply (Schiavon *et al.*, 2008). According to the research, a S deficit causes higher levels of oxidative stress. The concentration of TBARS was considerably higher under S deficiency than under the standard level of S. This was supported by research findings that demonstrated an antioxidative response generated by the additional S supply, which resulted in less GSH building up in plants treated with Ni (Matraszek-Gawron and Hawrylak-Nowak, 2019).

5.3.2. *Glutathione*

GSH has an important place in the endogenous defence system and is widely found in plant organelles, such as the endoplasmic reticulum, cytosol, mitochondria, chloroplast, vacuole and peroxisome (Kuloğlu *et al.*, 2022). GSH and perhaps its derivatives (phytochelatins) are synthesised in response to Ni in *Elodea* leaves (Tomaszewska and Mozia, 2002). Ni can cause sub-lethal oxidative stress in *Elodea* leaves at doses up to 10 μM. As a result, plants evolved detoxification strategies that included improved thiol-containing chemical production. Ni might efficiently accumulate and be sequestered in plant tissues under these circumstances. As a result, *E. canadensis* could be utilised to biologically remove up to 10 μM of Ni from contaminated water (Maleva *et al.*, 2009). In comparison to the control, GSH content increased in the 10, 50, 100 and 200 M Ni shoots as well. At 200 M Ni treatment, GSH contents in the shoot increased by 40.31%, whereas GSH contents in the Ni-treated rice roots increased by a marginally significant amount (Rizwan *et al.*, 2017). GSH levels dropped in all of the groups treated with Li_2CO_3; in particular, a notable reduction of 60.8% was found in the root meristematic tissue of *Allium cepa* treated with 100 mg L^{-1} Li_2CO_3 (Kuloğlu *et al.*, 2022).

The maximum levels of GSH, total polyphenols and total flavonoid contents were discovered in sweet potatoes after 60 mg L^{-1} Ni treatment. Additionally, the greatest increases of 344.2%, 312.7% and 273.5%, respectively, in the levels of GSH, total polyphenols and total flavonoids were found with 60 mg L^{-1} Ni treatment compared to the control (Kumar *et al.*, 2022). With increasing concentrations of Zn and Ni provided externally, the GR activity of two cultivars' roots and shoots rose and reported higher values when compared to their controls. In response to both metal treatments, LRG30, one of the two *pigeon pea* cultivars, recorded higher values of GR activity than ICPL87 (Madhava Rao and Sresty, 2000).

5.4. *Proteasome activity*

The intermediate Ni accumulators *Alyssum sibiricum* and *Alyssum caricum*, which are native to Turkey's serpentine soil, also have larger levels of Ni accumulation and higher proteasome activity compared to other low-accumulating *Festuca* or *Astragalus* species. *A. sibiricum* had higher proteasome activity on serpentine soil than on non-serpentine soil. The relatively minimal correlation between proteasome activity and Ni soil bioavailability or accumulation in leaf tissue in *A. caricum* suggests that proteasome function was not compromised in plants that accumulated the most Ni (Van Hoewyk *et al.*, 2018).

On serpentine soil, *Alyssum sibiricum* plants displayed almost a two-fold increase in proteasome activity. The amount of Ni in *Festuca valesiaca* was not affected by the kind of soil. Ni concentrations in *Alyssum sibiricum* were higher on serpentine soil than on non-serpentine soil, but they did not rise above the threshold of 1 ptt to signify hyperaccumulation (Van Hoewyk *et al.*, 2018).

Because they recycle the amino acids required to keep protein synthesis going, proteasomes are essential to metabolism (Suraweera *et al.*, 2012). Furthermore, proteasomal degradation of damaged proteins is essential for cellular survival under stress; however, it comes at a cost because it consumes a lot of energy. The slow growth rate of Ni hyperaccumulators may be due to the high costs of protein quality control, including high proteasome activity and *de novo* protein synthesis. In fact, in the Ankara Beynam Forest, *A. sibiricum* had higher proteasome activity than *Festuca valesiaca* and *Astragalus* sp. In comparison to the other species, the higher proteasome activity may offer protection from Ni stress. If this happens, higher proteasome activity in Ni accumulators may be responsible for their distinct ecophysiology and may also explain why they grow slowly (Van Hoewyk *et al.*, 2018).

5.5. *Other compounds*

Alongside the aforementioned substances, other compounds or reactions may exist. Increased protein nitration may be an effect at the protein level resulting from increased RNS signalling, as has already been shown in the case of excess elements (Feigl *et al.*, 2016). Ten nitrated protein bands were seen in untreated *Arabidopsis thaliana* seedlings, compared to eight protein bands in *Brassica juncea* that were immune positive for 3-nitrotyrosine (Kolbert *et al.*, 2020). These show that both plant species' healthy individuals have physiological nitroproteomes that are comparable to earlier findings (Kolbert *et al.*, 2017). From the foregoing, it may be inferred that excess Ni noticeably enhances protein nitration in the relatively sensitive *A. thaliana*, whereas Ni-induced protein tyrosine nitration is less severe in the comparatively resistant *B. juncea* (Molnár *et al.*, 2018).

6. Exogenous Applications Mediating Ni Tolerance

Exogenous injection of NO- or H_2 donors reduces Ni-induced toxicity in part because ROS levels and signalling are reduced (Hamdzah *et al.*, 2013; Kotapati *et al.*, 2017; Rizwan *et al.*, 2018). This is an intriguing connection between ROS, RNS and RSS.

6.1. *Plant growth regulators and plant growth promoting rhizobacteria*

Different examples of plant growth promoting rhizobacteria (PGPR) and their impacts on different plant species are illustrated in Table 1.

Table 1. Different applications used to overcome the nickel stress on many plants.

Application	Used plant	Action	Citation
Micrococcus luteus & Enterobacter cloacae	Tomato plants	Creation of phytohormones in addition to siderophores. Osmolality control and antioxidant capability.	(Badawy *et al.*, 2022)
Paecilomyces formosus LHL10	Soybean plants	Lowering the production of lipid peroxidation and the Building up of glutathione, peroxidase, linolenic acid, and Antioxidant enzymes. down-regulation of jasmonic acid and abscisic acid.	(Bilal *et al.*, 2017)
Rhizobium pisi PZHK2 and Ochrobacterium pseudogrignonense PZHK4	*Pongamia pinnata*	Promotion of non-enzymatic antioxidants to reduce nickel stress or shortages in nitrogen.	(Yu *et al.*, 2022)
Rhizobium pisi and Ochrobacterium pseudogrignonense	*Pongamia pinnata*	Inducing antioxidant capacity and increasing the nitrogen supplement.	(Shoaib *et al.*, 2021)
Pistachio husk biochar (PHB) and arbuscular mycorrhizal fungi (AMF)	Mung bean	Reduce oxidative stress and Ni distribution, and enhance the biological components in grain.	(Turan, 2021)
Brassinosteroid "28-homobrassinolide" (HBL)	*Vigna radiata*	Caused an increase in proline (an osmolyte) buildup and antioxidant enzyme overexpression.	(Yusuf *et al.*, 2014)
$CaCl_2$	*Cucurbita pepo L.*	Decreased Ni-induced buildup of hydrogen peroxide, malondialdehyde, and proline, altered cation homeostasis, enhanced peroxidase and catalase enzyme activity. In addition to deceased Ni- accumulation	(Valivand and Amooaghaie, 2021)

Table 1. (*Continued*).

Application	Used plant	Action	Citation
Citric acid	Sunflower	Increases the effectiveness of Ni phytoextraction and safeguards the plant against oxidative stress.	(Ijaz *et al.*, 2022)
Chitosan 2%		Effectively regulating Ni's bioavailability through immobilization.	(Turan, 2019)
Pistachio shell biochar 2%	Lettuce plants	Increased biomass, water content, photosynthetic capacity, biochemistry, nutritional potential, and antioxidant defence mechanism against oxidative damage.	(Turan, 2019)
Ethylene (Ethephon)	Mustard plants (*Brassica juncea*)	Increase the production of proline, glyoxalase systems, and antioxidant enzymes to resist oxidative stress.	(Khan *et al.*, 2020)
Ethylene (Ethephon)	Mustard plants (*Brassica juncea*)	Net photosynthesis is improved as a result of increased ascorbate peroxidase and glutathi (M. I. R. Khan & Khan, 2014).	(Khan and Khan, 2014)
Silicon (Si)	Young maize plants	Improving membrane stability, alleviating leaf water deficiency, and having an impact on both enzymatic (SOD, POX, CAT) and non-enzymatic (Pro, AsA) defence mechanisms.	(Fiala *et al.*, 2021)
Melatonin	Tomato seedlings	Recovered growth characteristics, increased genes involved in chlorophyll biosynthesis, reduced Ni uptake by plants, and decreased oxidative damage all helped to reduce the effects of Ni-induced phytotoxicity. Additionally, the quantity of secondary metabolites increased, which help to chelate heavy metals and prevent the creation of ROS.	(Jahan *et al.*, 2020)

Table 1. (*Continued*).

Application	Used plant	Action	Citation
Jasmonic acid (JA)	Glycine max L.	Favourably modifies different antioxidants while severely suppressing the stress-related mechanisms that lead to lipid peroxidation.	(Sirhindi *et al.*, 2015)
Overexpression of serine acetyltransferase (O-acetyl-L-serine)	From hyperaccumulator plant *Thlaspi goesingense*	Boosts nickel resistance in *Escherichia coli* because glutathione biosynthesis, a positive regulator of the cysteine regulon, and sulphur assimilation are overactivated.	(Freeman *et al.*, 2005)
Biochar (BC)	*Brassica napus*	Immobilize Ni in soil.	(Hannan *et al.*, 2021)
Rare-earth element yttrium	(Potamogeton crispus)	Low-dose Y can enhance the resistance to nickel stress via suppressing oxidative stress, boosting photosynthesis, and positively regulating polyamine transformation.	(Lyu *et al.*, 2019)
Combined applications of Polyamines (PAS) and arbuscular mycorrhiza (AM)	*Pigeonpea Cajanus cajan*	Increasing the production of phytoch (Saroy & Garg, n.d.) toxicity.	(Saroy & Garg, 2021)
Salicylic acid	Mustard plants	Restore growth, photosynthesis, and physio-biochemical characteristics while boosting the activity of enzymes connected to antioxidant systems, especially the ascorbate-glutathione (AsA-GSH) cycle and glyoxalase system, to lessen the negative impacts of Ni.	(Zaid *et al.*, 2019)
Salicylic acid	Chickpea (*Cicer arietinum L.*)	Activate nickel tolerance to control nickel toxicity.	(Çanakci and Dursun, 2011)

(*Continued*)

Table 1. (*Continued*).

Application	Used plant	Action	Citation
Seed priming with selenium or salicylic acid	Rice seedlings	Reduced ROS production, increased membrane integrity, a powerful antioxidant defence mechanism, and maintenance of mineral nutritional status.	(Khan *et al.*, 2020)
Seed priming with H_2S and Ca^{2+}	Zucchini seedlings	Trigger signal memory that promotes cross-adaptation against nickel stress, allowing these memory cells to be used later on to more powerfully improve the AsA-GSH cycle, redox homeostasis, and phytochelatin transcripts in leaves.	(Valivand *et al.*, 2019)
Silicon	*Oryza sativa L.*	Enhanced the majority of the glyoxalase and antioxidant defence systems, which enhanced MG detoxification and ROS scavenging.	(Hasanuzzaman *et al.*, 2019)
Terminalia arjuna bark extract	Rice seedlings	Decreasing lipid peroxidation, restoring the level of photosynthetic pigments and chloroplast ultrastructure, and regaining the levels of antioxidative enzymes. Suppressing the uptake of Ni and the production of reactive oxygen species.	(Rajpoot *et al.*, 2016)

6.1.1 *Ethylene*

Ethephon (2-chloroethyl phosphonic acid) is converted to ethylene in plant cells, where it is then used to control a variety of physiological processes (Khan and Khan, 2014). Ethylene is a gaseous phytohormone that may play a role in abiotic stress tolerance as well as developmental processes, from seed germination to senescence (Khan *et al.*, 2017). Ethephon treatment significantly boosted growth features as well as antioxidant enzyme and glyoxalase system production. Through improved photosynthetic efficiency and proline production, plants' water relations and cellular homeostasis were preserved by adding ethephon exogenously. At 30 DAS, ethephon application decreased TBARS concentration and electrolyte leakage by 49% and 62% in Ni-treated plants, and by 1.42 and 1.25 times at 60 DAS, respectively. In comparison to Ni-treated plants, the application of ethephon under Ni stress reduced methylglyoxal (MG) content by around 16% at 30 DAS and 12% at 60 DAS.

The application of ethephon to Ni-treated plants dramatically raised the activity of SOD, APX, GR and GPX by 99% at 30 DAS and by 105%, 195%, 163% and 104%, respectively, at 60 DAS. By promoting PS II activity, ribulose-1,5-bisphosphate carboxylase activity and photosynthetic nitrogen use efficiency (PNUE), ethylene safeguarded photosynthetic potential (Khan *et al.*, 2020b). Due to the induced activity of APX and GR, which results in increased production of GR, an application of 200 μL L^{-1} ethephon to Ni-grown plants significantly reduced toxicity and oxidative stress to a greater extent, along with improved net photosynthesis. By lowering the oxidative stress brought on by stress-induced ethylene production and maintaining an elevated GSH pool, ethylene creation brought about by ethephon treatment relieved Ni stress (Khan and Khan, 2014).

6.1.2. *Brassinosteroids*

The novel class of steroidal hormones known as brassinosteroids (BRs) exhibits a speedier response at very low concentrations. The majority of plants, including roots, contain BRs, which are structurally linked to steroid hormones found in animals and insects (Bajguz and Tretyn, 2003). Various biotic and abiotic stressors are ameliorated by BRs.

Exogenous administration of BRs improved the pool of antioxidant enzyme activity in Ni-stressed *R. sativus* L. and *Triticum aestivum* plants, reducing the harmful effects of Ni-metal (Yusuf *et al.*, 2014).

Brassica juncea L. plants were raised in pots with the goal of examining the function of BRs under high metal stress. Castasterone, Dolicholide, 24-Epibrassinolide (24-EBL) and Typhasterole are among the BRs isolated from the leaves of *Brassica* plants. It was shown that 24-EBL greatly reduced metal uptake in plants and boosted the activities of stress-relieving enzymes. Recent studies have shown that BRs give plants the ability to tolerate stress, including pathogen attack, high and low temperatures, salt stress, HM stress and water stress (Bajguz and Hayat, 2009).

Under Ni metal stress, foliar spray treatment with 24-EBL increased the activity of the antioxidative enzymes in the leaves of 60-day-old *B. juncea* plants, with the exception of APOX. The highest levels of CAT (121.74 mol UA mg^{-1} protein) and POD (194.47 m mol UA mg^{-1} protein) were discovered in the leaves of plants that had been sprayed with 109 M of 24-EBL and given a treatment of 2.0 mM of Ni metal. In plant leaves treated with 2.0 mM of Ni and sprayed with 1,011 M of 24-EBL, the superoxide radical-scavenging enzyme SOD displayed its highest level of activity (254.4 mol UA mg^{-1} protein). The application of 24-EBL as a therapy (foliar spraying) to Ni-challenged plants substantially boosted GR and DHAR activity. The plants that received a foliar application of 109 M of 24-EBL and a 2.0 mM metal treatment showed the greatest increases in the activities of GR (104.98 m mol UA mg^{-1} protein) and DHAR (286.46 m mol UA mg^{-1} protein). The plants treated with 2.0 mM of Ni metal and sprayed with 1,011 M of 24-EBL showed the highest activity of MDHAR (168.16 m mol UA mg^{-1} protein) (Bajguz and Hayat, 2009).

The findings of the above study supported (Mazorra *et al.*, 2002) finding that tomato's CAT activity was boosted when exposed to BRs, which showed that foliar spraying *B. juncea* plants with 24-EBL elevated antioxidant enzyme activities. Treatments with brassinolide could therefore reduce the ROS produced by HMs (Hayat *et al.*, 2007). In *Chorispora bungeana* under chilling stress, 24-EBL administration dramatically increased antioxidative enzyme activity and antioxidant content (Kanwar *et al.*, 2012).

6.1.3. *Melatonin application*

Melatonin, on the other hand, restored growth characteristics, improved photosynthesis efficiency (upregulation of genes involved in chlorophyll biosynthesis), improved mineral homeostasis (lower Ni uptake in plants) and decreased oxidative damage. Additionally, the amounts of secondary metabolites were increased in the melatonin-treated seedlings. These compounds primarily take part in the chelation of HMs and the suppression of ROS generation, which lessens the effects of Ni-induced growth inhibition. Overall, melatonin treatment effectively reduced the phytotoxicity caused by Ni and increased the resistance of tomato seedlings to Ni stress (Jahan *et al.*, 2020).

Plants that receive melatonin supplements also build up more GSH and phytochelatins, which help trap excess HMs in the cell (Cao *et al.*, 2018; Hasan *et al.*, 2015). Melatonin supplementation, as opposed to plants that simply experienced Ni stress, increased root activity by 21.2%. However, compared to tomato seedlings that had just been exposed to Ni stress, melatonin and Ni therapy showed a discernible drop in MDA concentration in leaves and roots of up to 20.9% and 24.3%, respectively. In addition, supplementation with melatonin reduced EL by as much as 21.6% and 17.9%, respectively, as compared to merely Ni stress, protecting the leaves and roots from cellular damage (Jahan *et al.*, 2020).

Furthermore, 100 M melatonin application combined with Ni treatment further increased the activity of SOD, APX, CAT and GR by 24.9/22.3%, 22.2/170.1%, 11.3/20.5% and 26.2/58.9%, respectively, in leaves/roots. Melatonin effectively reduced hydrogen peroxide (H_2O_2) and superoxide radical formation, boosted RBOH expression and restored cellular integrity (less MDA and electrolyte leakage). Notably, numerous defence genes (SOD, CAT, APX, GR, GST, MDHAR and DHAR), genes associated with melatonin production (TDC, T5S, SNAT and ASMT) and genes responsible for chlorophyll production were upregulated. Melatonin may reduce Ni-induced phytotoxicity in tomatoes by increasing biomass production, photosynthesis, nutrient uptake and redox balance, as well as improving mineral nutrient homeostasis and secondary metabolism. Melatonin treatment improved root architecture and organic acid anion exudation, which may inhibit excessive metal translocation and accumulation (Fashola *et al.*, 2016; Jahan *et al.*, 2020).

6.1.4. *Jasmonates*

Jasmonates (JAs) are widely distributed in plants and are a subclass of phytohormones generated from polyunsaturated fatty acids. According to Cheong and Choi (2003) and Creelman and Mullet (1995), JAs are signalling molecules that activate a number of signalling pathways involved in diverse functions in plants, including the control of plant growth and development, the induction of biotic and abiotic resistances, responses to various environmental signals and crosstalk with other phytohormones. According to (Wasternack and Hause, 2002), methyl jasmonate (MeJA) is an esterific derivative of JA that mimics stressors in plants and induces senescence-related genes. Garrido *et al.* (2003) presented the first evidence of a direct connection between MeJA – induced ROS and PODs’ *in vivo* activity in the roots of sunflower seedlings. Rice seedlings that received an exogenous application of MeJA exhibited lower water stress and chilling (Lee *et al.*, 1996).

The antioxidant defence mechanism of *Glycine max*, which is composed of SOD, POD, CAT, APOX and AsA, is triggered by the jasmonic acid treatments of plants. It further enhanced the POD activity to attain a maximum at both 1 M and 1 nM alone and in combination with metal JA, respectively, in mitigating Ni-induced oxidative stress by concurrently decreasing MDA and H_2O_2 content to significant levels. JA protects *G. max* seedlings from Ni poisoning by controlling the antioxidant system and safeguarding the DNA synthesis of total proteins. In *G. max* plants, adding JA before Ni treatment improved the chlorophyll contents compared to plants that did not receive JA, showing that JA may be inhibiting the interference of Ni with other metal ions needed by the photosynthetic machinery (Sirhindi *et al.*, 2015).

JA's regulatory reduction of MDA and H_2O_2 generation made plants more resilient to Ni^{2+} stress, which may be related to the ameliorative activity of the antioxidant enzyme system made up of SOD, POD, CAT and APOX. The AsA content of seedlings treated with Ni^{2+} and supplemented with various doses of JA showed manyfold improvements. The enhanced regulation of the transcriptional machinery by JA, which was also maintained by Ni^{2+} stress, was demonstrated by an increase in total protein content. JA was ineffective at protecting photosynthetic pigments; it significantly modified other plant machinery, including various antioxidants, while tightly inhibiting stress-related processes that lead to lipid peroxidation, making plants resistant to Ni^{2+} stress (Sirhindi *et al.*, 2015).

6.1.5. *PGPR*

The PGPR, which are bacterial phytohormones, control the hormonal and nutritional balance to increase stress tolerance in plants (Spence and Bais, 2015). In natural ecosystems, PGPR are essential for defending plants from a variety of stresses, such as excessive HM concentrations. In order to counteract the inhibitory effects of HM toxicity, PGPR has evolved a number of techniques (Tiwari and Lata, 2018). The five main strategies include HM biosorption/precipitation by the cell surface, HM efflux pumping out of the cell by the transporter system, HM binding in the cell vacuole and other intracellular compartments, the exclusion of HM chelates into the extracellular space and an enzymatic redox reaction via the conversion of HM cations into a less toxic state. Additionally, PGPR create a particular blend of volatile organic compounds (VOCs), which affect plant growth hormones and are crucial to their interactions with plants (Raza *et al.*, 2020). PGPR can secrete low-molecular-weight organic acids, which increase Ni and other HM bioavailability for plant uptake (Almaroai *et al.*, 2012). A number of organic acids, such as, citric, oxalic and malonic lactic, possess chelating properties. The salts formed from these organic acids with HMs enter the plants. For example, Ni–gluconate and Ni–citrate complexes were found to be present in cocoa (Peeters *et al.*, 2017).

The main Ni resistance mechanism in *Betula papyrifera* is the down-regulation of genes associated with translation (in the ribosome), binding and transporter activities (Theriault *et al.*, 2016). The activity of GR was significantly increased in pea plants (growing under Ni and Zn stresses) after inoculation with *Rhizobium* sp. RP5 (Wani *et al.*, 2008).

It was shown that the inoculation of rice with *Klebsiella variicola* F2, *Pseudomonas fluorescens* YX2 and *Raoultella planticola* YL2 led to the accumulation of N,N,N-trimethyl glycine (GB) and its precursor choline and improved water content in leaves (Gou *et al.*, 2015). GB *in vivo* is both an effective osmoprotectant and a compatible solute (Felitsky *et al.*, 2004). It was also found that GB increased under Ni stress (Sirhindi *et al.*, 2016).

The new metallophorpseudopaline from *Pseudomonas aeruginosa* is known to be more specific for the chelation of Ni and zinc (Lhospice *et al.*, 2017). Bacterial phytohormone PGPR regulate the nutritional and hormonal balance in plants and induce plant tolerance to stress (Spence and Bais, 2015).

Auxin-producing PGPR reduced the effect of HM stress on physiological processes in plants (Pishchik *et al.*, 2021). For instance, auxin-producing *B. megaterium* MCR-8 increased growth, the contents of total phenols, flavonoids and the activities of SOD, CAT, POD and APX in inoculated *Vinca rosea* plants under Ni stress (Khan *et al.*, 2017).

Abscisic acid (ABA) is a plant hormone that regulates water misbalance in plants by controlling stomatal closure and stress signal transduction pathways (Kudoyarova *et al.*, 2013). Therefore, we may suggest that ABA-producing PGPR can help plants overcome water misbalance caused by Ni stress.

Various plant–bacterial associations were explored by researchers: Ni hyperaccumulators *Alyssum bertolonii* (Mengoni *et al.*, 2001) and

Tlaspigoesingense with their Ni-resistant dominant bacteria from genera *Pseudomonas*, *Methylobacterium*, *Rhodococcus* and *Okibacterium*. According to (Theriault *et al.*, 2016), the principal Ni resistance mechanism in *Betula papyrifera* is the downregulation of genes linked to (ribosomal) binding, transporter and translation activities.

It was found that more than 80% of endophytic bacteria increased the production of siderophores in the presence of HMs (Ni, Co, Cr, Cu and Zn) and also reduced metal toxicity in their host plant, *Alyssum bertolonii* (Ma *et al.*, 2011a). Ni-resistant *Pseudomonas* sp. A3R3 increased plant biomass as well as Ni accumulation in *Brassica juncea* and *A. serpyllifolium* due to ACC-deaminase, siderophores and IAA activities (Ma *et al.*, 2011b). However, there also exists contradictory evidence that bacterial siderophores bound Ni cations (such as Pb and Zn), decreased Ni contents in plants and protected plants against HM toxicity (Tank and Saraf, 2009).

6.2. *Rare earth element yttrium*

As an exogenous additive, the low-dose rare earth element yttrium (REE Y) improved the DAO and SAMDC activities and further prevented oxidative stress. By increasing the total chlorophyll content and maximum/potential quantum efficiency of PS II in *P. crispus* under HM stress, Y also promoted photosynthesis in the macrophyte. The tolerance of *P. crispus* to Ni toxicity was physiological performance driven by Y rather than metal competition in the absorption processes. The exogenous Y treatment can significantly increase *P. crispus* tolerance to HMs (Lyu *et al.*, 2019).

6.3. *Bamboo biochar*

Due to its role in improving soil fertility and plant growth, using biochar (BC) as a soil amendment has become a promising practice. As a result of better NPK dynamics and BC amendment in the soil, *Fokienia hodginsii's* morphological traits increased, reaching a maximum increase in biomass of 29% (Tarin *et al.*, 2018). It was proved that BC application to contaminated soil immobilized HMs like (Zn, Pb, Cd and Cu) (Ali *et al.*, 2017). Ni can be immobilised in soil by immobilising chemicals, particularly biochar (BC). BC considerably improved plant biomass characteristics and total chlorophyll content, among other modifications. In both the absence and presence of Ni, amended soil drastically reduced the ROS content

(MDA, H_2O_2 and O^{2-}) of *B. napus* fresh leaves. Treatments with BC + Ni (50 mg), and BC + Ni (100 mg) reduced the levels of MDA by 51.73% and 58.1%, H_2O_2 by 24.7% and 30.4%, and O_2 by 78.14%, respectively, in comparison to the control. Overall, compared to the improved soil, the unamended soil under Ni stress had the highest ROS concentrations. During Ni stress, BC-treated soil considerably increased the activities of SOD, APX, CAT, GR, MDHAR and DHAR (176.27%, 207.37%, 117.93%, 104%, 94% and 74% in Ni 50 mg kg^{-1} and 149.76%, 249.28%, 137.33%, 217.74%, 171.76% and 55.91% in Ni 100 mg kg^{-1}, respectively). The amount of Ni in *Brassica napus* shoots and roots was dramatically reduced by the addition of BC. Whereas, in 100 mg kg^{-1} Ni, BC produced the largest reductions in root and shoot Ni concentrations compared to their respective controls. BC contained micro- and meso-holes, which contributed to the presence of more Ni adsorption sites. In BC, Ni mineral complexes were created that are less harmful and may provide nutrients to plants and soil enzymes (Pinto *et al.*, 2019).

6.4. *Microorganism inoculation*

Tsuruta (2005) investigated the accumulation of Li by microorganisms. Among the 70 strains of 63 species tested (20 bacteria, 18 actinomycetes, 18 fungi and 14 yeasts), *Arthrobacter nicotianae* (1.0 mg g^{-1} DW cells) and *Brevibacterium helvovolum* (0.7 mg g^{-1} DW cells) show a noticeable Li accumulation ability (Aral and Vecchio-Sadus, 2008). Yeast (*Saccharomyces cerevisiae*) was found to only take up a small quantity of Li in its early stage, and growth deterioration occurred at high levels (115–400 ppm) (Aral and Vecchio-Sadus, 2008).

It was revealed that the inoculation with endophytic bacteria *Enterobacter ludwigii* SAK5 and *Exiguobacterium indicum* SA22 led to increased Ni content in rice plants. However, the expression levels of stress-responsive genes, such as *OsGST* (glutathione-s transferase), *OsMTP1* (HM transporting) and *OsPCS1* (phytochelatin synthases), were lower in treated inoculated plants than in treated non-inoculated plants, which indicated a decrease in stress levels when inoculated with bacteria (Pishchik *et al.*, 2021).

Additionally, compared to plants under HM stresses, treatment with *Micrococcus luteus* and *Enterobacter cloacae* resulted in a significant increase in proline and phenol contents. When tomato plants were treated with *Micrococcus luteus* in comparison to stressed plants and extended to control plants, significant increases in proline and phenol contents were found (Badawy *et al.*, 2022). Endophytic bacterium *Caulobacter* sp. MN13 (alone and in combination with zeolite) reduced Ni uptake by sesame plants due to bacterial EPS and improved the biochemical and agronomic parameters of plants (Gulzar *et al.*, 2020).

6.5. *Arbuscular mycorrhizal fungi (AMF)*

According to (Turan, 2021), arbuscular mycorrhizal fungi (AMF) are among the most prevalent and important microbial species that have a major negative impact on the uptake of metals by plants through modifying metal bioavailability in the soil. AMF may successfully form symbioses with a variety of plant species, including lettuce and *Phaseolus vulgaris* (Aroca *et al.*, 2007; Jahromi *et al.*, 2008). In addition to providing the plant host with nutrients and water, AMF also confers resistance to a number of diseases and soil salinity (Begum *et al.*, 2019). More HMs can accumulate in spores, vesicles and mycelia (intraradical and extraradical) than in plant roots. Furthermore, according to (González-Guerrero *et al.*, 2009) and (Khan *et al.*, 2020a), the total glomalin (TG) released by AMF binds metal ions strongly and decreases their availability to plants.

6.6. *Fluoride*

Fluoride's presence during times of joint stress with Ni reduced Ni bioaccumulation to a point where it behaved more like a helpful micronutrient than an acute toxicant. However, because the elements were not accumulating as much during the combined Ni–fluoride stress, the physiology of the cells was practically unaffected (Ahmad and Ashraf, 2011).

Although the Hill activity in plants under combined Ni–fluoride stress was significantly higher than in plants treated with fluoride or with Ni, it was still 1.2 times lower than in the control sets of aromatic rice cultivar (Gobindobhog). This restriction of photosynthetic activity may be caused by the separate photoinhibitory effects of fluoride and Ni that accumulated to some degree in the joint-stressed seedlings (Hassan *et al.*, 2019; Yilmaz and Parlak, 2011).

The joint Ni–fluoride-treated plants retained substantial flavonoid sinks, which could possibly be attributed to the fact that the majority of the antioxidants were not used to reduce stress because of less ROS production and damage. Anthocyanin levels decreased after Pb, Ni and combined Pb–fluoride toxicity, possibly as a result of their potential use in reducing oxidative damage (Nakabayashi *et al.*, 2014). The interaction of more Ni with free PCs and the creation of bound complexes were the causes of the decreased PC level (Schat *et al.*, 2002). As a result of the synergistic coinhibitory absorption of Ni and fluoride, such a PC-induced reaction was not observed during simultaneous Ni–fluoride treatment, which greatly lowered overall Ni bioaccumulation. The seedlings of aromatic rice cultivar (Gobindobhog) accumulated the highest Fe and Cu contents in the aerial biomass to combat Ni toxicity, which allowed them to express seven of the eight SOD isozymes, including a new isoform. Given that Fe abundance reduces Ni-induced dysregulation of Cu homeostasis and related Ni-mediated toxicity in plants, the increased accumulation of Fe in the roots of aromatic rice cultivar (Gobindobhog, GB) treated with Ni–fluoride was likely the result of GB's natural defence mechanism. As a defence against potential cellular damage caused by existing Ni ions, the seedlings receiving combined Ni–fluoride treatment boosted the Fe content. The normal CAT enzyme activity was maintained under combined Ni–fluoride stress. The plants exposed to only Ni toxicity also exhibited the maximum Fe content in shoots among all the tested sets, which could also be a systemic response to abrogating Ni stress. Since they maintained the highest quantity of Fe in the aerial biomass, seedlings subjected to Ni poisoning showed the highest CAT activity. Fe is credited as a crucial cofactor and activator of CAT, the Fe–porphyrin enzyme (Gajewska and Skłodowska, 2010).

6.7 *Chelating agents*

Many trials were conducted with the aim of raising the Li accumulation capacity of hyperaccumulator plants. The application of exogenous chelating agents, such as ethylene diamine tetra acetic acid (EDTA) and ethylene diamine disuccinic acid (EDDS), to the soil near the maturity stage of some plants induces in them a greater accumulation capacity for Li. Kavanagh *et al.* (2018) observed a 5% mortality in *Brassica oleracea* and *Helianthus annus* while they accumulated 1,496 and 601 mg kg^{-1} of Li, respectively. They proved that EDTA addition improved Li phytoextraction by 300 and 210 mg kg^{-1} for *Brassica napus* and *Brassica oleracea*, respectively. Nevertheless, the mortality of *Brassica oleracea* amounted to >25%, *Helianthus annusto* to 25% and zero for that of *Brassica napus*.

Despite being less stable, EDDS may be superior to EDTA due to the biodegradable nature of the former. Comparing with EDTA, EDDS quenched the rate of mortality to >5% for both Brassica napus and Brassica oleracea; however, it enhanced Li accumulation to 2,590 and 3,091 mg kg^{-1} in the two species, respectively.

Although there is a lack of comprehensive knowledge about Li accumulating plants, it is intuitively deducible that these plants produce sufficient antioxidants to conserve a balance between the aggressive attack of the metals and the vital processes in their bodies. As previously mentioned by Kavanagh *et al.* (2018), some chelators offered a profitable privilege that has made the field of phyto-mining accessible. Some plant species exhibited accumulation levels of >4,000 mg kg^{-1}. Induced metal accumulation involves a considerable buildup of metals using non-hyperaccumulators with a huge biomass. The apoplast pathway rather than the symplast pathway is used to transport the solubilised metals into the plant (Nowack *et al.*, 2006). The effectiveness of utilising chelating chemicals to cause metal buildup in plants has been the subject of numerous studies (Kavanagh *et al.*, 2018).

6.8. *Genetic manipulations*

Li tolerance may be triggered in the presence of ethylene-insensitive mutants, such as ETR1-1 and EIN3-3. Retrograde ethylene production with 2-aminoisobutyrate inhibits Li toxicity (Bueso *et al.*, 2007). In an investigation of the consequences of a persistent rise in intracellular H_2O_2 onion homeostasis, (Bueso *et al.*, 2007) used a knock-out mutant of the Arabidopsis Catalase2 (CAT2, At4g35090) gene. In fact, K+ levels were lowered in the CAT2-1 mutant, whereas Li absorption was increased, as was the uptake of other hazardous cations, such as sodium and norspermidine. The most frequently expressed gene in leaves is CAT2; it is stimulated by some environmental cues, such as light and cold, and is controlled by the circadian clock, peaking in the early morning (McClung *et al.*, 1997).

According to gene expression research using microarrays, the CAT2-1 mutant's expression of genes involved in cation transport, ethylene production and perception was unaltered, which suggests that H_2O_2 regulates these processes at the protein level. These findings show how ethylene, cation homeostasis and oxidative stress interact (Bueso *et al.*, 2007). Although Li^+ perturbs a variety of plant processes, the induction of

1-aminocyclopropane-1-carboxylic acid (ACC) synthase (Liang *et al.*, 1996) and increased ethylene production (Naranjo *et al.*, 2003) are particularly intriguing. High concentrations of this hormone favour growth inhibition and senescence promotion (Bueso *et al.*, 2007). The stimulation of ACC synthase genes by Li is considered to be triggered by the suppression of inositol monophosphatase and the simultaneous decrease in cellular inositol and calcium. The exact method by which Li induces ethylene production is still mostly not understood (Liang *et al.*, 1996).

7. Conclusion

Metal contamination and toxicity date back hundreds of years. It was verified that Li and Ni toxicities to plants are dependent on dose, exposure time and plant species. The current review highlights the signs of Li/Ni oxidative stress, which are described in terms of the deterioration of membrane permeability and the alteration of metabolic behaviour as well as enzyme activities. Antioxidant enzymes, such as SOD, POD, CAT and Aps, show an ascending pattern to alleviate plant suffering. GSH, polyamines, proline and other non-enzymatic antioxidants exhibit trends of increasing concentration in response to increased stressful conditions to mitigate metal toxicity and preserve water balance and ion homeostasis. Several studies have shed light on the role of externally added antioxidants. Ethylene, jasmonate, PGPR and other phytohormone supplements in the growth medium improve plant health. Moreover, they boost the production of antioxidants, safeguard the photosynthetic process and apparatus, and upregulate or downregulate genes responsible for defensive compound production. Not only non-living but also living additives, such as mycorrhizal fungi and endophytic bacteria, can lessen metal uptake or sequester those metals away from plants. Furthermore, they aid the plant in nutrient and water intake after their inoculation and confer resistance to a number of diseases and soil salinity. Finally, extra efforts should be devoted to the investigation of toxic metal uptake, toxicity, interaction and remediation.

Acknowledgement

Mai S. Fouad would like to dedicate the current work to the soul of her father Prof. Dr. Sayed Fouad who was her first instructor and who set her on the right path.

References

Ahmad M.S.A., Ashraf M. (2011). Essential roles and hazardous effects of nickel in plants. *Rev. Environ. Contam. Toxicol.* 214:125–167.

Ali A., Guo D., Zhang Y., Sun X., Jiang S., Gu Z., Huang H., Liang W., Li R., Zhang Z. (2017). Using bamboo biochar with compost for the stabilization and phytotoxicity reduction of heavy metals in mine-contaminated soils of China. *Sci. Rep.* 7:2690. doi:10.1038/s41598-017-03045-9.

Almaroai Y.A., Usman A.R.A., Ahmad M., Kim K.R., Moon D.H., Lee S.S., Ok Y.S. (2012). Effects of synthetic chelators and low-molecular-weight organic acids on chromium, copper, and arsenic uptake and translocation in Maize (*Zea mays* L.). *Soil Sci.* 177:655–663.

Al-Thyabat S., Nakamura T., Shibata E., Iizuka A. (2013). Adaptation of minerals processing operations for lithium-ion (LiBs) and nickel metal hydride (NiMH) batteries recycling: Critical review. *Min. Eng.* 45:4–17.

Aral H., Vecchio-Sadus A. (2008). Toxicity of lithium to humans and the environment-A literature review. *Ecotoxicol. Environ. Saf.* 70:349–356.

Aroca R., Porcel R., Ruiz-Lozano J.M. (2007). How does arbuscular mycorrhizal symbiosis regulate root hydraulic properties and plasma membrane aquaporins in *Phaseolus vulgaris* under drought, cold or salinity stresses? *New Phytol.* 173:808–816.

Bach, Gallicchio. (1990). *Lithium and Cell Physiology*. Hardcover; Springer, New York, 1st edn; ISBN-13: 978-0387971285. Vol. 1, p.174.

Badawy I.H., Hmed A.A., Sofy M.R., Al-Mokadem A.Z. (2022). Alleviation of cadmium and nickel toxicity and phyto-stimulation of tomato plant L. by endophytic *Micrococcus luteus* and *Enterobacter cloacae*. *Plants* 11(15):2018–2045.

Bajguz A., Hayat S. (2009). Effects of brassinosteroids on the plant responses to environmental stresses. *Plant Physiol. Biochem.* 47:1–8.

Bajguz A., Tretyn A. (2003). The chemical characteristic and distribution of brassinosteroids in plants. www.elsevier.com/locate/phytochem.

Bakhat H.F., Rasu K., Farooq A.B.U., Zia Z., Natasha, Fahad S., Abbas S., Shah G.M., Rabbani F., Hammad H.M. (2020). Growth and physiological response of spinach to various lithium concentrations in soil. *Environ. Sci. Pollut. Res.* 27:39717–39725.

Bassi R., Sharma S.S. (1993). Proline accumulation in wheat seedlings exposed to zinc and copper. *Phytochemistry* 33:1339–1342.

Begum N., Qin C., Ahanger M.A., Raza S., Khan M.I., Ashraf M., Ahmed N., Zhang L. (2019). Role of Arbuscular Mycorrhizal Fungi in plant growth regulation: Implications in abiotic stress tolerance. *Front. Plant Sci.* 10: 1068–1082.

Bilal S., Khan A.L., Shahzad R., Asaf S., Kang S.M., Lee I.J. (2017). Endophytic paecilomyces formosus LHL10 Augments *Glycine max* L. Adaptation to ni-contamination through affecting endogenous phytohormones and oxidative stress. *Front. Plant Sci.* 8:870–887.

Bingham F.T., Bradford G.R., Page A.L. (1964). Toxicity of lithium to plants. *Calif. Agric.* 18(9):6–7.

Birch N.J., Troetel J., Nadelmann W.M. (1974). Enzymic and membrane transport effects of magnesium-dependent processes. *The Lancet* 19:965–966.

Birch-Hirschfeld, L. (1920). Investigation of the speed of diffusion of soluble dissolved substances in plants. *Jahrb. Wiss. Bot.* 59:170–262.

Borovik-Romanova T.F. (1965). The content of lithium in plants. In: Khitarov N. (ed.) *Problems of Geochemistry Akad.* Nauk SSSR Inst Geokhimii IAnaliticheskoi Khimii [in Russian]. pp. 675–682.

Bradley D.C., Stillings L.L., Jaskula B.W., Munk LeeAnn, McCauley A.D. (2017). Lithium. In *Critical Mineral Resources of the United States—Economic and Environmental Geology and Prospects for Future Supply*. US Geological Survey, pp. K1–K21.

Bradford G.R. (1966). Lithium. In: Chapman H.D. (ed.) *Diagnostic Criteria for Plants and Soil*. University of California, p. 793.

Bueso E., Alejandro S., Carbonell P., Perez-Amador M.A., Fayos J., Bellés J.M., Rodriguez P.L., Serrano R. (2007). The lithium tolerance of the Arabidopsis cat2 mutant reveals a cross-talk between oxidative stress and ethylene. *Plant J.* 52:1052–1065.

Çanakci S., Dursun B. (2011). Some physiological and biochemical responses to nickel in salicylic acid applied chickpea (*Cicer arietinum* L.) seedlings. *Acta Biol. Hung.* 62:279–289.

Cannon H., Harms T., Hamilton J. (1975). Lithium in unconsolidated sediments and plants of the Basin and Range Province, Southern California and Nevada. United States Geological Survey Professional Paper 918:23.

Cao S., Shao J., Shi L., Xu L., Shen Z., Chen W., Yang Z. (2018). Melatonin increases chilling tolerance in postharvest peach fruit by alleviating oxidative damage. *Sci. Rep.* 8:1.

Chares, A.E. (1990). Lithium in plants. In Bach and Gallicchio (Eds.), *Lithium and Cell Physiology* (1st edn), Springer-Verlag, New York Inc, pp. 25–46.

Chen C., Huang D., Liu J. (2009). Functions and toxicity of nickel in plants: Recent advances and future prospects. *Clean Soil Air Water* 37:304–313.

Chen R., Ma J., Luo D., Hou X., Ma F., Zhang Y., Meng Y., Zhang H., Guo W. (2019). CaMADS, a MADS-box transcription factor from pepper, plays an important role in the response to cold, salt, and osmotic stress. *Plant Sci.* 280:164–174.

Cheong J.J., Choi Y. (2003). Methyl jasmonate as a vital substance in plants. *Trend Genet*. 19:409–413.

Choi W.G., Miller G., Wallace I., Harper J., Mittler R., Gilroy S. (2017). Orchestrating rapid long-distance signaling in plants with Ca^{2+}, ROS and electrical signals. *Plant J.* 90:698–707.

Clemens S. (2006). Toxic metal accumulation, responses to exposure and mechanisms of tolerance in plants. *Biochimie* 88:1707–1719.

Corpas F.J., González-Gordo S., Palma J.M. (2021). Protein nitration: A connecting bridge between nitric oxide (NO) and plant stress. *Plant Stress* 2:100026.

Creelman R.A., Mullet J.E. (1995). Jasmonic acid distribution and action in plants: Regulation during development and response to biotic and abiotic stress. *Proceedings of the National Academy of Sciences of the United States of America*, 92(10): 4114–4119.

Dessemond C., Lajoie-Leroux F., Soucy G., Laroche N., Magnan J.F. (2019). Spodumene: The lithium market, resources and processes. *Minerals* 9(6):334.

Epstein E. (1960). Calcium-lithium competition in absorption by plant roots. *Nature* 185:705–706.

Evans H.J., Sorger G.J. (1966). Role of mineral elements with emphasis on the univalent cations. *Plant Physiol.* 17:47–76.

Fashola M.O., Ngole-Jeme V.M., Babalola O.O. (2016). Heavy metal pollution from gold mines: Environmental effects and bacterial strategies for resistance. *Int. J. Environ. Res. Public Health* 13:11.

Feigl G., Kolbert Z., Lehotai N., Molnár Á., Ördög A., Bordé Á., Laskay G., Erdei L. (2016). Different zinc sensitivity of Brassica organs is accompanied by distinct responses in protein nitration level and pattern. *Ecotoxicol. Environ. Saf.* 125:141–152.

Feigl G., Varga V., Molnár Á., Dimitrakopoulos P.G., Kolbert Z. (2020). Different nitro-oxidative response of *Odontarrhena lesbiaca* plants from geographically separated habitats to excess nickel. *Antioxidants* 9:1–3.

Felitsky D.J., Cannon J.G., Capp M.W., Hong J., Van Wynsberghe A.W., Anderson C.F., Record M.T. (2004). The exclusion of glycine betaine from anionic biopolymer surface: Why glycine betaine is an effective osmo protectant but also a compatible solute. *Biochemistry* 43:14732–14743.

Fiala R., Fialová I., Vaculík M., Luxová M. (2021). Effect of silicon on the young maize plants exposed to nickel stress. *Plant Physiol. Biochem.* 166:645–656.

Freeman J.L., Persans M.W., Nieman K., Salt D.E. (2005). Nickel and cobalt resistance engineered in Escherichia coli by overexpression of serine acetyltransferase from the nickel hyperaccumulator plant *Thlaspi goesingense. Appl. Environ. Microbiol.* 71:8627–8633.

Gajewska E., Skłodowska M. (2010). Differential effect of equal copper, cadmium and nickel concentration on biochemical reactions in wheat seedlings. *Ecotoxicol. Environ. Saf.* 73:996–1003.

Galardi F., Corrales I., Mengoni A., Pucci S., Barletti L., Barzanti R., Arnetoli M., Gabbrielli R., Gonnelli C. (2007). Intra-specific differences in nickel tolerance and accumulation in the Ni-hyperaccumulator *Alyssum bertolonii*. *Environ. Exp. Bot.* 60:377–384.

Garrido J.J., Giraud P., Carlier E., Fernandes F., Moussif A., Fache M.P., Debanne D., Dargent B. (2003). A targeting motif involved in sodium channel clustering at the axonal initial segment. *Science* 300(5628):2091–2094.

Gayathri N., Sailesh A.R., Srinivas N. (2022). Effect of lithium on seed germination and plant growth of *Amaranthus viridis*. *J. Appl. Nat. Sci.* 14: 133–139.

González-Guerrero M., Hong D., Argüello J.M. (2009). Chaperone-mediated Cu^+ delivery to Cu^+ transport ATPases. Requirement of nucleotide binding. *J. Biol. Chem.* 284:20804–20811.

Gou W., Tian L I., Ruan Z., Zheng P., Chen F., Zhang L., Cui Z., Zheng P., Li Z., Gao M., Shi W., Zhang L., Liu J., Hu J. (2015). Accumulation of choline and glycinebetaine and drought stress tolerance induced in maize (*Zea mays*) by three plant growth promoting rhizobacteria (pgpr) strains. *Pak. J. Bot.* 47:581–586.

Graham J.D., Rupp J.A., Brungard E. (2021). Lithium in the green energy transition: The quest for both sustainability and security. *Sustainability* 13(20):11274.

Gratão P.L., Polle A., Lea P.J., Azevedo R.A. (2005). Making the life of heavy metal-stressed plants a little easier. *Funct. Plant Biol.* 32:481–494.

Gulzar N., Hajam A.H., Khan S., Dar S.A., Bhat H.M. (2020). Nickel induced modulations in phosphorus metabolism Nickel induced modulations in phosphorus metabolism and other growth parameters of wheat. *Int. Res. J. Plant Sci.* 11:1–4.

Guo Y., George E., Marschner H. (1996). Contribution of an arbuscular mycorrhizal fungus to the uptake of cadmium and nickel in bean and maize plants. *Plant Soil* 184:195–205.

Gupta V., Jatav P.K., Verma R., Kothari S.L., Kachhwaha S. (2017). Nickel accumulation and its effect on growth, physiological and biochemical parameters in millets and oats. *Environ. Sci. Pollut. Res.* 24:23915–23925.

Habashi F. (2013). Extractive metallurgy and national policy. *Int. J. Non-Ferrous Met.* 2:31–34.

Hamdzah M., Nasef M., Hadibarata T., Olsson G., Hassan H., Rusli N. (2013). optimization of parameters affecting adsorption of Ni Zn and Pb on dowex 50 W resin using response surface methodology approach. In *Environmental Science and Technology* 6(3):106–118.

Handa A.K., Mattoo A.K. (2010). Differential and functional interactions emphasize the multiple roles of polyamines in plants. *Plant Physiol. Biochem.* 48:540–546.

Hannan F., Huang Q., Farooq M.A., Ayyaz A., Ma J., Zhang N., Ali B., Deyett E., Zhou W., Islam F. (2021). Organic and inorganic amendments for the remediation of nickel contaminated soil and its improvement on *Brassica napus* growth and oxidative defense. *J. Hazard Mater.* 416:125921.

Hasan S., Hétié P., Matunis E.L. (2015). Niche signaling promotes stem cell survival in the Drosophila testis via the JAK-STAT target DIAP1. *Dev. Biol.* 40:27–39.

Hassan M.U., Chattha M.U., Khan I., Chattha M.B., Aamer M., Nawaz M., Ali A., Khan M.A.U., Khan T.A. (2019). Nickel toxicity in plants: Reasons, toxic effects, tolerance mechanisms, and remediation possibilities — A review. *Environ. Sci. Pollut. Res.* 26:12673–12688.

Hasanuzzama M., Alam M.M., Nahar K., Mohsin S.M., Bhuyan M.H.M.B., Parvin K., Hawrylak-Nowak B., Fujita M. (2019). Silicon-induced antioxidant defense and methylglyoxal detoxification works coordinately in alleviating nickel toxicity in *Oryza sativa* L. *Ecotoxicology* 28: 261–276.

Hattab S., Hattab S., Flores-Casseres M.L., Boussetta H., Doumas P., Hernandez L.E., Banni M. (2016). Characterisation of lead-induced stress molecular biomarkers in *Medicago sativa* plants. *Environ. Exp. Bot.* 123:1–12.

Hawrylak-Nowak B., Kalinowska M., Szymańska M. (2012). A study on selected physiological parameters of plants grown under lithium supplementation. *Biol. Trace Elem. Res.* 149:425–430.

Hayat S., Ali B., Aiman Hasan S., Ahmad A. (2007). Brassinosteroid enhanced the level of antioxidants under cadmium stress in *Brassica juncea. Environ. Exp. Bot.* 60:33–41.

Hertel R.F., Maass T., Müller V.R. (1991). Nickel/published under the joint sponsorship of the United Nations Environment Programme, the International Labour Organisation, and the World Health Organization, & International Program on Chemical Safety.

Hinz U., Fischer H. (1976). Transport of lithium and cesium along the stolons of *Saxifraga sarmentosa* L.Z. *Pjlanzenphysiology* 78:283–292.

Hull S.L., Oty U.V., Mayes W.M. (2014). Rapid recovery of benthic invertebrates downstream of hyperalkaline steel slag discharges. *Hydrobiologia* 736: 83–97.

Ijaz M., Ansari M.U.R., Alafar H.A., Iqbal M., Alshaya D.S., Fiaz S., Ahmad H.M., Zubair M., Ramzani P.M.A., Iqbal J., Abushady A.M., Attia K. (2022). Citric acid assisted phytoextraction of nickel from soil helps to tolerate oxidative stress and expression profile of NRAMP genes in sunflower at different growth stages. *Front. Plant Sci.* 13:1072671.

Jahan M.S., Guo S., Baloch A.R., Sun J., Shu S., Wang Y., Ahammed G.J., Kabir K., Roy R. (2020). Melatonin alleviates nickel phytotoxicity by improving photosynthesis, secondary metabolism and oxidative stress tolerance in tomato seedlings. *Ecotoxicol. Environ. Saf.* 197:110593.

Jahromi L.B., Gulsrud A., Kasari C. (2008). Emotional competence in children with down syndrome: Negativity and regulation. In: *American Association on Intellectual and Developmental Disabilities*, 113:32–43.

Jiang L., Wang L., Zhang L., Tian C. (2018). Tolerance and accumulation of lithium in *Apocynum pictum* Schrenk. *PeerJ* 8:e5559.

Kabata-Pendias, A., Mukherjee, A.B. (2007). *Trace Elements from Soil to Human.* Springer, Berlin Heidelberg,1–550. doi:10.1007/978-3-540-32714-1, ISBN: 978-3-540-32713-4.

Kalinowska M., Hawrylak-Nowak B., Szymańska M. (2013). The influence of two lithium forms on the growth, L-ascorbic acid content and lithium accumulation in lettuce plants. *Biol. Trace Elem. Res.* 152:251–257.

Kanwar M.K., Bhardwaj R., Arora P., Chowdhary S.P., Sharma P., Kumar S. (2012). Plant steroid hormones produced under Ni stress are involved in the regulation of metal uptake and oxidative stress in *Brassica juncea* L. *Chemosphere* 86:41–49.

Kapoor D., Singh S., Kumar V., Romero R., Prasad R., Singh J. (2019). Antioxidant enzymes regulation in plants in reference to reactive oxygen species (ROS) and reactive nitrogen species (RNS). *Plant Gene* 19.

Kaul S., Sharma S.S., Mehta I.K. (2008). Free radical scavenging potential of L-proline: Evidence from *in vitro* assays. *Amino Acid* 34:315–320.

Kavanagh L., Keohane J., Cabellos G.G., Lloyd A., Cleary J. (2018). Induced plant accumulation of lithium. *Geosciences* 8:100182.

Kent N.L. (1941). Absorption, translocation, and ultimate fate of lithium in the wheat plant. *N. Phytol.* 40:291–298.

Khan F., Hussain S., Khan S., Geng M. (2020a). Seed priming improved antioxidant defense system and alleviated Ni-induced adversities in rice seedlings under N, P, or K deprivation. *Front. Plant Sci.* 11.

Khan M.I.R., Jahan B., AlAjmi M.F., Rehman M.T., Khan N.A. (2020b). Ethephon mitigates nickel stress by modulating antioxidant system, glyoxalase system and proline metabolism in Indian mustard. *Physiol. Mol. Biol. Plant* 26:1201–1213.

Khan M.I.R., Khan N.A. (2014). Ethylene reverses photosynthetic inhibition by nickel and zinc in mustard through changes in PS II activity, photosynthetic nitrogen use efficiency, and antioxidant metabolism. *Protoplasma* 251: 1007–1019.

Khan M.N., Mobin M., Abbas Z.K., Alamri S.A. (2017). Fertilizers and their contaminants in soils, surface and groundwater. *Encyclop. Anthrop.* 225–240.

Kiełczykowska M., Pasternak K., Musik I., Wrońiska J. (2004). The effect of lithium administration in a diet on the chosen parameters of the antioxidant barrier in rats. Annales Universitatis Mariae Curie-Sklodowska. *Sectio D: Medicina* 59:140–145.

Kljakovic M. (1984). Matters arising. *Family Pract.* 1:253–255.

Kolbert Z., Feigl G., Bordé Á., Molnár Á., Erdei L. (2017). Protein tyrosine nitration in plants: Present knowledge, computational prediction and future perspectives. *Plant Physiol. Biochem.* 113:56–63.

Kolbert Z., Oláh D., Molnár Á., Szőllősi R., Erdei L., Ördög A. (2020). Distinct redox signalling and nickel tolerance in *Brassica juncea* and *Arabidopsis thaliana. Ecotoxicol. Environ. Saf.* 189:109989.

Kotapati K.V., Palaka B.K., Ampasala D.R. (2017). Alleviation of nickel toxicity in finger millet (*Eleusine coracana* L.) germinating seedlings by exogenous application of salicylic acid and nitric oxide. *The Crop Journal* 5(3): 240–250.

Krämer U., Pickering I.J., Prince R.C., Raskin I., Salt D.E. (2000). Subcellular Localization and Speciation of Nickel in Hyperaccumulator and Non-Accumulator *Thlaspi* Sp. *Plant Physiol.* 122:1343–1354.

Kszos L.A., Stewart A.J. (2003). Review of lithium in the aquatic environment: Distribution in the United States, toxicity and case example of groundwater contamination. *Ecotoxicology* 12:439–447.

Kudoyarova G.R., Kholodova V.P., Veselov D.S. (2013). Current state of the problem of water relations in plants under water deficit. *Russ. J. Plant Physiol.* 60:165–175.

Kuloğlu S.S., Yalçin E., Çavuşoğlu K., Acar A. (2022). Dose-dependent toxicity profile and genotoxicity mechanism of lithium carbonate. *Sci. Rep.* 12:1.

Kumar S., Wang M., Liu Y., Fahad S., Qayyum A., Jadoon S.A., Chen Y., Zhu G. (2022). Nickel toxicity alters growth patterns and induces oxidative stress response in sweet potato. *Front. Plant Sci.* 13.

Lee J., Parthier B., Lijbler, M. (1996). Jasmonate signalling can be uncoupled from abscisic acid signalling in barley: Identification of jasmonate-regulated transcripts which are not induced by abscisic acid. In *Planta* 199(4): 625–632.

Lhospice S., Gomez N.O., Ouerdane L., Brutesco C., Ghssein G., Hajjar C., Liratni A., Wang S., Richaud P., Bleves S., Ball G., Borezée-Durant E., Lobinski R., Pignol D., Arnoux P., Voulhoux R. (2017). Pseudomonas aeruginosa zinc uptake in chelating environment is primarily mediated by the *Metallophore pseudopaline. Sci. Rep.* 7:1.

Liang X., Shen N.F., Theologis A. (1996). Li^+-regulated 1-aminocyclopropane-1-carboxylate synthase gene expression in *Arabidopsis thaliana. Plant J.* 10:1027–1036.

Liu D., Li T.Q., Jin X.F., Yang X.E., Islam E., Mahmood Q. (2008). Lead induced changes in the growth and antioxidant metabolism of the lead accumulating and non-accumulating ecotypes of *Sedum alfredii. J. Integr. Plant Biol.* 50:129–140.

Lu M.C., Kotelchuck M., Hogan V., Jones L., Wright K., Halfon N. (2010). Closing the Black-White gap in birth outcomes: A life-course approach. *Ethn. Dis.* 20:2019–2022.

Lyu K., Wang X., Wang L., Wang G. (2019). Rare-earth element yttrium enhances the tolerance of curly-leaf pondweed (*Potamogeton crispus*) to acute nickel toxicity. *Environ. Pollut.* 248:114–120.

Ma Y., Prasad M.N.V., Rajkumar M., Freitas H. (2011a). Plant growth promoting rhizobacteria and endophytes accelerate phytoremediation of metalliferous soils. *Biotechnol. Adv.* 29:248–258.

Ma Y., Rajkumar M., Luo Y.M., Freitas H. (2011b). Inoculation of endophytic bacteria on host and non-host plants-effects on plant growth and Ni uptake. *J. Hazard Mater.* 195:230–237.

Madhava Rao K.V., Sresty T.V.S. (2000). Antioxidative parameters in the seedlings of pigeonpea (*Cajanus cajan* (L.) Millspaugh) in response to Zn and Ni stresses. *Plant Sci.* 157:113–128.

Mahawar L., Kumar R., Shekhawat G.S. (2018). Evaluation of heme oxygenase 1 (HO^1) in Cd and Ni induced cytotoxicity and crosstalk with ROS quenching enzymes in two to four leaf stage seedlings of *Vigna radiata. Protoplasma* 252:527–545.

Maleva M.G., Nekrasova G.F., Malec P., Prasad M.N.V., Strzałka K. (2009). Ecophysiological tolerance of *Elodea canadensis* to nickel exposure. *Chemosphere* 77:392–398.

Matraszek-Gawron R., Hawrylak-Nowak B. (2019). Sulfur nutrition level modifies the growth, micronutrient status, and cadmium distribution in cadmium-exposed spring wheat. *Physiol. Mol. Biol. Plant* 25:421–432.

Mazorra L.M., Nunez M., Hechavarria M., Coll F., Sanchez-Blanco M.J. (2002). Influence of barassinosteroids on antioxidant enzymes activity in tomato under different temperature. *Biol. Planta* 45:593–596.

McClung A.M., Marchetti M.A., Webb B.D., Bollich C.N. (1997). Registration of 'Jefferson' *Rice. Crop Sci.* 37:629–630.

McStay N.G., Rodgers H.H., Anderson C.E. (1980). Effects of lithium on *Phaseolus vulgaris* L. *Sci. Total Environ.* 16:185–191.

Mengoni A., Barzanti R., Gonnelli C., Gabbrielli R., Bazzicalupo M. (2001). Characterization of nickel-resistant bacteria isolated from serpentine soil. *Environ. Microbiol.* 3:691–698.

Molas J. (2002). Changes of chloroplast ultrastructure and total chlorophyll concentration in cabbage leaves caused by excess of organic Ni(II) complexes. *Environ. Exp. Bot.* 47.

Molnár Á., Kolbert Z., Kéri K., Feigl G., Ördög A., Szőllősi R., Erdei L. (2018). Selenite-induced nitro-oxidative stress processes in *Arabidopsis thaliana* and *Brassica juncea. Ecotoxicol. Environ. Saf.* 148:664–674.

Nakabayashi R., Yonekura-Sakakibara K., Urano K., Suzuki M., Yamada Y., Nishizawa T., Matsuda F., Kojima M., Sakakibara H., Shinozaki K., Michael

A.J., Tohge T., Yamazaki M., Saito K. (2014). Enhancement of oxidative and drought tolerance in Arabidopsis by over accumulation of antioxidant flavonoids. *Plant J.* 77:367–379.

Naranjo M.A., Romero C., Bellés J.M., Montesinos C., Vicente O., Serrano R. (2003). Lithium treatment induces a hypersensitive-like response in tobacco. *Planta* 217:417–424.

Neumann P.M. (1986). Comparative phloem mobility of nickel in nonsenescent plants. *Plant Physiol.* 81:689–691.

Nishida S., Tsuzuki C., Kato A., Aisu A., Yoshida J., Mizuno T. (2011). AtIRT1, the primary iron uptake transporter in the root, mediates excess nickel accumulation in *Arabidopsis thaliana*. *Plant Cell Physiol.* 52:1433–1442.

Nowack B., Schulin R., Robinson B.H. (2006). Critical assessment of chelant-enhanced metal phytoextraction. *Environ. Sci. Technol.* 40:5225–5232.

Oktem F., Ozguner F., Mollaoglu H., Koyu A., Uz E. (2005). Oxidative damage in the kidney induced by 900-MHz-emitted mobile phone: Protection by melatonin. *Arch. Med. Res.* 36:350–355.

Peeters D., Sekeris E., Verschaffel L., Luwel K. (2017). Evaluating the effect of labeled benchmarks on children's number line estimation performance and strategy use. *Front. Psychol.* 8:1082.

Pietrini F., Iori V., Cheremisina A., Shevyakova N.I., Radyukina N., Kuznetsov V.V., Zacchini M. (2015). Evaluation of nickel tolerance in *Amaranthus paniculatus* l Plants by measuring photosynthesis oxidative status antioxidative response and metal-binding molecule content. *Environ. Sci. Pollut. Res.* 22:482–494.

Pinto M., Soares C., Pinto A.S., Fidalgo F. (2019). Phytotoxic effects of bulk and nano-sized Ni on *Lycium barbarum* L. grown *in vitro* — Oxidative damage and antioxidant response. *Chemosphere* 218:507–516.

Pishchik V., Mirskaya G., Chizhevskaya E., Chebotar V., Chakrabarty D. (2021). Nickel stress-tolerance in plant-bacterial associations. *PeerJ.* 9:e12230.

Rajpoot R., Rani A., Srivastava R.K., Pandey P., Dubey R.S. (2016). *Terminalia arjuna* bark extract alleviates nickel toxicity by suppressing its uptake and modulating antioxidative defence in rice seedlings. *Protoplasma* 253:1449–1462.

Rankin W.H. (1917). The penetration of foreign substances into trees. *Phytopathology* 7:5–13.

Rathor G., Chopra N., Adhikari T. (2014). Nickel as a pollutant and its management. *Int. Res. J. Environ. Sci.* 3:94–98.

Ravenna C., Maugini A. (1912). The behavior of plants toward lithium salts. *Atti. Accad. Naz. Lincei. Cl. Sci. Fis. Mat. Nat. Rend.* 21:292–298.

Raza W., Wang J., Jousset A., Friman V.P., Mei X., Wang S., Wei Z., Shen Q. (2020). Bacterial community richness shifts the balance between volatile organic compound-mediated microbe–pathogen and microbe–plant interactions. *Proc. R. Soc. B: Biol. Sci.* 287:1925.

Rizwan M., Imtiaz M., Dai Z., Mehmood S., Adeel M., Liu J., Tu S. (2017). Nickel stressed responses of rice in Ni subcellular distribution, antioxidant production, and osmolyte accumulation. *Environ. Sci. Pollut. Res.* 24: 20587–20598.

Rizwan M., Mostofa M.G., Ahmad M.Z., Imtiaz M., Mehmood S., Adeel M., Dai Z., Li Z., Aziz O., Zhang, Y., Tu, S. (2018). Nitric oxide induces rice tolerance to excessive nickel by regulating nickel uptake, reactive oxygen species detoxification and defense-related gene expression. *Chemosphere*, 191:23–35.

Sachan P., Lal N. (2017). An overview of nickel (Ni^{2+}) essentiality, toxicity and tolerance strategies in plants. *Asian J. Biol.* 2:1–15.

Salieri G., Vinci G., Antonelli M.L. (1995). Microcalorimetric study of the enzymatic hydrolysis of starch: An wamylase catalyzed reaction. *Anal. Chim Acta* 300:287–292.

Saroy K., Garg N. (2021). Relative effectiveness of arbuscular mycorrhiza and polyamines in modulating ROS generation and ascorbate-glutathione cycle in *Cajanus cajan* under nickel stress. *Environ. Sci. Pollut. Res.* 28: 48872–48889.

Schat H., Llugany M., Vooijs R., Hartley-Whitaker J., Bleeker P.M. (2002). The role of phytochelatins in constitutive and adaptive heavy metal tolerances in hyperaccumulator and non-hyperaccumulator metallophytes. *J. Exp. Bot.* 53:2381–2392.

Schat H., Sharma S.S., Vooijs R. (1997). Heavy metal-induced accumulation of free proline in a metal-tolerant and a nontolerant ecotype of *Silene vulgaris*. *Physiol. Planta* 101:477–482.

Schiavon M., Ertani A., Nardi S. (2008). Effects of an alfalfa protein hydrolysate on the gene expression and activity of enzymes of the tricarboxylic acid (TCA) cycle and nitrogen metabolism in *Zea mays* L. *J. Agric. Food Chem.* 56:11800–11808.

Schrauzer G.N. (2002). Lithium: Occurrence, dietary intakes, nutritional essentiality. *J. Am. Coll. Nutr.* 21:14–21.

Shacklette H.T., Boerngen J.G. (1984). Element concentrations in soils and other surficial materials of the conterminous United States. Geological Survey Professional Paper (USA). Vol. 1270, US Government Printing Office, Washington, DC.

Shahzad B., Tanveer M., Hassan W., Shah A.N., Anjum S.A., Cheema S.A., Ali I. (2016) Lithium toxicity in plants: Reasons, mechanisms and remediation possibilities — A review. *Plant Physiol. Biochem.* 107:104–115.

Shekhawat G.S., Verma K. (2010) Haem oxygenase (HO): An overlooked enzyme of plant metabolism and defence. *J. Exp. Bot.* 61:2255–2270.

Shevyakova N.I., Il'ina E.N., Stetsenko L.A., Kuznetsov V.V. (2011). Nickel accumulation in rape shoots (*Brassica napus* l.) increased by putrescine. *Int. J. Phytorem.* 13:345–356.

Shoaib M., Hussain S., Cheng X., Cui Y., Liu H., Chen Q., Ma M., Gu Y., Zhao K., Xiang Q., Zhou J., Liu J., Li S., Zou T., Yu X. (2021). Synergistic antioxidative effects of *Pongamia pinnata* against nickel mediated by *Rhizobium pisi* and *Ochrobacterium pseudogrignonense*. *Ecotoxicol. Environ. Saf.* 217:112244.

Singh R., Tripathi R.D., Dwivedi S., Kumar A., Trivedi P.K., Chakrabarty D. (2010). Lead bioaccumulation potential of an aquatic macrophyte *Najas indica* are related to antioxidant system. *Biores. Technol.* 101:3025–3032.

Sirhindi G., Mir M.A., Abd-Allah E.F., Ahmad P., Gucel S. (2016). Jasmonic acid modulates the physio-biochemical attributes, antioxidant enzyme activity, and gene expression in glycine max under nickel toxicity. *Front. Plant Sci.* 7:591.

Sirhindi G., Mir M.A., Sharma P., Gill S.S., Kaur H., Mushtaq R. (2015). Modulatory role of jasmonic acid on photosynthetic pigments, antioxidants and stress markers of *Glycine max* L. under nickel stress. *Physiol. Mol. Biol. Plant* 2:559–565.

Speirs J., Contestabile M., Houari Y., Gross R. (2014). The future of lithium availability for electric vehicle batteries. *Renew. Sust. Energy Rev.* 35:183–193.

Spence C., Bais H. (2015). Role of plant growth regulators as chemical signals in plant-microbe interactions: A double edged sword. *Curr. Opin. Plant Biol.* 27:52–58.

Sun, E.J., Wu, F.Y. (1998). Along-vein necrosis as indicator symptom on water spinach caused by nickel in water culture. *Botanical Bulletin of Academia Sinica*, 1–39.

Suraweera A., Münch C., Hanssum A., Bertolotti A. (2012). Failure of amino acid homeostasis causes cell death following proteasome inhibition. *Mol. Cell* 48:242–253.

Tank N., Saraf M. (2009). Enhancement of plant growth and decontamination of nickel-spiked soil using PGPR. *J. Basic Microbiol.* 49:195–204.

Tarin M.W.K., Fan L., Tayyab, M., Sarfraz, R., Chen, L., He, T., Rong, J., Chen, L., Zheng, Y. (2018). Effects of bamboo biochar amendment on the growth and physiological characteristics of *Fokienia hodginsii*. *Appl. Ecol. Environ. Res.* 16(6):8055–8074.

Thakur S., Sharma S.S. (2016). Characterization of seed germination, seedling growth, and associated metabolic responses of *Brassica juncea* L. cultivars to elevated nickel concentrations. *Protoplasma* 253:571–580.

Theriault G., Michael P., Nkongolo K. (2016). Decrypting the regulation and mechanism of nickel resistance in white birch (*Betula papyrifera*) using cross-species metal-resistance genes. *Gene Genom.* 38:341–350.

Tiwari S., Lata C. (2018). Heavy metal stress, signaling, and tolerance due to plant-associated microbes: An overview. *Front. Plant Sci.* 9:452.

Tomaszewska M., Mozia S. (2002). Removal of organic matter from water by PAC/UF system. *Water Res.* 36:4137–4143.

Tsuruta T. (2005). Removal and recovery of lithium using various microorganisms. *J. Biosci. Bioeng.* 100:562–566.

Turan V. (2019). Confident performance of chitosan and pistachio shell biochar on reducing Ni bioavailability in soil and plant plus improved the soil enzymatic activities, antioxidant defense system and nutritional quality of lettuce. *Ecotoxicol. Environ. Saf.* 183:109594.

Turan V. (2021). Arbuscular mycorrhizal fungi and pistachio husk biochar combination reduces Ni distribution in mung bean plant and improves plant antioxidants and soil enzymes. *Physiol. Planta* 173:418–429.

Valivand M., Amooaghaie R. (2021). Calcium signaling confers nickel tolerance in Cucurbita pepo L. *Int. J. Phytorem.* 23:362–373.

Valivand M., Amooaghaie R., Ahadi A. (2019). Seed priming with H_2S and Ca^{2+} trigger signal memory that induces cross-adaptation against nickel stress in zucchini seedlings. *Plant Physiol. Biochem.* 143:286–298.

Van Hoewyk D., Taskin M.B., Yaprak A.E., Turgay O.C., Ergul A. (2018). Profiling of proteasome activity in *Alyssum* Sp. on serpentine soils in Turkey reveals possible insight into nickel tolerance and accumulation. *Plant Physio. Biochem.* 124:184–189.

Vlasiuk P.A., Okhrimenko M.F. (1969). The effect of lithium on photochemical activity of chloroplasts of tomato and pepper leaves. *Acad. Sci. Ukran RSR Kiev. Ser. Geol. Geophy. Chem. Biol.* 4 [in Russian] UDK581.132:546.34.

Vlasyuk P.A., Kuz'menko L.M., Okhrimenko M.F. (1979). The role of lithium in protein-nucleic acid metabolism in plants. *Fiziol. Biokhim. Kul't. Rast.* 11:438–447.

Wani P.A., Khan M.S., Zaidi A. (2008). Chromium-reducing and plant growth-promoting Mesorhizobium improves chickpea growth in chromium-amended soil. *Biotechnol. Lett.* 30:159–163.

Wasternack C., Hause B. (2002). Jasmonates and octadecanoids: Signals in plant stress responses and development. *Prog. Nucleic Acid Res. Mol. Biol.* 72:166–206.

Wietelmann U., Bauer R.J. (2000). Lithium and lithium compounds. In: *Ullmann's Encyclopedia of Industrial Chemistry*. Wiley-VCH Verlag GmbH & Co. KGaA.

Yadav N., Sharma S. (2016). An account of nickel requirement, toxicity and oxidative stress in plants. *Biolog. Forum* 8:414–419.

Yilmaz D.D., Parlak K.U. (2011). Nickel-induced changes in lipid peroxidation, antioxidative enzymes, and metal accumulation in *Lemna gibba. Int. J. Phytorem.* 13:805–817.

Yu X., Shoaib M., Cheng X., Cui Y., Hussain S., Yan J., Zhou J., Chen Q., Gu Y., Zou L., Zhang X., Hao S., Zhao K., Ma M., Xiang Q., Li S., Zou T. (2022). Role of rhizobia in promoting non-enzymatic antioxidants to mitigate nitrogen-deficiency and nickel stresses in *Pongamia pinnata. Ecotoxicol. Environ. Saf.* 241:113789.

Yusuf M., Fariduddin Q., Ahmad I., Ahmad A. (2014). Brassinosteroid-mediated evaluation of antioxidant system and nitrogen metabolism in two contrasting cultivars of *Vigna radiata* under different levels of nickel. *Physiol. Mol. Biol. Plant* 20:449–460.

Yusuf M., Fariduddin Q., Hayat S., Ahmad A. (2011). Nickel: An overview of uptake essentiality and toxicity in plants. *Bull Environ Contam Toxico*, 86, 1–7. https://doi.org/10.1007/s00128-010-0171-1.

Zagorchev L., Seal C.E., Kranner I., Odjakova M. (2013). A central role for thiols in plant tolerance to abiotic stress. *Int. J. Mol. Sci.* 14:7405–7432.

Zaid A., Mohammad F., Wani S.H., Siddique K.M.H. (2019). Salicylic acid enhances nickel stress tolerance by up-regulating antioxidant defense and glyoxalase systems in mustard plants. *Ecotoxicol. Environ. Saf.* 180: 575–587.

https://doi.org/10.1142/9789811283123_0006

Chapter 6

Different Impacts of Ni^{2+} and Li^{+} to Plants and Mycorrhiza Systems: Roles of Bioligands Forming Metal-Binding Complexes in Cells and Apoplasts

Masahiro Inouhe[*,§], Yui Katsuta[*], Aki Kato[*], Hirotaka Kato[*], Yoh Sakuma[*], Clemens Walther[†], Veenu Joon[‡] and Dharmendra K. Gupta[‡]

[*]*Department of Biology, Faculty of Science, Ehime University, Matsuyama, 790-8577, Japan*

[†]*Gottfried Wilhelm Leibniz Universität Hannover, Institut für Radioökologie und Strahlenschutz (IRS), Herrenhäuser Str. 2, D-30419 Hannover, Germany*

[‡]*Ministry of Environment, Forest and Climate Change, Indira Paryavaran Bhavan, Jorbagh Road, Aliganj, New Delhi 110003, India*

[§]*inoue.masahiro.vb@ehime-u.ac.jp*

Abstract

Water uptake and nutritious mineral utilisation via roots from the soil are the critical requirements for terrestrial plants that have evolved as the major host organisms on the grove of Earth, autotrophically nourishing animals, fungi and many other heterotrophic organisms and continuing to play a crucial role in forming the landscapes of the planet's biosphere. Minerals taken up from the ground return to the earth via food chains and biome networks through the actions of consumers and decomposers. Through the cyclic usage of inorganic and organic resources, plants and their associates sustain and develop as whole communities. In this chapter, based on such a co-association, or co-evolution, concept, the important interactions between autotrophic plants and the coordinative or competitional fungi and microbes are discussed and emphasised, especially in relation to plant stresses caused by excess levels of Li^+ and Ni^{2+} ions in land environments.

Li and Ni are gathering more attention, as well as cobalt (Co), because of their importance as very useful elements in industrial activities and sustainable energy for people in the future; however, their use is also accompanied by the incidental threat and concern about their possible increases through release and contamination into land and other environments. Li belongs to the group of relatively abundant alkaline metals (light metals), and Ni is one of the rare and micro-elementary heavy metals. Although their differential and unique impacts on plants and fungi are known, their combinational effects are not well known, so it will be necessary to address the increasing risks in the near future. In this chapter, the existing evidence will be substantiated with details about their bindings and localisations with different bioligands and polymers in the cells of plants or fungi. Taken together, we discuss the importance of complex binding and detoxification systems in plants and the symbiotic associates under combined stresses caused by the two metals as well as others.

Keywords: Alkaline metal, apoplast, arbuscular mycorrhizal fungus (AMF), cell wall polymer, ectomycorrhizal fungus (EMF), glutathione (GSH), heavy metal (HM), histidine (His), hyperaccumulation, lithium ion (Li^+), membrane transport, metal-binding complex, metal-binding ligand (N-, O- and S-ligands), metallothionein (MT), mucigel, nickel ion (Ni^{2+}), mycorrhiza, organic acid (OA), pectin, phenolics, phytic acid, phosphate sugars, phytochelatin (PC), proline (Pro), rhizosphere, siderophore, silicate, symplast, symbiotic association, vacuole.

1. Introduction

The biomass of land plants (embryophytes) has been estimated at 450 Gt of carbon, which is 95% of the biomass in terrestrial areas (470 Gt) and 82% of total biomass (550 Gt) on Earth (Bar-On *et al.*, 2018). In the same terrestrial areas, fungi, bacteria, protists and animals (including us) exist with lower values of biomass at 12, 7, 1.6 and 0.5 Gt, respectively (Bar-On *et al.*, 2018). All these organisms co-exist dependently or independently of each other but are very closely associated with the primary host plants through co-evolutions on land (de Vries and Achibald, 2018). Water uptake and nutritional mineral utilisation from ground soil via roots are critical requirements for most vascular plants, which act as major hosts on the grove of Earth, autotrophically nourishing the other heterotrophic and aerobic organisms and continuing to successively form landscapes in the planet's biosphere (Whittaker, 1962, 1975). Here, minerals are taken up from the ground and returned to the earth via biome chains and networks through the actions of consumers and decomposers; therefore, the cyclic usage of inorganic and organic resources needs to be maintained by the plant associations to allow whole communities to sustain and develop.

Based on the concepts of co-association and co-evolution, the important interactions between autotrophic land plants and the coordinative or competitional fungi and bacteria are emphasised in this chapter, especially in response to excess levels of chemical stresses caused by Ni^{2+} and Li^{+} ions in environments. We also fully recognise that there are no aseptic conditions in the fields where plants grow and develop naturally or agronomically. In fact, more than 90% of species of land plants are estimated to be coexisting with arbuscular mycorrhizal fungi (AMFs) as their own partners, and much more, or perhaps almost all species, are affected or infected by symbiotic or pathogenic fungi, bacteria and other microbes in soils (Al-Karaki, 2000; Evelin *et al.*, 2009; Nadeem *et al.*, 2017; Begum *et al.*, 2019).

So far, the strong impacts of low levels of 'heavy metal' (HM) ions, such as the highly toxic Ni^{2+}, Cd^{2+}, Hg^{2+}, Pb^{2+} and metalloid arsenic (As), at micromolar order levels (>1 μM) have been studied using various plant species (Hossain *et al.*, 2012; Inouhe *et al.*, 2012, 2015; Gupta *et al.*, 2013a; Gupta and Chatterjee, 2017; Noor *et al.*, 2022). Concomitantly, the alkaline metal ions, such as Li^{+}, Na^{+} and Cs^{+}, which have moderate impacts at low levels but are stronger at excessive millimolar order levels

(>1 mM), have been widely studied in many plants (Hampton *et al.*, 2004; Flowers *et al.*, 2015; Shahzad *et al.*, 2016, 2017; Shakoor *et al.*, 2023). The same can be said for the so-called 'light metal' (LM) ions, such as the more toxic Al^{3+} (Panda *et al.*, 2009; Bhalerao and Prabhu, 2013; Shetty *et al.*, 2021). In general, their effects are variable, depending on the species of metals and plants, as well as the conditions of the external medium solutions, pH and matrixes in the soils (Gupta and Walther, 2014). Moreover, differences in uptake and extrusion, extra- and intracellular binding and complexation, long-distance transport between root and shoot, and detoxification and compartmentation mechanisms can serve as clues for understanding the modes through which metals impact plants.

Among these, the most notable features for Ni and Li are selected as a common subject of focus in this chapter. They are delivered from different resources, i.e., either Ni-rich ores from mines and grounds or Li-rich continental brines, salty lakes and some ores. Even though their resources are geologically separated from each other, their industrial utilisation and subsequent formation of by-products may increase the risks of their complicated contamination of the environments and the exposure of their toxic ionic forms to plants and other organisms in lands and waters. Here, the existing evidence is substantiated with details about their extra- and intracellular distributions and bindings to major biological ligands or polymers, both in plant roots and fungi. Taken together, we finally discuss the importance of complex binding and detoxification systems in plant associates under the combined stresses of both metals.

2. Plants and Mycorrhiza Systems: Overview

Plants and the associated fungi interact at the cellular level to obtain nutrition and devise survival strategies in stressful soil environments. In terrestrial plants, about 90% of species have been reported to be in symbiotic interrelationships with endophytic AM fungi (Frey-Klett *et al.*, 2007; Bonfante and Genre, 2010; Nadeem *et al.*, 2017; Evelin *et al.*, 2019). Usually, the typical mycorrhizal associations are categorised into seven groups, as shown in Table 1. Plants are also unique in that they have associations with other types of endophytes, such as bacteria, to establish well-balanced symbiotic interactions (Maillet *et al.*, 2011). AM fungi are endomycorrhizal fungi that belong to the phylum Glomeromycota. *Glomus mosseae* and other species have been demonstrated to provide

Table 1. The symbiotic association between typical plants and fungi.

Mycorrhiza types	**Fungi**	**Plants**	**Benefits***
Arbuscular mycorrhiza (AM)	Glomeromycota	Angiosperms, Gymnosperms Pteridophyta, Bryophyta	H_2O, P, N, K, etc., organics, tolerance
Ectomycorrhiza (EM)	Basidiomycota Ascomycota Zygomycota	Angiosperms, Gymnosperms (in forests)	H_2O, P, N, K, etc., tolerance
Ectendomycorrhiza (EEM)	Basidiomycota Ascomycota	Angiosperms, Gymnosperms	H_2O, P, N, K, etc.,
Arbutoid mycorrhiza	Basidiomycota	Ericales (Angiosperms)	H_2O, P, N, K, etc.
Monotropoid mycorrhiza	Basidiomycota	Monotropoideae (Ericaceae)	H_2O, P, N, K, etc., organics, germination
Ericoid mycorrhiza	Ascomycota	Ericales	H_2O, P, N, K, etc., tolerance
Orchidaceous mycorrhiza	Basidiomycota	Orchids (Angiosperms)	H_2O, P, N, K, etc., most organics, germination

Note: *Major nutritional benefits available to host plants in symbiosis. For more information about mycorrhizae, refer to Bonfante and Genre (2010), Figueiredo *et al.* (2021) and other articles cited in the text.

various merits through their arbuscules, vesicles and hyphal coils penetrated into the root cells for plant growth and development, including increases in root biomass, basic physiological and metabolic functions and tolerance/resistance characteristics to various abiotic and biotic stresses (Al-Karaki, 2000; Martínez-Madina *et al.*, 2009; Jung *et al.*, 2012; Sharifi *et al.*, 2007; Nadeem *et al.*, 2017; Singh *et al.*, 2022). Among these, AM fungi-enhanced uptakes of water and nutrients, such as N, P and K, contribute to the improved photosynthetic efficiency and other metabolic functions of plants in terrestrial woodlands and grasslands. In turn, the plants are estimated to provide 10–20% of carbon as sugars to the fungi as a cost for the benefits (Allen *et al.*, 2003; Allen, 2011; Farrar *et al.*, 2003). Beside the endophytic AM fungi, ecophytic fungi are also known to be present at the surfaces of or around plant roots in soils. They develop thick hyphal sheaths on root surfaces and hyphal networks called 'Hartig nets' in the intercellular spaces of cortical cells, which offer several merits, including water absorption, mineral retention, the immobilisation of harmful chemicals and the protection of invasion xenobiotics to the host root systems, named 'ectomycorrhizae' (EM) (Anderson and Cairney, 2007). This type of symbiotic system has been demonstrated to play a central role in biomass maintenance in woody plant species (though it accounts for only 3% of the total seed plant species, Spermatophyta), constructing various types of forest ecosystems in the world. Furthermore, different types of symbiotic roots are found mainly in angiosperms, with some of them restricted to close and specific symbiotic relationships. Fungi and bacteria occupy about 95% of the biomass of soil microbes (Beare, 1997; Wardle, 2002; Wardle *et al.*, 2004). Non-symbiotic fungi and bacteria are estimated to be present in the order of 100 g m^{-2} in fields (Hoorman and Islam, 2010; Bhattarai *et al.*, 2015). They also affect plant development and health in a positive or negative manner, depending on the local habitats and vegetation and/or the whole ecosystem conditions.

3. Biological and Chemical Properties of Li^+ and Ni^{2+} Ions: An Overview

It is well known that plants require 17 essential elements (nine macro- and eight micro-elements) and use some other beneficial elements for growth. Of these, more than 11 are metallic elements (Table 2). Fungi and microbes have similar requirements; however, the degree of dependency and levels of micro-elements are different from plants and vary with their species and/or microhabitats. Ni has been demonstrated to be essential for plants and fungi (Mulrooney and Hausinger, 2003; Chen *et al.*, 2009). However, the beneficial roles of Li are not fully validated, especially in fungi and bacteria, as compared to plantae (Shahzad *et al.*, 2016, 2017).

Table 2. Essential, favourable and not useful elements to plants.

Essential macroelements (9)	C H O N Mg Ca K S P
Essential microelements (8)	Fe Mn Mo B Zn Cu Cl Ni
Favourable (Beneficial) useful elements (5)	Si Na Co (Li) (Se)
Not useful or harmful elements (>8)	Cs Al Ba I Cd Hg Pb As

Although information about their ubiquitous functions is limited, their nutritional functions may also be linked to the availability of other related nutritional cations, such as K^{+}, Na^{+}, Mg^{2+}, Ca^{2+} and/or Fe^{2+}, Fe^{3+}, Zn^{2+}, Cu^{2+} and Mo^{2+} ions, to the organisms, either in a competitive or compensatory manner (Behr and Zörb, 2023; Chen *et al.*, 2009). Furthermore, plants and these microbes strongly suffer from toxic HMs, such as Cd, Hg, Pb and As (metalloid), and LMs, such as Al; however, they have constitutive or inducible potentials to develop their own unique tolerance and/or resistance mechanisms against the respective ions' toxicities. These and further information gathered with respect to the usefulness or non-usefulness of metal ions can aid our understanding of the biological roles and impacts of Ni^{2+} and Li^{+} ions.

In general, the ionisation tendencies of metals are greater and follow the order of **Li** > K > Ca > Na > Mg > Al > Zn > Fe > **Ni** > Sn > Pb > (H_2) > Cu > Hg > Ag > Pt > Au. This order means that Li can be ionised most easily in water and that Ni is moderately and more easily ionisable, like Fe or Zn, than the other HMs. Once ionised in solutions, they become cations, with their charges varying from monovalent to tervalent (M^{+} – M^{3+}) in general, and exert their distinct properties in liquid or solid phases. The metal cations display a variety of configurations, depending on their size and electronic valency (Persson, 2010). For example, the atomic radii follow the order of Li, Na, K, Rb and Cs, and thus, conversely, their hydrate radii as monovalent ions are larger in the order of Li^{+}, (H^{+}), Na^{+}, K^{+}, Rb^{+} and Cs^{+}, according to their degree of hydration. The hydrate radii of cations affect their absorption capacities for polyanionic resins with cation-exchange capacity (CEC) and for the negatively charged surface of soil matrixes. Affinities for absorption by anionic resins are greater in the order of Cs^{+}, Rb^{+}, K^{+}, Na^{+}, (H^{+}) and Li^{+}. Similarly, HMs, such as Ni, Cd, Cu, Co and Zn have lower atomic radii and tend to become divalent cations in solution, and the resulting HM^{2+} ions have much higher affinities for the polyanionic resins than do Cs^{+} or other alkaline metal ions. As a simple example, the binding affinities for the strong-acid exchanger resins with sulphonic acid groups ($–SO_3^{-}$) are known to occur in the following order: Al^{3+}> Ba^{2+}> Pb^{2+}> Sr^{2+}> Ca^{2+}> $\mathbf{Ni^{2+}}$> Cd^{2+}> Cu^{2+}> Co^{2+}> Zn^{2+}> Mg^{2+}> Ag^{+}> Cs^{+}> Rb^{+}> K^{+}> NH_4^{+}> Na^{+}> (H^{+})> Li^{+}. Here, Ni^{2+} has the greatest affinity for the resins among the HM^{2+} ions (except Pb^{2+}), whereas Li^{+} has the smallest affinity among all the metal, H^{+} or NH_4^{+} ions. These characteristics become relevant when discussing the interactions between Ni^{2+} and Li^{+} ions and within the same group of elements for binding to matrixes or polymers in soils. As well as Al^{3+} ions, the oxidised-form Fe^{3+} ions can bind to the resions more strongly than the reduced-form Fe^{2+} ions. Such a relationship between redox states and binding affinities is also valid for other cations with multivalent electric charges in liquid phases.

4. Metal Bindings and Complexations in Soils: Solid and Liquid Phases

Under many field conditions, clays and humus are the major components that absorb metal ions in soils. They have a higher negative charge than sands and retain the metal ions, such as Al^{3+}, Ca^{2+}, Ni^{2+}, Mg^{2+}, K^+, Na^+ and Li^+, more strongly in that order as major exchangeable cations in soils under neutral or high pH conditions (pH > 5.5). However, as the pH is decreased (pH < 5) by acid rainfall or water flooding, the metal ions release more readily in the reversed order, and Al^{3+} becomes the most dominant metal ion adsorbed in the soils. These provide a background for us to evaluate Ni^{2+} and Li^+ impacts in natural soils and vegetation.

The influence of soil–solid matrixes on the availability of metal ions to plant roots can be simply estimated by comparing the difference between the effects or changes in plants grown in the same liquid medium supplemented with or without the solid culture matrices, such as commercially available vermiculite. Vermiculite (a hydrated magnesium aluminium silicate mineral) is made of sterile inorganic matrices, for example, 75% silicate (SiO_2), 13 % aluminate (Al_2O_3) and certain other oxidised minerals (MgO, K_2O, Na_2O, Fe_2O_3, etc.), but contains little organic or inorganic free ions. Vermiculite has a high CEC of 100–150 meq per 100 g and shows higher affinities for cations with bigger valency numbers and lesser hydration radii, as mentioned above. Ni^{2+} ions strongly bind to the solid matrices as do other HM^{2+} ions, or LM^{2+} ions such as Ca^{2+} and Mg^{2+}. In contrast, Li^+ ions weakly bind to solid matrices, such as those of H^+. This means that excess Li^+ may affect the binding of the other metals to the matrices and enhance some parts of the releases of metals into the liquid phases, e.g., increased levels of H^+ (i.e., acidification), Na^+ (high salinity), and NH_4^+ or K^+ (high fertilisations) in the soils. Acid rain has been shown to cause damage to plants and forests by increasing Al^{3+} ions released from soils. Excessive Li^+ at mM order levels may mimic this scenario in part, contributing to the toxic mode of action in land plants.

Although soil matrix compositions have diverse origins and undergo different weathering processes, their basal properties for binding are similar to each other and to vermiculite. However, actual soils contain many other organic and inorganic components, depending on their sources and origins, as shown earlier. These make the existence states of cations in soils and thus their availability to plant roots and the associated organisms more complex.

The liquid phase in soils contains many nutritional cations, anions and molecules in their dissolved forms in water. Their concentrations and compositions largely fluctuate on a daily basis under the influences of drought and watering cycles, through interaction with chemical and organic fertilisers and by additional contamination with toxic metal ions or other xenobiotic chemicals. Ionic bindings and/or complex formations in soil-water solution may also strongly affect the speed and extent of metal contamination in plant roots, making them critical environmental factors. Ionic bindings and/or complex formations with counteracting pairs of anions and organic or inorganic ligands present in the soil-water solutions are also critical. The behaviour of the former is pH-dependent, being weaker at lower pH, and that of the latter depends on the ligand–metal affinity. Inorganic ligands, such as H_2O, NH_3, OH^- and S^{2-}, can coordinate with metal ions and form various metal complexes in soluble or insoluble forms in the solution. Other inorganic acid anions, such as phosphate, carbonate, sulphate, nitrate and chloride, also affect metal complex formation and ionic bindings in the soil. In addition, various types of organic anions and bioligands are released from plant root cells, as well as from fungal and bacterial cells. Their interactions in the solid and liquid phases of soils strongly influence the metal mobilisation into and utilisation by plant and microbe cells, as described in the following section.

5. Metal Bindings and Complexations with Cell Wall Polymers: Plants and Fungi

Plant cells and fungal cells are surrounded by cell wall structures. The plant root cells have been traditionally divided into two compartments: the *apoplast* (cell wall compartment outside the cell membrane) and the *symplast* (cytoplasmic compartment inside the cell membrane). Both extra- and intercellular spaces are included in the apoplast. The apoplasts of plant cells consist of two distinguishable sites, called water free space (WFS) and Donnan free space (DFS). WFS refers to the space where solute materials can move freely with water and therefore, includes inter and extracellular spaces within liquids and solutions. DFS is an electrically charged space that can absorb charged materials and ions, especially cations. The properties of DFS can be demonstrated by those of the cell wall polymers.

Plant cell walls contain various polysaccharides, glycoproteins, phenolics and some inorganic components. The structures and compositions of primary cell walls are different between dicotyledonous plants and some monocotyledonous plants (cereal plants) and are classified as Type-I and Type-II cell walls (Yokoyama and Nishitani, 2004). In young tissues, both types of cell walls contain cellulosic β-1,4-glucans (approximately 25–30% of total cell-wall polysaccharides) as major cellulose microfibrils, where the former is richer in pectin (~30%) and glycoproteins and the latter in non-cellulosic β-1,3:1,4-glucans (~30%) and arabinoxylans. Pectin consists of poly-galacturonic acids (PGAs) with many carboxyl groups. Thus, the PGAs and peptides in glycoproteins can be the major DFS sites with CEC in the roots. Cereal roots are poor in pectin (less than 5%) and glycoproteins; however, β-1,3:1,4-glucans and other components provide the DFS sites (Yokoyama and Nishitani, 2004). For example, cell wall polysaccharides, such as β-1,3:1,4-glucans, form bonds with silicates through coordination and can thus act as sources for CEC in the roots (Kobayashi *et al.*, 1996; Kido *et al.*, 2015). Silicone also increases the strength of cell walls by binding to β-1,3:1,4-glucans in rice (Kido *et al.*, 2015) and inhibits Na^+ influx in rice root by absorption into the cell wall (Gong *et al.*, 2006). In mature cell walls (secondary cell walls) of root tissues, pectin and non-cellulosic polysaccharides decrease, but cellulosic glucans increases to near 60%, and phenolics, such as lignin, are newly deposited and become providers of more –OH groups. These groups act as direct *O*-ligands to bind cations and also form coordination bonds with phosphates, borates and silicates, which also provide secondary sites with CEC on the cell surfaces (Guo *et al.*, 2008). Therefore, pectin, glycoproteins and most glycans and phenolics associated with inorganic anions or oxides can be considered the major parts of cell walls with CEC in plant root tissues. Here, hydrophobicity and site-specific localisations of the secondary cell walls are distanced temporally or spatially from the primary cell walls of growing young root tissues, which should be considered while investigating the roles of cell wall bindings, especially in whole rhizosphere systems.

Fungi have long and massive filamentous hyphae through which they make contact and communicate with the host plant cells. The structure and composition of the cell walls are quite different from those of plants, as shown in Fig. 1. Fungal cell walls are typically composed of chitin or chitosan (instead of cellulose), glucan, mannan, glycoproteins and some pigments (Coelho *et al.*, 2017, 2018; Gow *et al.*, 2017). Of these polymers, the inner layers of chitin and chitosan with amino groups and the outermost layers of glycoproteins (mannoproteins) with carboxyl groups are the sites highly capable of binding to metal ions. The CEC of the fungal cell walls depends on the total amount of mycelium biomass; therefore, the exchange can occur more effectively in EM fungi covering the surface of roots with developed hyphal networks than in AM fungi. In turn, AM fungi may have more intrinsic endogenous roles within the root tissues or cells of the host plants.

Bacteria have different cell wall structures and components between Gram-positive and Gram-negative groups. The cell walls of Gram-positive bacteria consist of a layer of peptidoglycan (PG) with teichoic acid (TeA). The carboxyl groups of PGs and phosphate groups of TeAs bind metal ions. On the other hand, the cell walls of Gram-negative bacteria have another layer consisting of lipopolysaccharide (LPS), phospholipid (PL) and protein surrounding the thinner peptidoglycan layer. Phosphate groups of LPS and PL and carboxyl groups of proteins are involved in the binding of metal ions. These chemical compositions and conformational structures of bacterial cell walls are involved in metal biosorption by bacteria (Johnson *et al.*, 2006; Moriwaki *et al.*, 2013; Ogi *et al.*, 2012, 2016). Bacteria develop various types of biofilms using cell wall components such as LPSs and other extracellular polymeric substances (EPSs), such as exopolysaccharides and proteins, or other small organics secreted from the cells (Casterton *et al.*, 1995). The biofilms have important functions in plant growth, nutrition and biotic and abiotic stress tolerances, including salinity and HM tolerances (Velmourougane *et al.*, 2017; Sharma, 2022). Fungi also produce unique EPSs and biofilms (Breitenbach *et al.*, 2019, 2022).

Among protistic microbes living in terrestrial habitats, some groups, such as green algae, have cell wall structures that are rich in mannan and other different polysaccharides and glycoproteins from the other organisms. Groups of diatoms have frustule structures comprising silicates on

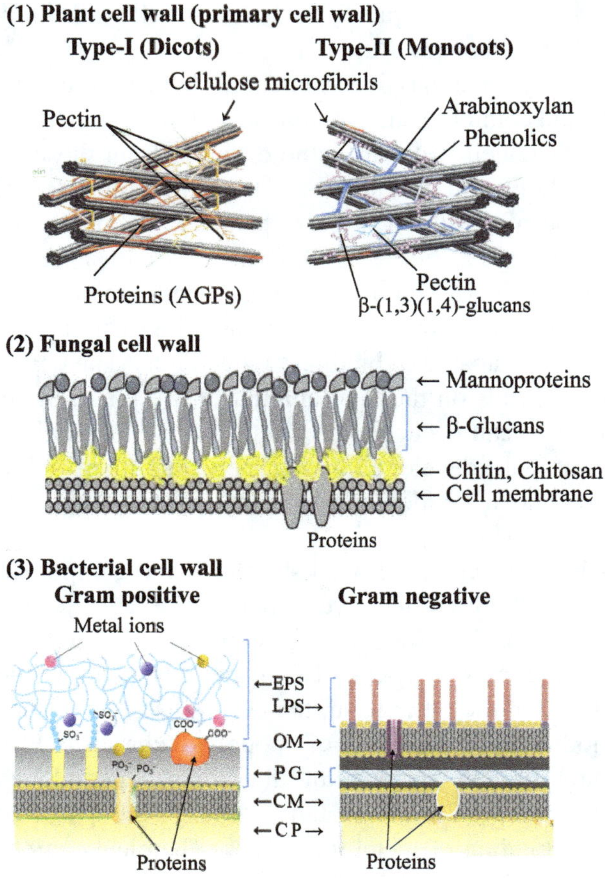

Figure 1. Comparisons of cell wall structures and compositions among plants, fungi and bacteria: Structures of primary cell walls are different between dicotyledonous plants and some monocotyledonous plants (cereal plants) and are classified as Type-I and Type-II cell walls (1). Fungal cell walls are composed of chitin, chitosan, glucan, mannan, glycoproteins and some pigments (2). Bacteria have different cell wall structures and components between Gram-positive and Gram-negative groups (3). HMs and LMs are adsorbed by ionic and co-ordinational bindings to the many polymeric bioligands of the different polysaccharides and glycoproteins of cell walls in plants, fungi and bacteria (see text for more details). CP: cytoplasm; CM: cell membrane; OM: outer membrane; PG, peptidoglycan; LPS: lipopolysaccharide; EPS: extracellular polymeric substance. Figures adapted from Yokoyama and Nishitani (2004) for (1), Vega and Kalkum (2012) for (2) and Mwandira *et al.* (2020) for (3), respectively. The images of the Gram-positive and Gram-negative bacteria are from the websites https://bre.eng.hokudai.ac.jp/en/research/biosorption.html and https://togotv.dbcls.jp/en/togopic.2022.444.html, respectively.

the surfaces (Hoffmann, 1989; Wilkinson *et al.*, 2012). However, their roles in symbiotic interactions with plants require further verification.

The DFS sites of plant apoplasts absorb nutritional cations, such as K^{+}, Mg^{2+}, Ca^{2+}, Fe^{2+}, Mn^{2+}, Mo^{2+}, Zn^{2+} and Cu^{2+} efficiently, as they do for Ni^{2+} and Li^{+} ions too. Excessive cations may inhibit the enzymes and other functional proteins in the cell walls or on cell membranes as direct inhibitors or through substitutions of the original metal components in the proteins. Ni^{2+} strongly binds to the cell walls and thus competes with the other major cations, especially mono- and divalent metal ions. On the other hand, excessive Li^{+} may cause the substitution of the accumulated cations in the cell wall and the solubilisation of the basic and acidic proteins that ionically bind to the cell wall polymers (Inouhe and Nevins 1991, 1997; Inouhe *et al.*, 2000). Plant roots have a unique cell layer between cortex and stele called endodermis, or *Casparian strips*, that has thick and hydrophobic cell walls consisting of suberin, lignin and proteins. Here, the passive movements of metal ions in water are controlled and restricted only through symplastic movement with plasmodesmata. These diffusion barriers give dominant and specific roles to the cell membrane transporters to carry nutrients and metals from the outside to the internal xylem tissues. Mycorrhizal systems are well established in their own styles, as mentioned above; however, their functions express themselves outside the endodermis or stele. Therefore, more detailed information on the trance-gate mechanisms for metals is needed, especially in many mycorrhizal systems.

6. Metal Complexations with Cytoplasmic Ligands: In Apoplast and Symplast

Various types of bioligands are synthesised within the cells of plants and microbes and play a multitude of functions for metal homeostasis in the organisms and the symbiotic habitats. According to the HSAB (hard and soft acids and bases) principle (Pearson, 1968a, 1968b), hard-acid metals, such as Li, Na, K, Cs, Mg, Ca or Al, prefer the *O*-donor ligands (–COOH, $-H_2PO_4$, –OH, –CHO, etc.), and soft-acid metals, such as Ag, Hg, Au and Pt, prefer the *S*-donor (–SH, –SS–, etc.) and *N*-donor ligands ($-NH_2$, =NH, etc.) for bindings. The intermediate metals, such as Ni, Cu, Zn, Mn, Fe, Co, Mo and Cd, are capable of binding to each type of those ligands (Pearson, 1968a, 1968b; Woolhouse, 1983). Representative bioligands that might be involved in Li or Ni bindings are listed in Table 3, and some

Table 3. List for major organic and inorganic ligands and biopolymers that are involved in metal bindings and complexations in plants and fungi and other microbes associated with the rhizosphere.

Types	Ligands and group	Metal choice favourite	Plant sites*
Thiol peptides	GSH, PCs (–SH, –COOH)	Ni (Cd, Zn, Cu)	Sy, Va, Ph
Thiol proteins	MTs (–SH, –COOH, $-NH_2$)	Ni (Cd, Zn, Cu)	Sy
Amino acids	Cysteine (–SH, –COOH, $-NH_2$)	Ni (Cd, Zn, Cu)	Sy, Va
	Histidine (=NH, –NH, –COOH)	Ni (Fe, Zn)	Sy, Va
	Proline (=NH, –COOH)	Ni (Cd, Zn), Li (Na, K)	Sy, Va
	Glu, Asp (–COOH, $-NH_2$)	Ni (Fe, Zn), Li (Na, K)	Ap, Sy, Va, Xy. Ph
Phytosiderophore	MuA (=NH, $-NH_2$, –COOH, –OH)	Ni (Fe, Zn)	Ap, Sy, Va, Xy, Ph
	NiA (–NH, $-NH_2$, –COOH)	Ni (Fe, Zn)	Sy, Va, Xy, Ph
Siderophore, Fungi	Ferricrocin (=NH =COOH –OH)	Ni, Fe	Ap, Sy
Siderophore, Bacteria	Enterobactin (–OH =NH –COOH) and many other kinds	Ni (Fe, Co)	Ap, Sy
Organic acids	CiA, MaA (–COOH, –OH)	Ni (Zn, Fe), Li (Na, K), Al	Ap, Sy, Va, Xy, Ph
	OxA, LaA, AcA, etc.(–COOH)	Ni (Zn, Fe), Li (Na, K)	Ap, Sy, Va, Xy, Ph
Polysaccharides	Pectin (–COOH –OH) β-1,3:1,4-glucan, –OH ($-SiO_2$) Cellulose, Xylan, Mannan (–OH) Chitin, Chitosan, ($-NH_2$, –OH) Lipopolysaccharide (–OH, $-H_2PO_3$)	Ni (Ca, Mg), Li (Na, K)	Ap
Glycoproteins	AGP, HRP, etc. (–COOH, –OH etc.) Peptidoglycans, lipoproteins, EPS	Ni (Ca, Mg), Li (Na, K)	Ap
Polyphenols	Lignin, Lignan (–OH)	Ni, Li, and others	Ap, Sy, Va, Xy
Polyols (sugars)	Sugars, Sugar alcohols (–OH)	Ni, Li and others	Ap, Sy, Va
Polyphosphates	Polyphosphate ($-H_2PO_4$)	Ni (Fe, Mg)	Ap, Sy, Va, So
	Phytins ($-H_2PO_4$)	Ni (Zn, Cd, Fe), Li	Ap, Sy, Va

Note: *Ap: apoplast; Sy: symplast; Va: vacuole; Xy: xylem; Ph: phloem; So: soil matrix.

of the structures are shown in Fig. 2. These bioligands have roles in metal bindings and complexations inside or outside the cells, as well as in metals for membrane transports, compartmentations, immobilisations or precipitations and long-distant transports from root to shoot or from source to sink in plants, as discussed later.

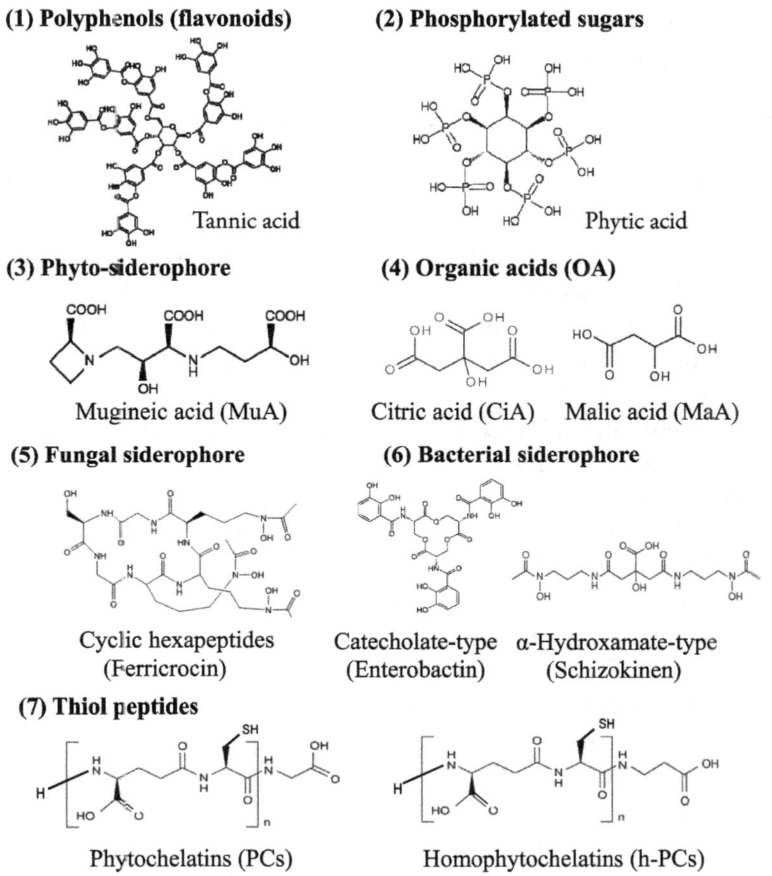

Figure 2. Representative bioligands produced by plants and microbes: Polyphenols and phytic acids produced by plant roots are typical *O*-bioligands that make stable bindings to various metal ions on the surface or inside the cells, as well as pectic substances with carboxyl groups. Plant roots and microbes (fungi and bacteria) produce various types of organic acids (OAs) and amino acids in the cells and release some of them into the apoplasts as common bioligands. Besides, they release more unique bioligands called siderophores into soils to solubilise and acquire Fe^{3+} strongly coordinated with cray matrixes under Fe-deficient and alkaline pH conditions. Plants produce phyto-siderophores, such as mugineic acid (MuA) or deoxy-MuA. Microbes produce much more complex and diverse types of fungal or bacterial siderophores. These and other types of bioligands, along with protons (H^{+} ions) associated with the anionic ligands, may promote the dissociation of many metals, including Al^{3+}, Fe^{3+}, Fe^{2+}, Mg^{2+}, Mn^{2+}, Ni^{2+}, Cu^{2+} and Zn^{2+}, from solid clay minerals in rhizosphere sites. Phytochelatins (PCs) and homo-phytochelatins (h-PCs) synthesised from glutathione or homo-glutathione have important roles in HM-detoxification in the cytoplasm, such as metallothioneins (MTs). However, their roles in binding or transport of Ni or Li ions are still unknown.

6.1. *Bioligands released from cytoplasm to apoplastic compartment*

Plant roots have been reported to extrude several kinds of organic bioligand substances into soils to chelate and solubilise metals tightly combined within the soil matrixes. The typical amounts of secretions have been estimated at 20–30% of total organics synthesised by the plant (Hinsinger, 2001). Of the root exudates, the well-known ones are organic acids (OAs), such as citric acid (CiA), malic acid (MaA), acetic acid (AcA), lactic acid (LaA), oxaloacetic acid (OAA), succinic acid (SuA) and malonic acid, released from the roots of many plants (Hinsinger, 2001; Fan *et al.*, 2012; Sun *et al.*, 2021). CiA is an α-hydroxy acid with one hydroxyl group and three carboxyl groups in the molecule and is used as a very effective OA ligand that provides a soluble form of the CiA–Fe–phosphate complex when these nutrients are deficient in the soils as free forms (Gardner *et al.*, 1983). It is also used for the solubilisation and utilisation of Ca^{2+} and for the complexation and detoxification of Al^{3+} in soils (Ma, 2000). Mugineic acid (MuA) and deoxy-MuA are released more specifically from cereal plant roots, especially under Fe^{3+}-deficient conditions (Takagi, 1976; Ma *et al.*, 1993; Ma, 2000). They are called "*phytosiderophores*" because of their functional resemblance to "*siderophores*" that are produced and released from many kinds of bacterial and fungal species (Haas, 2014; Saha *et al.*, 2016; Boiteau *et al.*, 2019). Plant cells also produce unique bioligands, such as phyotochelatin (PC) as *S*-ligands, phytic acid and various polyphenols as *O*-ligands, and common amino acids, including histidine (His) and proline (Pro), as *N*- and *O*-ligands. Of these, PC peptides synthesised from glutathione (GSH) in the cytoplasm appear to be retained in the site or transported to vacuoles and may not release into the apoplast (Inouhe, 2005). Furthermore, plant roots produce and secrete well-hydrated viscous substances called *mucigel* (or mucilage) composed of pectin and some other polysaccharides, which are known to protect young root tips from physical and chemical damages and to facilitate the symbiotic association with microbes (Iijima *et al.*, 2004; Canarini *et al.*, 2019). These polymeric and monomeric ligands produced by roots can interact with each other to exchange metals bonded onto the solid surfaces of soil particles through ligand–ligand substitution reactions and then to mobilise them into the liquid phase (Gardner *et al.*, 1983; Otani *et al.*, 1996; Romheld and Awad, 2000; Hinsinger, 2001; Sun *et al.*, 2021).

Fungi and bacteria also produce and secrete various types of OAs, including CiA, AcA, SuA, LaA, oxalic acid (OxA), lactobionic acid, itaconic acid, gluconic acid, fumaric acid and propionic acid (Scervino *et al.*, 2010; Gadd *et al.*, 2012). Therefore, the different combinations of OAs in the apoplasts may contribute to the efficiency and diversity of the utilisation of nutritional metal ions or detoxification of toxic ones in plants and the associated microbes. Organic ligands released from Ectomycota stimulate the solubilisations of many nutritional elements that are also available to plants and not only to their own mycelium cells. Similar contributions from fungi to plant growth and nutrition, especially under some poor soil vegetation, have been consistently demonstrated (Mohammadi *et al.*, 2011, for review). Fungi produce different types of siderophore ligands into cell wall spaces in response to changes in the levels of HM ions in both cases of excessive and deficient metals (Renshaw *et al.*, 2002). This means that their bioligands have roles in the binding and detoxification of excessive and toxic levels of metal ions, such as Ni^{2+} and Li^{+}, in the apoplasts. Furthermore, it is known that a greater number of species of bacteria produce and release stronger chelator ligands into soils and catch up with the Fe^{3+} ions to be usable by the roots of neighbourhood plants. Sympatric actions due to mixed types of bioligands (OAAs, MuA and siderophores) are very useful for capturing Ni^{2+} and can be applied for Ni detoxification in the apoplasts. Symbiosis with fungi and bacteria brings a number of benefits to the host plant through association. Boiteau *et al.* (2018) identified both plant and fungal metallophores, revealing compound-specific patterns of chelation to biologically essential metals. They showed that numerous metabolites typically implicated in plant Fe acquisition and homeostasis, including MuA, deoxy-MuA, NiA and hydroxy-NiA, dominated the speciation of divalent metals, such as Ni^{2+}, Cu^{2+} and Zn^{2+}, while the fungal siderophore ferricrocin was specific for Fe^{3+}. MuA and deoxi-MuA have high affinities for Fe^{3+} ions but are also effective in the solubilisation and mobilisation of Ni^{2+} and Zn^{2+} (Boiteau *et al.*, 2018). Cereal plants and other angiosperm plants in dry grasslands selectively exudate bioligands to chelate metal ions, i.e., MuA for Fe^{3+} and Ni^{2+} and OAs (CiA, MaA, etc.) for Ca^{2+}, Mg^{2+}, Mn^{2+}, etc. As shown in this typical case, competence and interaction among these bioligands mixed up in s liquid phase may become critical for the metal-binding and/or immobilising processes after hydration under the influences of the other polymeric bioligands in the cell walls and the surface ligands of clay particles or organic humus in the soils. The quantity and quality (variety) of the bioligands secreted from plants and microbes may be important for the maintenance of balance and diversity of minerals in the rhizosphere.

6.2. *Metal-bindings and complexations in the symplast and vacuole*

The symplast consists of the cell membrane, cytosol and different organelles, including the nucleus. Furthermore, plant cells have well-developed vacuoles surrounded by tonoplasts within the symplastic compartment. Land plants produce various organic compounds necessary for their own growth and development via photosynthesis and accumulate those in excess in the vacuoles. These unique abilities of plants allow them to produce multiple types of bioligands involved in metal bindings, detoxifications, movements and compartmentations. The representative bioligands in the cytoplasm or vacuoles are shown in Table 3. Among them, the best characterised are phytochelatins (PCs) and the related thiol-peptides. Their biosynthesis, derivatives, functions and specificities have been studied in detail (Rauser, 1990, 1995, 1999; Zenk, 1996; Cobbett and Goldsbrough, 2002; Inouhe, 2005). PCs have the general structure of (γ-glutamyl-cysteinyl)$_n$-glycine ($n = 2$–11), and they are synthesised from glutathione (Grill *et al.*, 1989; Rauser, 1995). The PC–metal conjugates are formed in the cytoplasm and transported to the vacuole to produce more stable complex mixtures with inorganic sulphur (S^{2-}). By contrast, little evidence supports the idea that PCs have a central role in xylem transport or immobilisation in the shoots of heavy metals. PCs and GSH can make a stable complex with Ni^{2+}, as with Cd^{2+}, Zn^{2+} and other HM^{2+} or As, but not with Li^{+} or other metal ions in the Class A group (Inouhe *et al.*, 2015; Gupta *et al.*, 2013b). OAs have been proposed as important cellular ligands for Zn, Cd and Ni (Salt *et al.*, 1999; Kupper *et al.*, 2004). The carboxylic acids known to be present in high concentrations in the cell vacuoles of photosynthetic tissues include citrate (CiA), malic (MaA), isocitric, oxalic, tartaric, malonic and aconitic acids (Callahan *et al.*, 2006). Many studies have implied that these acids play a role in hyperaccumulation (Rauser, 1999; Salt *et al.*, 1999; Chiang *et al.*, 2006; Panchal *et al.*, 2021). CiA is the predominant ligand for Zn in the leaves of *Thlaspi caerulescens* (Salt *et al.*, 1999), and Ni-CiA accounted for one-quarter of the Ni species in the leaves of the Ni hyperaccumulator *T. goesingense* and in the related non-accumulator *T. arvense*

(Kramer *et al.*, 2000). The vacuole was the major subcellular compartment for the accumulations of Zn, Cd and Ni, and the favouring metal complexes were demonstrated to be made with CiA and NiA as the only ligands for these metals within vacuoles (Kramer *et al.*, 2000; Ma, 2005). NiA and its derivatives are synthesised in all plants through the condensation of three *S*-adenosyl-methionine molecules through the activity of NiA synthase (NAS). It is involved in the chelation of metals, such as Fe, Cu and Zn, for their enhanced extraction by roots and/or transport to shoots, especially under mineral-deficient conditions (Takahashi *et al.*, 2003; Mari *et al.*, 2006). The idea that NiA plays a role in heavy-metal tolerance and hyperaccumulation is strengthened by its tolerating and accumulating functions for Zn, Cu and Fe (Weber *et al.*, 2004) and conferring resistance against Ni (Pianelli *et al.*, 2005).

Plant cells contain many other small organic ligands with variable functional groups, including amino acids, polyamines, nucleotides, phytates and other phosphate sugars. Of these, polyamines appear to act as a messenger or a molecule to stabilise or protect the cell membranes rather than as direct binding ligands to toxic heavy metals (Sharma and Dietz, 2006). Nucleotides, phytates and sugar phosphates can conjugate to Ca, Mn, Mg, Al and other metals through their *O*-bonds. The particular importance of phytates in the coordination and storage of phosphates and metals, such as Zn, Mg, and K, in the vacuole and cytoplasm and also in the detoxification of Cd has been widely suggested (Van Steveninck *et al.*, 1992; Hayden and Cobbett, 2007a, 2007b). Amino acids are the most abundant amphoteric ions with variable forms and residues, existing in the 10–100 mM range of concentration and serving multiple functions in plant cells. Cysteine (Cys) is a thiol compound that has a S-donor residue equivalent to a GSH molecule. However, its internal level does not usually exceed that of GSH or PCs, probably because of the restricted supply of total S available for it and its quick turnover and utilisation for the other thiol ligands and proteins. Acidic amino acids, glutamic acid (Glu) and aspartic acid (Asp) provide an extra carboxyl group (–COOH), and the amides, glutamine (Gln) and asparagine (Asn) provide an acid amide group consisting of both *O*- and *N*-donors ($-CO-NH_2$). All of these are generally rich in phloem sap, for example, at near 300 mM in cereals and 50 mM in some dicotyledonous plants (Ohshima *et al.*, 1990; Winter *et al.*, 1992), and can be potential ligands for metal translocations to sink organs. Histidine (His) is the most characterised imidazole (=NH), containing an amino acid that plays a central role in binding to and the transport of Ni, especially in Ni-hyper-accumulating plants (Kramer *et al.*, 2000; Callahan *et al.*, 2006). Two His molecules can make up a stable complex by chelating to one Ni (Callahan *et al.*, 2006). Furthermore, proline (Pro) has been most extensively studied for its unique and important function as a compatible solute in many plants affected by water-deficit and salinity stresses; however, interestingly, heavy metals, such as Cu, Cd, Zn or Pb, also significantly stimulate the accumulation and/or biosynthesis of Pro in many plants (Sharma and Dietz, 2006). The possible roles of Pro as a direct N-donor ligand conjugating to HMs or LMs are not established as yet, but will be more attractive in combination with its role as an osmotic protectant or antioxidant under complex conditions, including drought-salinity stresses and Ni^{2+} or Li^{+} contaminations.

As mentioned above, there are possible interactions between different soluble organic ligands and different metals in the cytoplasm and vacuole, as well as other compartments in the shoots and roots of plants. These solutions also contain inorganic anions, such as sulphate, phosphate, nitrate, borate, carbonate, chloride and silicate. These inorganic anions and their counterpart cations affect the organic ligand's interactions with metals in each site at different but almost constant pH conditions (Callahan *et al.*, 2006). Some bindings between metals and ligands are not specific and not stable, especially under varied pH and ion-strength conditions. Conversely, well-regulated conditions or mechanisms promise a unique and established mechanism for metal transport and binding systems in land plants. Actually, specific bioligands and transporters function in the metal accumulations into the cytoplasm or vacuoles by coupling with active H^{+} transporters (ATPases), and the long-distance transportations of metals and ligands from root to shoot operate in the developed vascular bundle systems in the steles (central cylinders) of stems and roots (Inouhe *et al.*, 2012; Gupta *et al.*, 2013a, 2013b; 2018).

6.3. *Bioligands for long-distance transports to shoots*

Detailed mechanisms for the short-distance transports of respective HM and LM ions have been disclosed both in plants and microbes (for reviews, González-Guerrero *et al.*, 2016; Yang *et al.*, 2022). Here, information for Li^{+} transport is limited, though it is supposed to follow the modes of well-known Na^{+} transports through the cell membrane (CM) and the vacuolar membrane, i.e., the tonoplast (TP), which are also linked to the transports of H^{+}, K^{+} and other LM ions. Membrane transports for

Ni^{2+} are already known to be mediated by several types of transporters, including the bioligand NiA-dependent and/or independent transporters (YSL, ZIP and HMA at the PM and MTP at the TP, etc.).

During long-distant transports, the co-transports of Li^{+} ions with counter anions, such as Cl^{-} and NO_3^{-}, and other hard-base anionic *O*-ligands, such as OAs (CiA and MaA), phosphate, silicate and borate, via the xylem in the transpiration stream from root to shoot are likely. The bindings are not tight and are variable, depending on the solute compositions and concentrations, pH (H^{+} concentrations) and Li^{+} concentrations of the apoplast solution in xylem vessels (xylem sap), which may be affected by those in the extracellular liquid phase (i.e., medium) of the roots and by characteristics that vary with plant species. Ni^{2+} appears to bind and coordinate more tightly with some specific organic and inorganic ligands, such as OAs (especially CiA), NiA, MuA, His and Pro in the xylem. The uptake of Ni^{2+} ions in plant roots is dependent on the transporter activities directed toward a Ni complex with MuA secreted from plant roots into the apoplast and the conversions to a Ni complex with NiA and others. Therefore, the Ni-complex movements from the apoplast to root cells and further to shoot cells may incur time and costs for organic ligand synthesis, secretions and collections in a competing manner with those for other essential cations, such as Fe, Zn and Cu. This bears some inevitable risks of deficiency for the essential elements in exchange for Ni acquisitions (Mizuno, 2024).

If excessively applied to soils or media, as long as Ni^{2+} ions are absorbed into the soil matrices or extracellular polymeric ligands, acute toxic symptoms due to Ni might not occur. However, acute damage to the root cells apparently occurs, and this is possibly due to their binding to the metal-sensitive sites on the cell membrane, either or both of phospholipids and proteins. In this case, total functional damage to roots causes an overall inhibition in the growth and functions of the aboveground parts of the plants. Therefore, ligand–bond transports and ligand–ligand exchange reactions are the bases for Ni utilisation and homeostasis in plants. On the other hand, Li-specific bioligands or transporters have not yet been identified. It is necessary to establish these systems in selective Li-specific hyperaccumulators. Here, as a summary, possible bindings, movements and interactions of Ni^{2+} and Li^{+} ions in rhizospheres and plants are shown as a simple illustration in Fig. 3, which may be useful as an overview of the presented topics and for more issues or prospects that are not fully covered in this chapter.

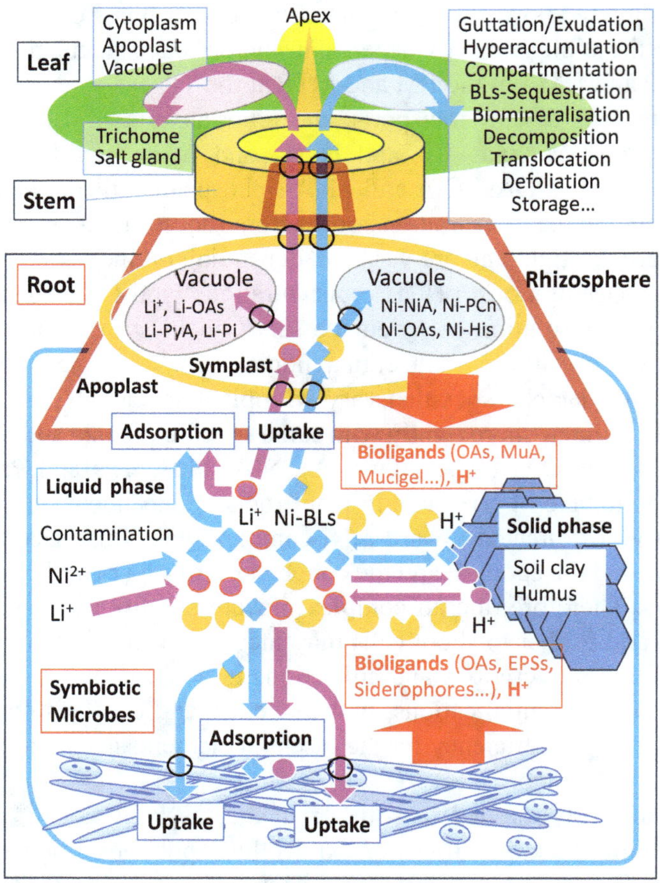

Figure 3. Possible bindings, movements and interactions of Ni^{2+} and Li^{+} ions contaminated into the liquid phases of rhizosphere systems under supposed common field conditions (BLs: bioligands).

7. Conclusions

The general points of view in this chapter that need to be stressed are as follows:

(1) The diversity and biomass of rhizosphere microbes provide benefits to plants through their abilities for rapid absorption and complex formation of multivalent cations, such as Ni^{2+}.

(2) They also contribute to the mitigation of salt-induced damages and osmotic stresses due to higher levels of monovalent cations, such as Li^+, through their quick absorption and dilution of the free cations.
(3) In both cases, the cations equip themselves with a duality of functions: either as nutrients or as inhibitors, depending on their concentrations.
(4) Of particular importance is that their constitutive cell wall properties (chitin, peptidoglycan, LPS and EPS) contribute to the absorbance of Ni^{2+} and other HM ions.
(5) Further, the inducible transporters of the cell membrane and tonoplast contribute to the uptake of Li^+ and other metal ions.
(6) Plants also possess their own virtues, including: the intra-, extra-, and inter-cellular specificities of the roots and the nutritional and developmental interrelationships between different organs and in the whole individual system.
(7) In addition to these points, the support of soil microorganisms offers great advantages (adaptability and robustness) to plants and coexisting systems. This surpasses the cost paid by plants to the fungi and other microbes in the rhizosphere.

References

Al-Karaki G.N. (2000). Growth of mycorrhizal tomato and mineral acquisition under salt stress. *Mycorrhiza* 10:51–54.

Allen M.F. (2011). Linking water and nutrients through the vadose zone: A fungal interface between the soil and plant systems. *J. Arid. Land* 3:155–163.

Allen M.F., Swenson W., Querejeta J.I., Egerton-Warburton L.M., Treseder K.K. (2003). Ecology of mycorrhizae: A conceptual framework for complex interactions among plants and fungi. *Annu. Rev. Phytopathol.* 41:271–303.

Anderson I.C., Cairney J.W.C. (2007). Ectomycorrhizal fungi: Exploring the mycelial frontier. *FEMS Microbiol. Rev.* 31:388–406.

Bar-On Y.M., Phillips R., Milo R. (2018). The biomass distribution on earth. *Pro. Natl. Acad. Sci. USA* 115:6506–6511.

Beare M.H. (1997). Fungal and bacterial pathways of organic matter decomposition and nitrogen mineralization in arable soils. *Soil Ecol. Sustainable Agric. Syst.* 37–70.

Begum N., Qin C., Ahanger M.A., Raza S., Khan M.I., Ashraf M., Ahmed N., Zhang L. (2019). Role of arbuscular mycorrhizal fungi in plant growth regulation: Implications in abiotic stress tolerance. *Front. Plant Sci.* 10:1068.

Behr J.G., Zörb C. (2023). Interactive effect of lithium on concentration of alkali cations in sugar beet (*Beta vulgaris* L.) under saline conditions *J. Plant Nutr. Soil Sci.* 186:38–49.

Bhalerao S.A., Prabhu D.V. (2013). Aluminium toxicity in plants — A review. *J. Appl. Chem.* 2:447–474.

Bhattarai A., Bhattarai B., Pandey S. (2015). Variation of soil microbial population in different soil horizons. *J. Microbiol. Exp.* 2:75–78.

Boiteau R.M., Shaw J.B., Pasa-Tolic L., Koppenaal D.W., Jansson J.K. (2018). Micronutrient metal speciation is controlled by competitive organic chelation in grassland soils. *Soil Bio. Biochem.* 120:283–291.

Boiteau R.M., Fansler S.J., Farris Y., Shaw J.B., Koppenaal D.W., Pasa-Tolic L., Jansson J.K. (2019). Siderophore profiling of co-habitating soil bacteria by ultra-high resolution mass spectrometry. *Metallomics* 11:166–175.

Bonfante P., Genre A. (2010). Mechanisms underlying beneficial plant–fungus interactions in mycorrhizal symbiosis. *Nat. Commun.* 1:48.

Breitenbach R., Silbernagl D., Toepel J., Sturm H., Broughton W.J., Sassaki G.L., Gorbushina A.A. (2019). Corrosive extracellular polysaccharides of the rock-inhabiting model fungus Knufia petricola. *Extremophiles* 22:165–175.

Breitenbach R., Gerrits R., Dementyeva P., Knabe N., Schumacher J., Feldmann I., Radnik J., Ryo M., Gorbushina A.A. (2022). The role of extracellular polymeric substances of fungal biofilms in mineral attachment and weathering. *Mater. Degrad.* 6:42.

Callahan D.L., Baker A.J.M., Kolev S.D., Wedd A.G. (2006). Metal ion ligands in hyperaccumulating plants. *J. Biol. Inorg. Chem.* 11:2–12.

Canarini A., Kaiser C., Merchant A., Richter A., Wanek W. (2019). Root exudation of primary metabolites: Mechanisms and their roles in plant responses to environmental stimuli. *Front. Plant Sci.* 10:157.

Casterton J.W., Lewandowski Z., Caldwell D.E., Korber D.R., Lappin-Scott H.M. (1995). Microbial biofilms. *Annu. Rev. Microbiol.* 49:711–745.

Chen C., Huang D., Liu J. (2009). Functions and toxicity of nickel in plants: Recent advances and future prospects. *Clean* 37:304–313.

Chiang P.N., Wang M.K., Chiu C.Y., Chou S.Y. (2006). Effects of cadmium amendments on low-molecular-weight organic acid exudates in rhizosphere soils of tobacco and sunflower. *Environ. Toxicol.* 21:479–488.

Cobbett C., Goldsbrough P. (2002). Phytochelatins and metallothioneins and role in heavy metal detoxification and homeostasis. *Annu. Rev. Plant Biol.* 53:159–182.

Coelho L.C.B.B., Silva P.M.S., Lima, V.L.M., Pontual E.V., Paiva P.M.G., Napoleao T.H., Correia M.T.S. (2017). Lectins, interconnecting proteins with biotechnological/pharmacological and therapeutic applications. *J. Evid. Based Complement Alternat. Med.* 2017:1–22.

Coelho L.C.B.B., Silva P.M.S., Oliveira W.F., Moura M.C., Pontual E.V., Gomes F.S., Paiva P.M.G., Napoleao T.H., Correia M.T.S. (2018). Lectins as antimicrobial agents. *J. Appl. Microbiol.* 125:1238–1252.

de Vries J., Achibald J.M. (2018). Plant evolution: Landmarks on the path to terrestrial life. *New Phytol.* 217:1428–1434.

Evelin H., Kapoor R., Giri B. (2009). Arbuscular mycorrhizal fungi in alleviation of salt stress: A review. *Ann. Bot.* 104:1263–1281.

Fan B., Carvalhais L.C., Becker A., Fedoseyenko D., von Wirén N., Borriss R. (2012). Transcriptomic profiling of *Bacillus amyloliquefaciens* FZB42 in response to maize root exudates. *BMC Microbiol.* 12:116.

Farrar J., Hawes M., Jones D., Lindow S. (2003). How roots control the flux of carbon to the rhizosphere. *Ecology* 84:827–837.

Figueiredo A.F., Boy J., Guggenberger G. (2021). Common mycorrhizae network: A review of the theories and mechanisms behind underground interactions. *Front. Fungal Biol.* 2:735299.

Flowers T.J., Munns R., Colmer T. (2015). Sodium chloride toxicity and the cellular basis of salt tolerance in halophytes. *Ann. Bot.* 115:419–431.

Frey-Klett P., Garbaye J., Takka M. (2007). The mycorrhiza helper bacteria revisited. *New Phytol.* 176:22–36.

Gadd G.M., Rhee Y.J., Stephenson K., Wei Z. (2012). Geomycology: Metals, actinides and biominerals. *Environ. Microbiol. Rep.* 4:270–296.

Gardner W.K., Garder W.K., Barber D.A., Parbery D.G. (1983). The acquisition of phosphorus by *Lupinus albus* L. III. The probable mechanism by which phosphorus movement in the soil/root interface is enhanced. *Plant Soil* 70:107–124.

González-Guerrero M., Escudero V., Saéz Á., Tejada-Jiménez M. (2016). Transition metal transport in plants and associated endosymbionts: Arbuscular mycorrhizal fungi and rhizobia. *Front. Plant Sci.* 7:1088.

Gow N.A.R., Latge J.P., Munro C.A. (2017). The fungal cell wall: Structure, biosynthesis, and function. *Microbiol. Spect.* 5. doi: 10.1128/microbiolspec. FUNK-0035-2016.

Grill E., Winnacker E.L., Zenk M.H. (1989). Phytochelatins, a class of heavy-metal-binding peptides from plants, are functionally analogous to metallothioneins. *Pro. Natl. Acad. Sci. USA* 84:439–443.

Guo X., Zhang S., Shan X. (2008). Adsorption of metal ions on lignin. *J. Haz. Mat.* 151(1):134–42.

Gupta D.K., Huang H.G., Corpas F.J. (2013a). Lead tolerance in plants: Strategies for phytoremediation. *Environ. Sci. Pollut. Res.* 20:2150–2161.

Gupta D.K., Vandenhove H., Inouhe M. (2013b). Role of phytochelatin in heavy metal stress and detoxification mechanisms in plants. In: Gupta D.K., Corpas F.J., Palma J.M. (eds.) *Heavy Metal Stress in Plants.* Springer, Berlin, Heidelberg. pp. 73–94.

Gupta D.K., Walther C. (2014). *Radionuclide Contamination and Remediation through Plants*. Springer-Cham, Switzerland.

Gupta D.K., Chatterjee S. (2017). *Arsenic Contamination in the Environment: The issues and Solutions*. Springer-Cham, Switzerland.

Gupta D.K., Palma J.M., Corpas F.J. (2018). Generation and scavenging of reactive oxygen species (ROS) in plant cells: An overview. In: Gupta D.K., Palma J.M., Corpas F.J. (eds.) *Antioxidants and Antioxidant Enzymes in Higher Plants*. Springer-Cham, Switzerland.

Haas H. (2014). Fungal siderophore metabolism with a focus on *Aspergillus fumigatus*. *Nat. Prod. Rep.* 31:1266–1276.

Hampton C.R., Bowen H.C., Broadley M.R., Hammond J.P., Mead A., Payne K.A., Pritchard J., White P.J. (2004). Cesium toxicity in Arabidopsis. *Plant Physiol.* 136:3824–3837.

Haydon M.J., Cobbette C.S. (2007a). A novel major facilitator superfamily protein at the tonoplast influences zinc tolerance and accumulation in Arabidopsis. *Plant Physiol.* 143:1705–1719.

Haydon M.J., Cobbett C.S. (2007b). Transporters of ligands for essential metal ions in plants. *New Phytol.* 174:499–506.

Hinsinger P. (2001). Bioavailability of soil inorganic P in the rhizosphere as affected by root-induced chemical changes: A review. *Plant Soil* 237: 173–195.

Hoffmann L. (1989). Algae of terrestrial habitats. *Bot. Rev.* 55:77–105.

Hoorman J.J., Islam R. (2010). Understanding soil microbes and nutrient recycling. *Agric. Nat. Res.* 16(10):SAG–16.

Hossain M.A., Piyatida P., da Silva J.A.T., Fujita M. (2012). Molecular mechanism of heavy metal toxicity and tolerance in plants: Central role of glutathione in detoxification of reactive oxygen species and methylglyoxal and in heavy metal chelation. *J. Bot.* 2012:1–37.

Iijima M., Higuchi T., Barlow P.W. (2004). Contribution of root cap mucilage and presence of an intact root cap in maize (*Zea mays*) to the reduction. *Annal. Bot.* 94:473–477.

Inouhe M., Nevins D.J. (1991). Inhibition of auxin-induced cell elongation of maize coleoptiles by antibodies specific for cell wall glucanases. *Plant Physiol.* 96:426–431.

Inouhe M., Nevins D.J. (1997). Regulation of cell wall glucanase activities by non-enzymic proteins in maize coleoptiles. *Int. J. Biol. Macromol.* 21:15–20.

Inouhe M., Hayashi K., Thomas B.R., Nevins D.J. (2000). Exo- and endoglucanases of maize coleoptile cell walls: Their interaction and possible regulation. *Int. J. Biol. Macromol.* 27:157–162.

Inouhe M. (2005). Phytochlatins. *Braz. J. Plant Physiol.* 17:65–78.

Inouhe M., Huang H., Chaudhary S.K., Gupta D.K. (2012). Heavy metal bindings and their interactions with thiol peptides and other biological ligands in plant cells. In: Gupta D.K., Sandalio L.M. (eds.) *Metal Toxicity in Plants: Perception, Signaling and Remediation.* Springer, Berlin, Heidelberg, pp. 1–21.

Inouhe M., Sakuma Y., Chatterjee S., Datta S., Jagetiya B.L., Voronina A.V., Walther C., Gupta D.K. (2015). General roles of phytochelatins and other peptides in plant defense mechanisms against oxidative stress/primary and secondary damages induced by heavy metals. In: Gupta D.K., Palma J.M., Corpas F.J. (eds.) *Reactive Oxygen Species and Oxidative Damage in Plants Under Stress.* Springer, Berlin, Heidelberg, pp. 219–245.

Johnson K.J., Cygan R.T., Fein J.B. (2006). Molecular simulations of metal adsorption to bacterial surfaces. *Geochim. Cosmochim. Acta* 70:5075–5088.

Jung S.C., Martinez-Medina A., Lopez-Raez J.A., Pozo M.J. (2012). Mycorrhiza-induced resistance and priming of plant defenses. *J. Chem. Ecol.* **38**: 651–664.

Kido N., Yokoyama R., Yamamoto T., Furukawa J., Iwai H., Satoh S., Nishitani K. (2015). The matrix polysaccharide (1;3,1;4)-β-D-glucan is involved in silicon-dependent strengthening of rice cell wall. *Plant Cell Physiol.* 56:268–276.

Kobayashi M., Matoh T., Azuma J. (1996). Two chains of rhamnogalacturonan II are cross-linked by borate-diol ester bonds in higher plant cell walls. *Plant Physiol.* 110:1017–1020.

Kramer U., Pickering I.J., Price R.C., Raskin I., Salt D.E. (2000). Subcellular localization and speciation of nickel in hyperaccumulator and non-accumulator *Thlaspi* species. *Plant Physiol.* 122:1343–1354.

Kupper H., Mijovilovich A., Meyer-Klaucke W., Kroneck P.M.K. (2004). Tissue- and age-dependent differences in the complexation of cadmium and zinc in the cadmium/zinc hyperaccumulator *Thlaspi caerulescens* (Ganges ecotype) revealed by X-ray absorption spectroscopy. *Plant Physiol.* 134:748–757.

Ma J.F. (2000). Role of organic acids in detoxification of aluminum in higher plants. *Plant Cell Physiol.* 41:383–390.

Ma J.F., Kusano G., Kimura S., Nomoto K. (1993). Specific recognition of mugineic acid-ferric complex by barley roots. *Phytochemistry* 34:599–603.

Ma J.F. (2005). Plant root responses to three abundant soil minerals: Silicon, aluminum and iron. *Crit. Rev. Plant Sci.* 24:267–281.

Maillet F., Poinsot V., André O., Puech-Pagès V., Haouy A., Gueunier M., Cromer L., Giraudet D., Formey D., Niebel A., Martinez E.A., Driguez H., Bécard G., Dénarié J. (2011). Fungal lipochitooligo saccharide symbiotic signals in arbuscular mycorrhiza. *Nature* 469:58–63.

Mari S., Gendre D., Pianelli K., Ouerdane L., Lobinski R., Briat J.-F., Lebrun M., Czernic P. (2006). Root-to-shoot long-distance circulation of nicotianamine and nicotianamine–nickel chelates in the metal hyperaccumulator *Thlaspi caerulescens*. *J. Exp. Bot.* 57:4111–4122.

Martínez-Medina A., Pascual J.A., Lloret E., Roldán A. (2009). Interactions between arbuscular mycorrhizal fungi and *Trichoderma harzianum* and their effects on *Fusarium wilt* in melon plants grown in seedling nurseries. *J. Sci. Food Agric.* 89:1843–1850.

Mizuno T. (2024). Nickel Disorder in Plants on Ultramafic Soils. In: (Inouhe M., Walther C., Gupta D.K. (eds.) *Lithium and Nickel Contamination in Plants and Environment*. World Scientific Publishing, Singapore, pp. 171–183.

Mohammadi K., Khalesro S., Sohrabi Y., Heidari G. (2011). A review: Beneficial effects of the mycorrhizal fungi for plant growth. *J. Appl. Environ. Biol. Sci.* 1:310–319.

Moriwaki H., Koide R., Yoshikawa R., Warabino Y., Yamamoto H. (2013). Adsorption of rare earth ions onto the cell walls of wild-type and lipoteichoic acid-defective strains of *Bacillus subtilis*. *Appl. Microbiol. Biotechnol.* 97:3721–3728.

Mulrooney S.B., Hausinger R.P. (2003). Nickel uptake and utilization by micro-organisms. *FEMS Microbiol. Rev.* 27:239–261.

Mwandira W., Nakashima K., Kawasaki S., Arabelo A., Banda K., Nyambe I., Chirwa M., Ito M., Sato T., Igarashi T., Nakata H., Nakayama S., Ishizuka M. (2020). Biosorption of Pb (II) and Zn (II) from aqueous solution by *Oceanobacillus profundus* isolated from an abandoned mine. *Sci. Rep.* 10:21189. And a figure was adopted from the web site of https://bre.eng.hokudai.ac.jp/en/research/biosorption.html.

Nadeem S.M., Khan M.Y., Waqas M.R., Binyamin R., Akhtar S., Zahir Z.A. (2017). Arbuscular mycorrhizas: An overview. In: Wu Q.S. (ed.) *Arbuscular Mycorrhizas and Stress Tolerance of Plants*. Springer, Singapore, pp. 1–24.

Noor I., Sohail H., Sun J., Nawaz M.A., Li G., Hasanuzzaman M., Liu J. (2022). Heavy metal and metalloid toxicity in horticultural plants: Tolerance mechanism and remediation strategies. *Chemosphere* 303:135196.

Ogi T., Tamaoki K., Saitoh N., Higashi A., Konishi Y. (2012). Recovery of indium from aqueous solutions by the Gram-negative bacterium *Shewanella algae*. *Biochem. Eng. J.* 63:129–133.

Ogi T., Makino T., Iskandar F., Tanabe E., Okuyama K. (2016). Heat-treated *Escherichia coli* as a high-capacity biosorbent for tungsten anions. *Biores. Technol.* 218:140–145.

Ohshima T., Hayashi H., Chino M. (1990). Collection and chemical composition of pure phloem sap from *Zea mays* L. *Plant Cell Physiol.* 31: 735–737.

Otani T., Ae N., Tanaka H. (1996). Phosphorus (P) Uptake mechanisms of crops grown in soils with low P status II. Significance of organic acids in root exudates of pigeon pea. *Soil Sci. Plant Nutr.* 42:553–560.

Panchal P., Miller A.J., Giri J. (2021). Organic acids: Versatile stress-response roles in plants. *J. Exp. Bot.* 72:4038–4052.

Panda S.K., Baluska F., Matsumoto H. (2009). Aluminum stress signaling in plants. *Plant Signal Behav.* 4:592–597.

Pearson R.G. (1968a). Hard and soft acids and bases, *HSAB*, part I: Fundamental principles. *J. Chem. Edu.* 45:581–587.

Pearson R.G. (1968b). Hard and soft acids and bases, *HSAB*, part II: Underlying theories. *J. Chem. Edu.* 45:643–648.

Persson I. (2010). Hydrated metal ions in aqueous solution: How regular are their structures? *Pure Appl. Chem.* 82:1901–1917.

Rauser W.E. (1990). Phytochelatins. *Ann. Rev. Biochem.* 59:61–86.

Rauser W.E. (1995). Phytochelatins and related peptides: Structure, biosynthesis, and function. *Plant Physiol.* 109:1141–1149.

Rauser W.E. (1999). Structure and function of metal chelators produced by plants — The case for organic acids, amino acids, phytin, and metallothioneins. *Cell Biochem. Biophys.* 31:19–48.

Renshaw J.C., Robson G.D., Trinci A.P.J., Wiebe M.G., Livens F.R., Collison D., Taylor R.J. (2002). Fungal siderophores: Structures, functions and applications. *Mycol. Res.* 106:1123–1142.

Romheld V., Awad F. (2000). Significance of root exudates in acquisition of heavy metals from a contaminated calcareous soil by graminaceous species. *J. Plant Nutr.* 23:1857–1866.

Saha M., Sarkar S., Sarkar B., Sharma B.K., Bhattacharjee S., Tribedi P. (2016). Microbial siderophores and their potential applications: A review. *Environ. Sci. Pollut. Res.* 23:3984–3999.

Scervino J.M., Mesa M.P., Mónica I.D., Recchi M., Moreno N.S., Godeas A. (2010). Soil fungal isolates produce different organic acid patterns involved in phosphate salts solubilization. *Biol. Fertil. Soils* 46:755–763.

Shahzad B., Tabveer M., Hassan W., Shah A.N., Anjum S.A., Cheema S.A., Ali I. (2016). Lithium toxicity in plants: Reasons, mechanisms and remediation possibilities — A review. *Plant Physiol. Biochem.* 107:104–115.

Shahzad B., Mughal M.N., Tanveer M., Gupta D., Abbas G. (2017). Is lithium bio-logically an important or toxic element to living organisms? An overview. *Environ. Sci. Pollut Res.* 24:103e115.

Shakoor N., Adeel M., Ahmad M.A., Zain M., Waheed U., Javaid R.A., Haider F.U., Azeem I., Zhou P., Li Y., Jilani G., Xu M., Rinklebe J., Rui R. (2023). Reimagining safe lithium applications in the living environment and its impacts on human, animal, and plant system. *Environ. Sci. Ecotechnol.* 15:100252.

Sharma P. (2022). Role and significance of biofilm-forming microbes in phytoremediation — A review. *Environ. Technol. Innovation* 25:102182.

Sharma S.S., Dietz K.J. (2006). The significance of amino acids and amino acid-derived molecules in plant responses and adaptation to heavy metal stress. *J. Exp. Bot.* 57:711–726.

Sharifi M., Ghorbanli M., Ebrahimzadeh H. (2007). Improved growth of salinity-stressed soybean after inoculation with salt pre-treated mycorrhizal fungi. *J. Plant Physiol.* 164:1144–1151.

Shetty R., Vidya C.S.N., Prakash N.B., Lux A., Vaculík M. (2021). Aluminum toxicity in plants and its possible mitigation in acid soils by biochar: A review. *Sci. Total Environ.* 765:142744.

Singh S.K., Wu X., Shao C., Zhang H. (2022). Microbial enhancement of plant nutrient acquisition. *Stress Biol.* 2:3.

Sun H., Jiang S., Jiang C., Wu C., Gao M., Wang S. (2021). A review of root exudates and rhizosphere microbiome for crop production. *Environ. Sci. Pollut. Res.* 28:54497–54510.

Takagi S. (1976). Naturally occurring iron-chelating compounds in oat- and rice-root washings. I. Activity measurement and preliminary characterization. *Soil Sci. Plant Nutr.* 22:423–433.

Takahashi M., Terada Y., Nakai I., Nakanishi H., Yoshimura E., Mori S., Nishizawa N.K. (2003). Role of nicotianamine in the intracellular delivery of metals and plant reproductive development. *Plant Cell* 15:1263–1280.

Van Steveninck R.F.M., Van Stevenick M.E., Fernando D.R. (1992). Heavy-metal (Zn, Cd) tolerance in selected clones of duck weed (*Lemna minor*). *Plant Soil* 146:271–280.

Vega K., Kalkum M. (2012). Chitin, chitinase responses, and invasive fungal infections. *Int. J. Microbiol.* 2012:1–11.

Velmourougane K., Prasanna R., Saxena A.K. (2017). Agriculturally important microbial biofilms: Present status and future prospects. *J. Basic Microbiol.* 57:548–573.

Wardle D.A. (2002). *Communities and Ecosystems: Linking the Aboveground and Belowground Components*. Princeton University Press, NJ, USA.

Wardle D.A., Bardgett R.D., Klironomos J.N., Setala H., van der Putten W.H., Wall D.H. (2004). Ecological linkages between aboveground and below-ground biota. *Science* 304:1629–1633.

Weber M., Harada E., Vess C., Roepenack-lahaye E.V., Clemens S. (2004). Comparative microarray analysis of *Arabidopsis thaliana* and *Arabidopsis helleri* roots identifies nicotianamine synthase, a ZIP transporter and other genes as potential metal hyperaccumulation factors. *Plant J.* 37:269–281.

Whittaker R.H. (1962). Classification of natural communities. *Bot. Rev.* 28: 1–239.

Whittaker R.H. (1975). *Communities and Ecosystems*. MacMillan Publishing, New York. ISBN 13:9780024273901, ISBN 10: 0024273902.

Wilkinson D.M., Creevy A.L., Valentine J. (2012). The past, present and future of soil protist. *Ecol. Acta Protozool.* 51:189–199.

Winter H., Lohaus G., Heldt H.W. (1992). Phloem transport of amino acids in relation to their cytosolic levels in barley leaves. *Plant Physiol.* 99: 996–1004.

Woolhouse H.W. (1983). Toxicity and tolerance in the responses of plants to metals. In: Lange O.L, Osmond L.B., Ziegler H. (eds.) *Encyclopedia of Plant Physiology*. Springer, Berlin, Heidelberg, pp. 245–300.

Yang Z., Yang F., Liu J.L., Wu H.T., Yang H., Shi Y., Liu J., Zhang Y.F., Luo Y.R., Chen K.M. (2022). Heavy metal transporters: Functional mechanisms, regulation, and application in phytoremediation. *Sci. Total Environ.* 809:151099.

Yokoyama R., Nishitani K. (2004). Genomic basis for cell-wall diversity in plants. A comparative approach to gene families in rice and *Arabidopsis. Plant Cell Physiol.* 45:1111–1121.

Zenk M.H. (1996). Heavy metal detoxification in higher plants — A review. Heavy metal bindings and their interactions with thiol peptides. *Gene* 179:21–30.

https://doi.org/10.1142/9789811283123_0007

Chapter 7

Nickel Disorder in Plants on Ultramafic Soils

Takafumi Mizuno

Laboratory of Soil Science and Plant Nutrition,
Graduate School of Bioresources,
Mie University, 1577 Kriyamachya-cho, Tsu, Mie, Japan
tmizuno@bio.mie-u.ac.jp

Abstract

Symptoms of plant nickel (Ni) disorders occurring on ultramafic soils resemble those of iron (Fe) deficiency, such as chlorosis. Since plants absorb Ni via root iron transporters, it is expected that Ni would compete with Fe acquisition on the root surface in soils with high nickel concentrations, resulting in Fe deficiency in plants. However, experiments using *Arabidopsis thaliana* confirmed that the addition of Ni to a hydroponic solution induced an Fe starvation response instead. It has also been observed that Fe content in plant roots increases with Ni content, and in fact, plants from ultramafic soils tend to have higher Fe contents than those from other soil types. This phenomenon, in which Fe deficiency occurs even though Fe is acquired sufficiently, can be attributed to the disruption of the Fe transport system in plants. Nicotianamine (NA) is a biological compound that chelates Fe and is responsible for its transport in plants; however, NA has a higher affinity for Ni than Fe. Because of that, it is predicted that NA, which should be used for Fe transport, is instead bound by Ni, resulting in the inhibition of chlorophyll synthesis in plants under high Ni conditions. Experimentally, plants that overexpress NA genes are more tolerant to high levels of heavy metals, especially Ni.

Soil Ni strongly inhibits plant growth during or immediately after germination. This may be because Ni inhibits the supply of Fe, which is essential for chlorophyll synthesis, and thus prevents the initiation of photosynthesis. An adequate supply of NA may be an essential condition for plants to adapt to ultramafic soils.

Keywords: Iron deficiency, IRT1, nickel toxicity, nicotianamine, serpentine, ultramafic soil

1. Introduction

Ultramafic soil (also known as serpentine soil), which is formed through the metamorphism of peridotite, is characterised by high magnesium (Mg) content and is mainly composed of serpentine ($Mg_3Si_2O_5(OH)_4$). Since ultramafic soil originates from the weathering of serpentinite, it naturally reflects the chemical properties of serpentinite in its inorganic composition, containing not only Mg but also heavy metals, such as nickel (Ni) and chromium (Cr). In addition, essential elements for plants, such as phosphorus (P), potassium (K) and calcium (Ca), are present in very low concentrations, and the high pH resulting from the high Mg content inhibits the solubilisation and absorption of essential metallic elements. The low levels of essential elements, such as nitrogen, phosphorus, and potassium (often referred to as NPK), limit the number of plant species that can grow on ultramafic soils and strongly suppress the growth of those plants that can survive. Plants that grow in ultramafic soil are believed to adapt to the harsh soil conditions, have few competitors for resources such as sunlight and form unique plant communities, despite often being susceptible to disturbances.

The unique vegetation formed on ultramafic soil has drawn attention from botanists around the world and has been investigated as a model of plant ecology, adaptation and evolution (Brooks, 1987). In addition, plants capable of accumulating high levels of metals, such as Ni and manganese (Mn), can be found in ultramafic soils all over the world. Those plant species that can accumulate metals or metalloids beyond specified concentrations in their shoots are called 'hyperaccumulators' (Reeves *et al.*, 2018). This unique characteristic of hyperaccumulators have been the subject of the studies of plant physiology and plant nutrition, and used on new scientific techniques, such as phytoremediation and phytomining (Chaney, 1983; van der Ent *et al.*, 2015).

Meanwhile, the stunted growth of crops caused by ultramafic soil has been a concern, and investigations into its cause have been conducted primarily at agricultural research stations. It was established early that the main cause was Ni toxicity, with one of the contributing factors being iron (Fe) deficiency (Agurla *et al.*, 1955). However, it took a long time to uncover the differences in Ni tolerance between crop plants and the mechanisms that induce Fe deficiency in them. On the other hand, transporters involved in the absorption and distribution of essential elements in plants were identified, and genetical research on Ni toxicity in plants began to be reported.

In this chapter, the causes of Ni impairment in plants growing on ultramafic soils are explained in relation to the mechanisms of Fe acquisition and transport, especially with the use of nicotianamine (NA). It is hypothesised that the inhibition of seed germination by Ni may be involved in the formation of specialised vegetation often found on ultramafic soils.

2. Relationship between Ni Toxicity and Plant Fe Content

Among crops, oat (*Avena sativa*) has low Ni tolerance and has been used as a model plant for Ni toxicity studies for some time (Hunter and Vergnano, 1953). Under excess Ni conditions, this plant shows chlorosis (whitening) with horizontal striping in leaves immediately after germination, which later becomes vertical striping as growth progresses (Fig. 1). Additionally, as Ni toxicity progresses, necrotic black spots (necrosis) occur in the leaves, a condition that is commonly found in *Brassicaceae* vegetables when tissues contain a high level of Ni. Crooke and Knight (1955) showed that Fe concentrations are low in oats with Ni toxicity, especially in parts where necrosis occurs, suggesting that Fe deficiency is an inducing factor for excess Ni. In that report, it was also noted that Ni toxicity was induced by a decrease in the Fe/Ni ratio in the leaves.

Figure 1. Ni toxicity symptoms in crops: (a) young oats with chlorosis, (b) grown oats with chlorosis, (c) Japanese white radish leaves with chlorosis, (d) cabbage leaves with tissue necrosis.

Damage to oats is not strongly related to total soil Ni content, but rather shows a strong correlation with exchangeable soil Ni content. Various studies, such as that by Mizuno and Kobayashi (1971), have demonstrated a clear correlation between Ni concentration in a plant and exchangeable Ni content in soil. They also reported that the application of calcium carbonate can control crop growth inhibition because it increases soil pH and suppresses exchangeable Ni concentration.

In these studies, it was suggested that Ni damage to crops (inhibition of growth) differs greatly depending on crop species because of the suppression of internal Ni concentration in relation to the exchangeable Ni concentration in the soil. Crops, such as oats, that increase internal Ni concentration in proportion to the increase in exchangeable Ni in soil are more prone to Ni toxicity, while crops, such as potatoes, that suppress internal Ni concentration even with an increase in exchangeable soil Ni have less Ni toxicity (Mizuno, 1968). From these results, it was suggested that a Fe/Ni ratio of 5–10 or less in the plant body is one indicator of Ni toxicity in crops.

3. Ni is Taken Up into the Body through Fe Transporters

The average concentration of Ni in the soil is very low, approximately 40 mg kg^{-1}, and it is unlikely to compete with Fe, which is present at about 38,000 mg kg^{-1}. However, in most soils, Fe exists mainly as trivalent ion compounds (iron oxide and iron hydroxide), and the solubility of Fe (III) at a pH of 7 is very low (10^{-17} mol L^{-1}). Therefore, in ultramafic soil with high levels of Ni, Ni can compete with Fe for absorption. On Sugashima Island in Toba, Japan, the total Ni concentration in the soil is in the range of 1,000–4,000 mg kg^{-1}, while in ultramafic soil areas in Hokkaido, Japan, it is 600–3,000 mg kg^{-1}, with an average of around 2,000 mg kg^{-1} in both. In Hokkaido, the amount of exchangeable Ni is on average about 7 mg kg^{-1} (Horie, 2002; Mizuno and Kirihata, 2015).

For Ni to induce Fe deficiency in plants, it should be taken up by them through the Fe uptake pathway, or Ni should inhibit the transport of Fe within them. Plant Fe uptake mechanisms are known as Strategy-I for non-graminaceous plants and Strategy-II for graminaceous plants (Marschner and Römheld, 1994). It is also known that Fe taken up by a plant is mainly chelated by NA for transport to tissues (Gayomba *et al.*, 2015). On the other hand, while Ni is considered an essential element for plant urea metabolism, there is little research done on the Ni uptake mechanism due to its low requirement volume, and information on its relationship with Fe transport is also limited.

Our group conducted research on Japanese pennycress (*Noccaea japonicum*), a plant species that exhibits hyperaccumulation of Ni, and confirmed that proteins involved in Fe transport (ZIP and NRAMP transporter groups) are capable of transporting Ni (Mizuno *et al.*, 2005). This suggested that pathways for Ni and Fe absorptions may overlap. To further demonstrate this in a model plant, we obtained *Arabidopsis thaliana* lacking IRT1, a protein involved in Fe absorption, and conducted experiments on it. Results revealed that expression of IRT1 in yeast enhances Ni sensitivity and that accumulation of Ni in the roots of *A. thaliana* is suppressed without IRT1, indicating that Ni is transported into plants via Fe transporters, at least in *A. thaliana* (Nishida *et al.*, 2011). Furthermore, subsequent studies have suggested that some zinc transporters are also involved in Ni transport (Nishida *et al.*, 2015).

4. What Really is Ni Toxicity?

On confirming that Ni enters plant roots through Fe transporters, I initially believed that the inhibition of Fe uptake due to competition with Ni was the cause of Ni toxicity. However, as the amount of Ni added to the hydroponic solution increased, the concentration of Fe in roots actually increased, and Fe uptake itself was not inhibited (Nishida *et al.*, 2011). In fact, in a survey of 75 wild plant species found on the ultramafic soil of Sugashima Island, the Fe concentration in the plants was recorded to be about three times (Mizuno and Kirihata, 2015) that in plants found in typical Japanese soil (114 mg kg^{-1}) (Takada *et al.*, 1994). In plants growing on ultramafic soils in Hokkaido, the Fe concentration was double its level in general plants and three times higher than that in species endemic to ultramafic soil (Horie, 2002). This indicated that plants from ultramafic soil areas are actually promoting Fe uptake due to the presence of Ni.

Despite sufficient absorption of Fe, why does the presence of Ni cause Fe-deficiency symptoms in plants? This can be explained by the response of the Fe-deficiency response gene group induced by Ni addition. The genes and transcription factors that respond to Fe deficiency in plants, such as *A. thaliana* and rice, have been elucidated (Kobayashi *et al.*, 2010). When Ni was applied to *A. thaliana* grown in a hydroponic solution with sufficient Fe, the expression of genes, such as Fe transporter IRT1, Fe reductase FRO2 and transcription factor FIT, which positively induce the expression of these genes during Fe deficiency, was increased (Nishida *et al.*, 2012). This means that, even if sufficient Fe is provided, the presence of Ni alone can induce the Fe deficiency response in *A. thaliana*.

Taking into account the fact that Ni and Fe transports occur in the same pathway in the body of *A. thaliana*, it is expected that a vicious cycle will occur as follows: I: Ni is taken up into the body → II: Ni accumulation is detected as 'iron deficiency' → III: Various genes required for Fe acquisition, including IRT1, are expressed → IV: More Ni is absorbed into the plant through IRT1 → V: Even stronger Fe starvation is detected → VI: Fe transporters are further expressed, leading to the uptake of even more Ni (Fig. 2). If Ni flows into the body in large amounts due to this feedback loop, what will happen inside a plant?

NA is a precursor of mugineic acids that chelate soil Fe in the Fe uptake mechanism of Strategy-II in graminaceous plants. It is also involved in the transport of Fe in a plant, playing an important role in the

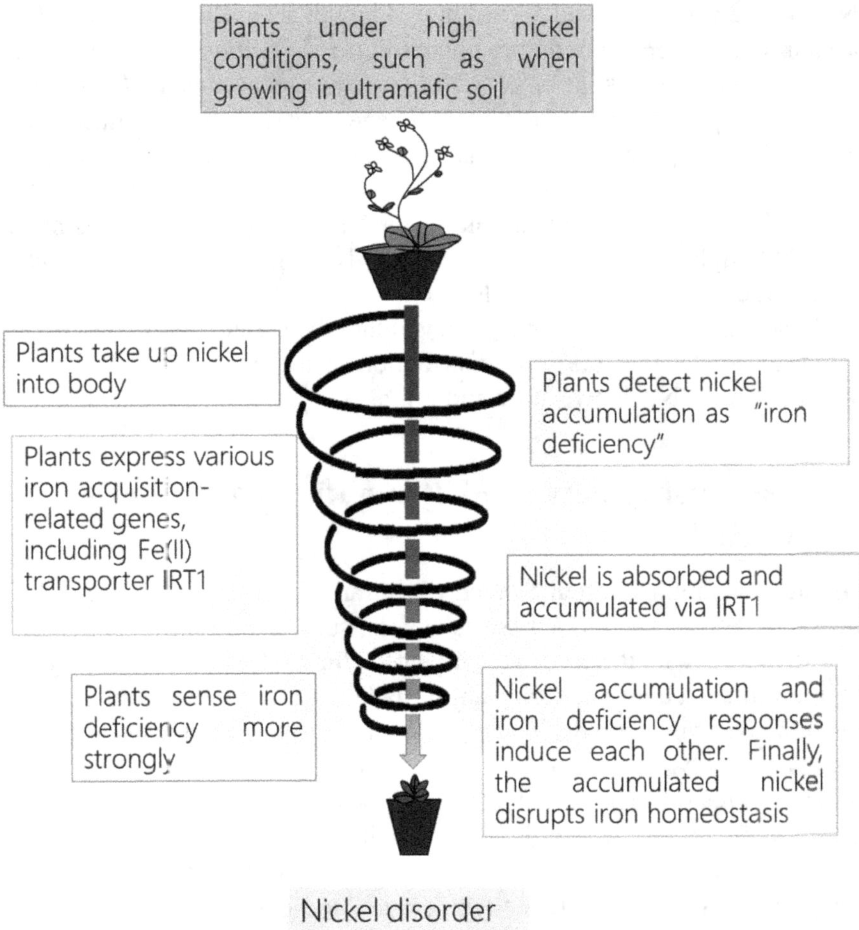

Figure 2. Correlation model of nickel absorption and iron utilisation deficiency.

supply of Fe for chlorophyll synthesis in chloroplasts (Takahashi *et al.*, 2003). On the other hand, NA can chelate Ni ions (Ni^{2+}) much more strongly than divalent Fe ions (Fe^{2+}) (Beneš *et al.*, 1983). This suggests that the Fe transport system in plants can be disturbed easily by Ni when growing on high Ni soil. While it has been shown in studies of plants with Ni accumulation, such as *Noccaea caerulescens* infield conditions, that NA can strongly bind to Ni and may be involved in Ni transport in plants (Mari *et al.*, 2006), it was also reported that the nicotianamine synthase

(NAS) gene was expressed at extremely high levels in *Arabidopsis helleri* (a cadmium hyperaccumulator) compared to *A. thaliana* (Weber *et al.*, 2004). Furthermore, Kim *et al.* (2005) created recombinant *A. thaliana* and tobacco plants expressing the NAS gene of barley and examined their metal tolerance. They revealed that, among heavy metals, Ni damage in particular was alleviated, and tobacco plants overexpressing the NAS gene did not show growth inhibition by Ni when grown on ultramafic soils. Although indirect, these studies strongly suggest that when Ni enters a plant, it competes with Fe for NA.

Based on these reports, it is believed that the essence of Ni toxicity is the failure of Fe transport and distribution in the plant body due to the competition for NA between Fe and Ni.

5. Relationship between Ultramafic Vegetation and Ni Toxicity

Unique vegetation is often formed on ultramafic soils, and various soil factors have been mentioned as the causes, such as nutrient deficiencies, excessive heavy metals and high pH levels due to the presence of Mg. In addition, the accumulation of plant-derived organic matter from decomposition takes time, and the development of granular structures is difficult, leading to low water retention and a tendency to dry easily. In the ultramafic soil area of Sugashima Island, very slow-growing tree species, such as boxwood (*Buxus microphylla*) and Ubame oak (*Quercus phillyraeoides*) constitute the main vegetation, and the trees do not grow large. In Japan, where there is moderate temperature and good rainfall, such poor vegetation can be clearly distinguished from that on other soils; therefore, mountains with ultramafic vegetation are sometimes called 'bald mountains'. On the other hand, under similar meteorological conditions, vegetation similar to that on overseas ultramafic soil may be formed. For example, species such as boxwood can be found growing naturally in many ultramafic soils around the world (Fig. 3). Furthermore, since available P content is low, the plants of the *Brassicaceae* family are rarely seen on Sugashima Island because they are unable to form a symbiotic relationship with mycorrhizal fungi to acquire P. Similarly, each factor that contributes to the stressful nature of ultramafic soils inhibits the growth of various plants, and it is believed that the limited number of plant species that can reproduce in such soils is strictly constrained by the compounding of these factors.

Figure 3. Similarity of ultramafic vegetation: boxwood (a) and thyme (b) on Sugashima Island, Japan; similar vegetation is shown from an ultramafic area in Pogradec, Albania (c and d).

Plants synthesise chlorophyll through various intermediate products from glutamic acid, and Fe is an essential element in this process. Fe deficiency causes leaf whitening (chlorosis) due to the inhibition of chlorophyll synthesis. Therefore, even at the stage when roots are immature immediately after seed germination, plants store and use Fe in vacuoles to enable chlorophyll synthesis (Lanquar *et al.*, 2005). However, when considering germination in ultramafic soils, plants absorb Ni immediately after germination, and Fe usage is inhibited from that time onwards, suggesting that these young and weak plants will die before they can synthesise the necessary amount of chlorophyll. In fact, the report by Kim *et al.* (2005) suggests that, although various heavy metals can inhibit the growth of *A. thaliana*, Ni is the only element that can almost completely stop growth immediately after germination. Similarly, in pot experiments using a buckwheat cover crop with high lead accumulation conducted by our team, only Ni caused strong growth inhibition (Fig. 4).

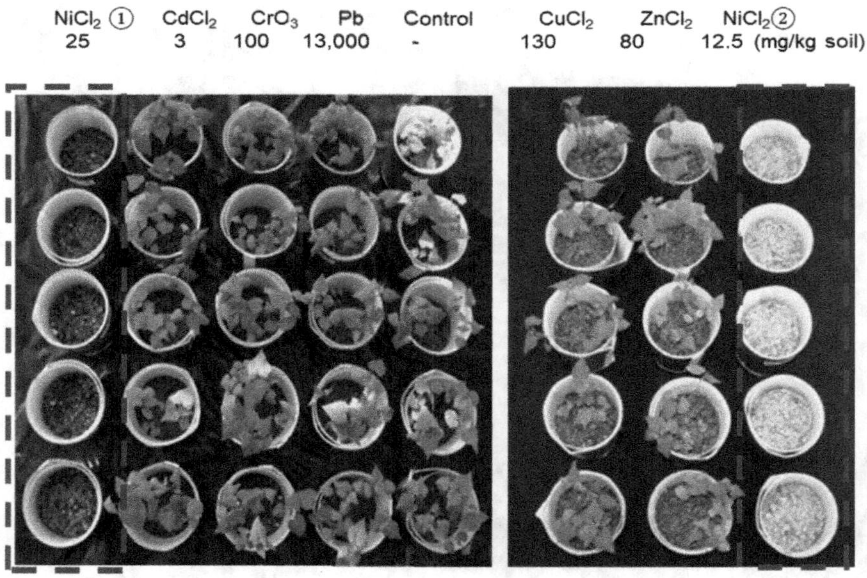

Figure 4. Suppression of seed germination by nickel.

It is believed that the strong inhibitory effect of Ni on early growth could create a strong bias in plant species selection on ultramafic soil. It may be hypothesised that only plant species that can overcome this germination inhibition in ultramafic soils can survive and reproduce, thus contributing to the unique vegetation on ultramafic soil (Fig. 5). To test this hypothesis, it is necessary to investigate whether plants on ultramafic soil have a higher NA synthesis ability than those on normal soil, or whether there are differences in NA contents in seeds or in germination rates under high Ni conditions. These issues need to be addressed in future research.

6. Summary

Ni disorder in plants on ultramafic soils is caused not by the competition between Ni and Fe uptake in the roots but rather by the failure of Fe transport and distribution in the plant due to the binding of NA with Ni. Ni strongly suppresses the growth of plants immediately after germination; however, this can be avoided by enhancing the rate of NA synthesis.

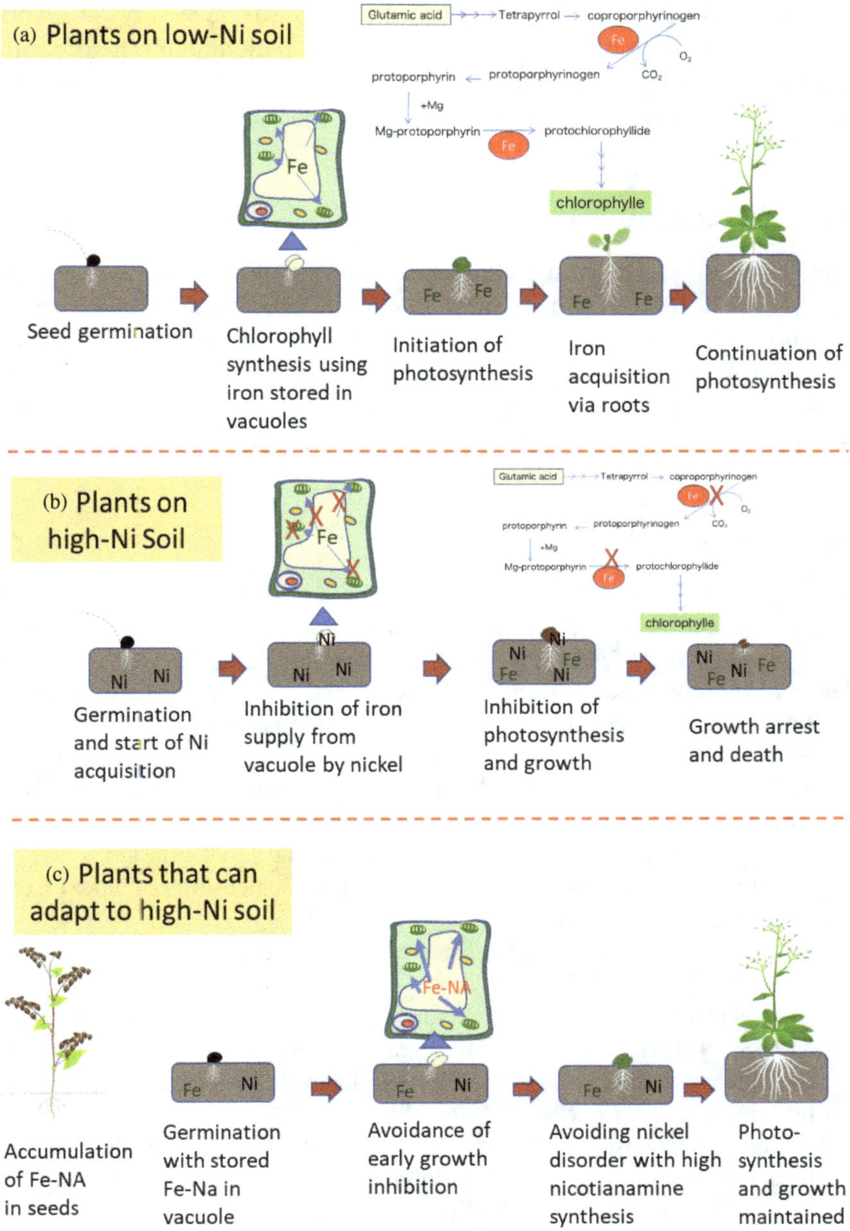

Figure 5. A hypothesis about plant adaptation under high nickel conditions (ultramafic soil): (a) plants on low-Ni soil, (b) plants on high-Ni soil, (c) plants that can adapt to high-Ni soil.

This suggests that a high ability to synthesise NA is necessary for plants to germinate and grow in high-Ni environments, including ultramafic soil, and that unique vegetation on this soil may be formed by plants selected by their NA synthesis capability.

Acknowledgements

I am grateful to Dr. Kenji Horie (the curator of the Asahikawa City Northern Wild Plant Garden) for valuable information and advice about serpentine plants in Japan. I would like to thank Prof. Robert S. Boyd of Auburn University, USA, for English language review. This work was supported, in part, by Grant-in-Aid for Young Scientists (B) 21780054 (T. Mizuno) from the Japan Society for the Promotion of Science.

References

Agurla S., Gayatri G., Crooke W.M., Knight A.H. (1955). The relationship between nickel-toxicity symptoms and the absorption of iron and nickel. *Ann. Appl. Biol.* 43:454–464.

Beneš I., Schreiber K., Ripperger H., Kircheiss A. (1983). Metal complex formation by nicotianamine, a possible phytosiderophore. *Experientia* 39:261–262.

Brooks R.R. (1987). *Serpentine and Its Vegetation, a Multidisciplinary Approach.* Croom Helm, London, pp. 1–454.

Gayomba S.R., Zhai Z., Jung H., Vatamaniuk O.K. (2015). Local and systemic signaling of iron status and its interactions with homeostasis of other essential elements. *Front. Plant Sci.* 14:716.

Hunter J.G., Vergnano O. (1953). Trace-element toxicities in oat plants. *Ann. Appl. Biol.* 40:761–777.

Kim S., Takahashi M., Higuchi K., Tsunoda K., Nakanishi H., Yoshimura E., Mori S., Nishizawa N.K. (2005). Increased nicotianamine biosynthesis confers enhanced tolerance of high levels of metals, in particular nickel, to plants. *Plant Cell Physiol.* 46:1809–1818.

Kobayashi T., Nakanishi H., Nishizawa N.K. (2010). Recent insights into iron homeostasis and their application in graminaceous crops. *Proc. Jpn. Acad. Ser. B Phys. Biol. Sci.* 86:900–9013.

Horie K. (2002). Studies on the chemical characteristics of the ultramafic plants in Hokkaido. *J. Rakuno Gakuen Univ. Nat. Sci.* 26:155–264 [in Japanese].

Lanquar V., Lelièvre F., Bolte S., Hamès C., Alcon C., Neumann D., Vansuyt G., Curie C., Schröder A., Krämer U., Barbier-Brygoo H., Thomine S. (2005).

Mobilization of vacuolar iron by AtNRAMP3 and AtNRAMP4 is essential for seed germination on low iron. *EMBO J.* 24:4041–4051.

Mari S., Gendre D., Pianelli K., Ouerdane L., Lobinski R., Briat J.F., Lebrun M., Czernic P. (2006). Root-to-shoot long-distance circulation of nicotianamine and nicotianamine-nickel chelates in the metal hyperaccumulator *Thlaspi caerulescens*. *J. Exp. Bot.* 57:4111–4122.

Marschner H., Römheld V. (1994). Strategies of plants for acquisition of iron. *Plant Soil* 165:261–274.

Mizuno N. (1968). Interaction between iron and nickel and copper and nickel in various plant species. *Nature* 219:1271–1272.

Mizuno N., Kobayashi S. (1971). Several trace elements in serpentine paddy soil. Effects of Ni on rice plant growth with reference to contents of Mn, Fe, Cu, Zn and Mo in rice and soil. *Jpn. J. Soil Sci. Plant Nutr.* 42:214–220 [In Japanese].

Mizuno T., Kirihata Y. (2015). Elemental composition of plants from the serpentine soil of Sugashima Island, Japan. *Aust. J. Bot.* 63:252–260.

Mizuno T., Usui K., Horie K., Nosaka S., Mizuno N., Obata H. (2005). Cloning of three ZIP/Nramp transporter genes from a Ni hyperaccumulator plant *Thlaspi japonicum* and their Ni^{2+}-transport abilities. *Plant Physiol. Biochem.* 43:793–801.

Nishida S., Aisu A., Mizuno T. (2012). Induction of IRT1 by the nickel-induced iron-deficient response in *Arabidopsis*. *Plant Signal Behav.* 7:329–331.

Nishida S., Kato A., Tsuzuki C., Yoshida J., Mizuno T. (2015). Induction of nickel accumulation in response to zinc deficiency in *Arabidopsis thaliana*. *Int. J. Mol. Sci.* 16:9420–9430.

Nishida S., Tsuzuki T., Kato A., Aisu A., Yoshida J., Mizuno T. (2011). AtIRT1, the primary iron uptake transporter in the root, mediates excess nickel accumulation in *Arabidopsis thaliana*. *Plant Cell Physiol.* 52:1433–1442.

Reeves R.D., Baker A.J.M., Jaffré T., Erskine P.D., Echevarria G., van der Ent A. (2018). A global database for plants that hyperaccumulate metal and metalloid trace elements. *New Phytol.* 218:407–411.

Takada J., Takamatsu T., Satake K., Sase H. (1994). *Data on Elemental Concentration in Land Plants by Neutron Activation Analysis* (No. 1). National Institute for Environmental Studies, Tsukuba, Japan, pp. 154–167. [In Japanese]. https://www.nies.go.jp/kanko/gyomu/pdf/f058-1993.pdf.

Takahashi M., Terada Y., Nakai I., Nakanishi H., Yoshimura E., Mori S., Nishizawa N.K. (2003). Role of nicotianamine in the intracellular delivery of metals and plant reproductive development. *Plant Cell* 15:1263–1280.

Weber M., Harada E., Vess C., Roepenack-Lahaye E.V., Clemens S. (2004). Comparative microarray analysis of *Arabidopsis thaliana* and *Arabidopsis halleri* roots identifies nicotianamine synthase, a ZIP transporter and other genes as potential metal hyperaccumulation factors. *Plant J.* 37:269–281.

https://doi.org/10.1142/9789811283123_0008

Chapter 8

Ecotoxicity, Oxidative Stress and Phytoremediation of Nickel on Aquatic Plants

Walter Darío Di Marzio[*,†,‡,§], **María Elena Sáenz**[*,†] **and Ricardo Santiago Martinez**[*]

[*]*Program Research in Ecotoxicology (PRIET), Basic Sciences Department, National University of Luján, Luján, Argentina*

[†]*National Council on Scientific and Technical Research CONICET, Godoy Cruz 2290, Buenos Aires, Argentina*

[‡]*ecotoxicologia@aae.org.ar*

[§]*wdimarzio@gmail.com*

Abstract

The search for environmentally friendly practices and new technologies in order to reduce the ecological impact associated with nickel disposal and pollution remains open. In this sense, phytoremediation stands out as a high-efficiency, eco-friendly green technology for removing pollution from contaminated soil, water and air. It is a form of bioremediation and an economical and sustainable alternative to traditional chemical and physical pollution remediation methods. High metal bioaccumulation efficiency and tolerance are requisites for plants employed in phytoremediation techniques; thus, investigating new aquatic plants with such characteristics is a priority. The antioxidant defence system plays a significant role in providing resistance to plants by protecting labile macromolecules against the attack of free radicals. This ability made possible the use of aquatic plants in technological applications for the restoration of contaminated sediments, solid wastes and organic contaminants. Accordingly, results have demonstrated that *Lemna gibba* and *Ceratophyllum demersum* would have the capacity to tolerate and, therefore, resist nickel stress. APX would be a more efficient enzyme in the detoxification of hydrogen peroxide, as its activity was stimulated at higher exposure concentrations. This reinforces the proposal of selecting APX as a good biomarker for the evaluation of the welfare of plants and growth performance. It is important to keep in mind that plant stress tolerance may be improved by the enhancement of the levels of antioxidant enzymes, in particular those related to the application of aquatic plants in the phytoremediation of nickel from wastewaters and industrial effluents.

Keywords: Aquatic plants, reactive oxygen species, biomarkers, phytoremediation, industrial effluents

1. Introduction

Nickel (Ni) is naturally present in the Earth's crust; it belongs to the transition metal group of elements and is extensively distributed in the environment (air, water and soil). Despite existing in several oxidative states, the Ni^{+2} oxidation state is predominant in the environment and in biological systems (Genchi *et al.*, 2020). Plants can uptake the ionic form Ni^{+2} relatively easily, while higher-molecular-weight chelated compounds are less available (Harasim and Filipek, 2015). Ni is of major importance in terms of environmental mobility and availability in acidic organic-rich soils and can be quite mobile; consequently, Ni contamination may pose a risk to natural waters as leaching of the soil profile increases (Harasim and Filipek, 2015; Rodriguez-Jimenez *et al.*, 2016). Additionally, in aquatic ecosystems, metals are not easily contained because they can spread with the water flow and, depending on the physicochemical properties of the system, their state at the sediment–water interphase can undergo changes (Nguyen *et al.*, 2021). In surface waters, the natural level of Ni is generally less than 0.005 mg L^{-1}; however, in recent decades, surface water pollution has considerably increased, becoming a severe problem (Song *et al.*, 2018). Ni^{2+} in natural waters (pH of 5–9) occurs mostly as the hexaquonickel ion $[Ni(H_2O)_6]^{2+}$ when no strong complexing agents are present, and to a minor degree, complexes with common ligands, such as HCO_3^-, Cl^-, OH^-, NH_3 and SO_4^{2-}, are formed (EFSA, 2015).

In addition to the natural presence of this metal in the environment, different anthropogenic sources contribute to the release of this metal. The widespread use of Ni in industrial and commercial fields contributed to extensive release of this metal into the ecosystem, leading to contamination issues (Ding *et al.*, 2022). Different Ni compounds have many industrial and commercial uses (Gautam *et al.*, 2017). Nickel acetate, carbonate, hydroxide and oxides are commonly used in industrial processes and are destined for accumulation in the environment, where they can be absorbed by plants (Ghazanfar *et al.*, 2021). Some human activities, such as the burning of fossil fuels, fertilisers and organic manure application, coal combustion, disposal of household, municipal and industrial wastes and vehicle emissions, contribute to Ni release (Rathor *et al.*, 2014; Chen *et al.*, 2019; Hassan *et al.*, 2019). The main industries that are sources of Ni contamination carry out activities such as mining and a broad variety of metallurgical processes, including alloy production, smelting, iron and steel production, stainless steel production, electroplating and catalysts in the chemical and food industries (Rathor *et al.*, 2014; Harasim and Filipek, 2015; Rodriguez-Jimenez *et al.*, 2016; Elshkaki *et al.*, 2017; Chen *et al.*, 2019; Hassan *et al.*, 2019; Borah *et al.*, 2023). In modern technology, it is also an essential component of rechargeable batteries, such as NiCd (Ameen *et al.*, 2019; Genchi *et al.*, 2020) and lithium-ion batteries, where Ni in different combinations is part of the cathode materials (Held *et al.*, 2022). Owing to the wide variety of Ni applications, based on its corrosion resistance, strength and high-temperature stability, Ni contamination in soil and water has reached undesirable levels in the past decades, becoming a serious environmental and health risk (Ameen *et al.*, 2019). Elshkaki *et al.* (2017) estimated that total Ni demand by 2050 will be 210–315% higher than that in 2010.

Passive diffusion and active transport are the principal ways of Ni absorption by plant roots; the ratio between the two depends on plant species, Ni form and Ni content in soil or nutrient solution (Ferreira-Barreto and Cruz, 2021). Generally, metal ion uptake occurs through secondary transporters, such as channel proteins and/or H^+-coupled carrier proteins, and no specific Ni-uptake system has been characterised in higher plants

(Mishra and Dubey, 2006; Mishra *et al.*, 2006). It was mentioned that poorly selective cation transporters, such as members of the ZRT/IRT-like (ZIP) family, appear to be primarily responsible for Ni uptake (Van der Pas and Ingle, 2019). This protein family of transporters is also present in macrophytes; Lyu *et al.* (2021) mentioned that, due to the thin cuticle, especially of submerged species, metal ions could enter the plants through plasma membrane transporters, such as the ZRT/IRT-like protein family members. Nishida *et al.* (2015) demonstrated using *Arabidopsis thaliana* as a model that Ni is absorbed by iron-regulated transporter 1 (IRT1), the primary Fe uptake transporter in roots; nevertheless, they concluded that additional pathways are involved in Ni accumulation in roots, suggesting that the Mg^{2+} uptake system is also engaged.

Ni is required in trace amounts by all higher plants and is, therefore, a part of the group of elements considered essential micronutrients (Baran, 2021). The metal tolerance proteins (MTPs) are described as metal efflux transporters from the cytoplasm, transporting mainly Zn^{2+} but also Mn^{2+}, Fe^{2+}, Cd^{2+}, Co^{2+} and Ni^{2+}. To avoid cellular damage, heavy metals are generally chelated by low-molecular-weight compounds, sequestered into organelles or expelled to the extracellular space by specific transporters. One class of metal transporters involved in these functions is the cation diffusion facilitator (CDF) family, also known as MTPs in plants (Ricachen *et al.*, 2013). On the other hand, natural resistance-associated macrophage proteins (NRAMPs) are transporters involved in the transport of toxic metals. Data on three rice NRAMPs, known as OsNramp1, OsNramp2 and OsNramp3, showed their presence in various parts, playing a major role in transporting metal ions. The role of NRAMP is mainly related to the transport of Fe and, to a lesser extent, manganese. Further, overexpression of AtNRAMP1 increases plant resistance to toxic Fe concentrations. The roles of MTPs or CDFs are responsible for the transfer of zinc, nickel, cadmium and cobalt and the sequestration of these ions away from the cytoplasm into the vacuole (Jain *et al.*, 2018). The corresponding gene is expressed in roots as well as shoots, and when overexpressed in *Arabidopsis*, it helps in zinc tolerance. It is necessary for normal plant growth and development, as it plays important roles in a wide range of physiological processes (Shahzad *et al.*, 2018). It has also been demonstrated that Ni is an important constituent of plant enzymes, such as urease and glyoxalase I (Fabiano *et al.*, 2015; Ameen *et al.*, 2019). Glyoxalase I has been related to aging processes or oxidative stress because it is important for the degradation process of methylglyoxal, a cytotoxic byproduct of cellular metabolism capable of forming adducts with nucleic acids, leading to DNA, RNA or protein modifications (Fabiano *et al.*, 2015; Baran, 2021). Urease is a Ni metalloenzyme ubiquitously found in plants; it catalyses the hydrolysis of urea into ammonia and carbamate, which spontaneously react with water to produce bicarbonate and a second ammonia molecule, thus being important in nitrogen metabolism (Baran, 2021; Myrach *et al.*, 2017). This enzyme has a major role in plants to limit the toxic accumulation of urea, and the N from the ammonia may be reused for different metabolic pathways, such as amino acid synthesis, polyamines and other nitrogen compounds, thus being important for nitrogen recycling (Ghazanfar *et al.*, 2021).

Despite Ni's essentiality in proper plant functioning, its excess has a detrimental influence on plant growth and development, affecting nutrient and water uptake, photosynthesis, membrane permeability, protein structure and content and reducing CO_2 uptake (Song *et al.*, 2018; Malea *et al.*, 2021). Ni also induces reactive oxygen species (ROS), which cause oxidative damage to different cellular components (Gautam *et al.*, 2017). As Ni is a redox-inactive metal, it is not able to directly generate ROS; however, the production of a highly toxic oxygen species has been found in plants after Ni exposure (Maheshwari and Dubey, 2009; Gautam *et al.*, 2017). Hao *et al.* (2006), studying the roots of wheat (*Triticum durum* D.), observed that Ni-induced enhancement in H_2O_2 and $O_2^{\cdot -}$ was mainly originated from plasma-membrane NADPH oxidase. Induced ROS production causes membrane destabilisation and DNA damage, as these species can react with lipids, proteins and nucleic acids; furthermore, its production reduces electron flow in the photosynthesis electron transport chain by disrupting the reaction centres P680 for PSII and P700 for PSI and also retards the biosynthesis of chlorophyll molecules (Zaid *et al.*, 2019). One way Ni toxicity in plants is related to alteration in uptake and competition with essential elements, such as iron (Fe^{+2}) (Ameen *et al.*, 2019; Merlot, 2020). Nevertheless, Ni can lead to responses at physiological and transcriptomic levels that are distinct from simple iron deficiency (Merlot, 2020). Working with gene expression microarrays in *Arabidopsis thaliana*, it was demonstrated that Ni at 100 μM significantly changed (by more than twofold) the expression of 235 genes, of which only 25% were related to iron starvation, while 75% are not known to respond to iron deficiency; most of these differentially expressed genes are connected with the regulation of hydrogen peroxide (H_2O_2), metabolism, nitrate transport and response and plant cell wall structure (Lešková *et al.*, 2020).

An excess of Ni concentration causes inhibition of root growth (Ghazanfar *et al.*, 2021) to a greater extent than inhibition of shoot growth (Hassan *et al.*, 2019). The observed effects in plant roots after excess Ni exposure may be related to alterations in the distribution of auxins, which are plant hormones relevant to the regulation of root cell elongation and tropic growth. Different plasma membrane transporters are responsible for auxin transport in the root apical meristem. PIN2, localised on the apical side, towards the shoot, of the epidermal cells of root tips is the main PIN-type transporter responsible for shootward auxin flows in roots (Lešková *et al.*, 2020).

These authors demonstrated that different auxin transporters were affected by Ni, where the strong and rapid inhibition of PIN2 protein accumulation and the alteration of polar distribution in plasma membranes may explain the lateral root phenotype of Ni-exposed roots they observed. They concluded that the strong repression of this transporter, mainly at the post-transcriptional level, could explain the decrease in auxin response detected in the outer cells of the root meristem. Besides, they studied the orientation and integrity of cortical microtubules in cells of the elongation zone and observed in Ni-exposed roots (after two days of exposure to 150 μM Ni) that a significant proportion of microtubules underwent random, oblique and longitudinal realignments with respect to the control and also a decreased density, suggesting that Ni could induce microtubule depolymerisation (Fig. 1).

The search for environmentally friendly practices and new technologies in order to reduce the ecological impact associated with Ni disposal and pollution remains open. In this sense, phytoremediation stands out as a high-efficiency, eco-friendly green technology for removing pollution from contaminated soil, water and air. It is a form of bioremediation and an economical and sustainable alternative to traditional chemical and physical pollution remediation methods (Shmaefsky, 2020).

High metal bioaccumulation efficiency and tolerance are requisites for plants employed in phytoremediation techniques; thus, investigating new aquatic plants with such characteristics is a priority (Song *et al.*, 2018). These authors reviewed some macrophytes that have been considered for the remediation of Ni-contaminated water: *Lemna minor*, *Lemna gibba*, *Spirodela polyrhiza*, *Elodea canadensis*, *Nasturtium officinale*, *Pistia stratiotes*, *Ceratophyllum demersum*, *Salvinia minima*, *Myriophyllum aquaticum* and *Sesuvium portulacastrum*.

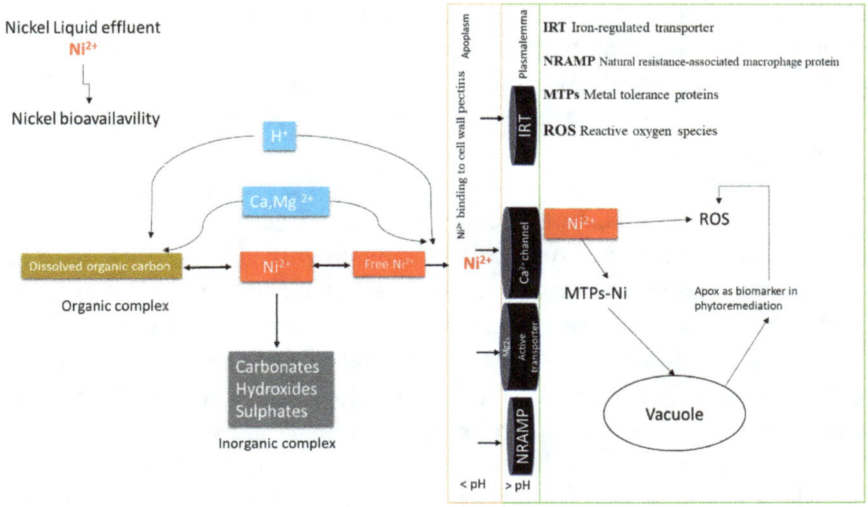

Figure 1. Schematic representations of Ni bioaccumulation process, bioavailability and cellular process as target in the phytoremediation process.

Aquatic plants remove pollutants via submerged roots and/or absorption from the water column through leaves. In submerged plant species in particular, it was mentioned that leaves are the main site of mineral uptake, where metal absorption occurs by passive movement through the thin cuticle and cell walls because of the inward flow favoured by the negative charges of pectin and cuticle polymers (Dhir *et al.*, 2009). Aside from the thinner cuticle of submerged leaves, they have more stomatal pores than air-exposed ones, which can facilitate dissolved and nanoparticulate metal uptake (Nguyen *et al.*, 2021). Besides, the same authors mentioned that macrophyte leaves can remove metals from surrounding water by biosorption to different parts of their cell walls via complexation, ion exchange, physical adsorption and precipitation—processes that can be followed by metal accumulation. All these characteristics make aquatic plants more than interesting organisms to continue considering in the development of phytoremediation processes.

2. Effects of Nickel in Aquatic Plants

Ni can have both positive and negative effects on aquatic plants, depending on the concentration of the metal and the specific plant species.

The effects of Ni on aquatic plant growth also depend on other environmental factors, such as water temperature, pH and nutrient levels. For example, some studies have found that Ni toxicity is more severe at low pH levels, while others have found that it is less severe in the presence of certain nutrients.

Ni is a micronutrient that is essential for the growth and development of some aquatic plants. At low concentrations, it can enhance growth, as it plays a role in nitrogen metabolism and is involved in the biosynthesis of certain enzymes and proteins. However, at higher concentrations, Ni can be toxic to aquatic plants and inhibit their growth. Ni toxicity can lead to chlorosis (yellowing of leaves), necrosis (tissue death) and reduced photosynthesis. The extent of the toxicity depends on the plant species, with some species being more tolerant of Ni than others.

Studies conducted on the effects of Ni on *Ceratophyllum demersum* have demonstrated a significant decrease in growth rate (μ) at exposure to 3.42 mg L^{-1} and higher (Table 1). It is relevant to indicate that the growth rate based on main shoot length was the most inhibited, compared to those estimated based on fresh weight and total shoot length (Table 1). The exposure of plants of *C. demersum* to a range of Ni concentrations for 14 days following the OECD (2014b) protocol resulted in an EC50 of 6.51 (4.5–9.2) mg L^{-1} (Martinez, 2019).

There is experimental evidence that Ni also affects the growth of *Lemna gibba* (Fam Araceae, Subfam Lemnoidae, APG IV, 2016), known as duckweed, an aquatic plant very common in freshwater environments. Studies have shown that the estimated growth rate (μ) based on the

Table 1. Average specific growth rate values (μ) and percent inhibition of growth rate (% I_r) of *C. demersum* exposed to nickel (mg L^{-1}) based on (a) main shoot length, (b) fresh weight and (c) total shoot length.

	μ (a)	% I_r (a)	μ (b)	% I_r (b)	μ (c)	% I_r (c)
0	0.0048		0.022		0.122	
0.85	0.0041	14.5	0.022	1.8	0.0113	7.85
1.71	0.0041	14.5	0.0178	20.2	0.0092	25.26
3.42	0.0024*	49.6*	0.0133*	40.3*	0.0064*	47.71*
6.84	0.0028*	42.2*	0.0126*	43.5*	0.0060*	51.02*
13.69	0,0019*	61.1*	0.0131*	41.3*	0.0055*	55.39*

Note: (*) Significant, ANOVA–Dunnet's test, $\alpha < 0.05$.

Table 2. Average specific growth rate values (μ), percent inhibition of growth rate (% I_r) and percent inhibition of yield (%I_y) of *L. gibba* exposed to different concentrations of nickel (mg L^{-1}).

	0	**0.18**	**0.37**	**0.92**	**1.46**	**3.61**	**6**	**11.82**
μ	0.200	0.207	0.187	0.113*	0.090*	0.058*	0.041*	0*
%I_r		3.1**	6.60	43.5*	56.1*	71.1*	80.1*	100*
%I_y		4.7**	12.7	61.1*	71.7*	83.9*	89.3*	100*

Notes: * Significant, ANOVA-Dunnet's test, $\alpha < 0.05$.
** Percentage of stimulation.

number of fronds decreased significantly as the exposure concentrations increased from a concentration of 0.92 mg Ni L^{-1}. On the other hand, a stimulation of growth was observed in plants exposed at concentrations of 0.18 mg L^{-1}, in accordance with the positive effects on growth of this metal mentioned earlier (Table 2).

Ni can also produce morphological changes, altering the characteristic shape and size of aquatic plants. These changes can include reduced leaf area and leaf thickness, decreased leaf axis, both transversal and longitudinal, as well as root and shoot lengths. Ni toxicity can also lead to the development of necrotic lesions on the leaves and, in severe cases, can cause the death of the plant. Studies about the morphological changes of fronds of *L. gibba* have been performed, showing that Ni can alter the area and longitudinal and transverse axes of the frond.

Martinez (2019) has reported that the area of the fronds was significantly reduced at exposure concentrations of 0.37 mg L^{-1} and above, while the lowest Ni exposure concentration of 0.18 mg L^{-1} did not have significant effects on the total area compared to the controls. A concentration-dependent response was observed, where the total area value decreased with increasing exposure concentrations.

Regarding the effect on the longitudinal and transverse axes of the fronds, a significant decrease was observed starting at 0.92 mg L^{-1}. As with the area of the fronds, a concentration-dependent response was observed. A decrease in these parameters was observed at the two lower exposure concentrations of 0.18 and 0.37 mg L^{-1}; however, these differences were not significant. Among the morphometric characters evaluated, the frond area was the one that showed adverse effects at a lower Ni exposure concentration. The study demonstrated that the total area

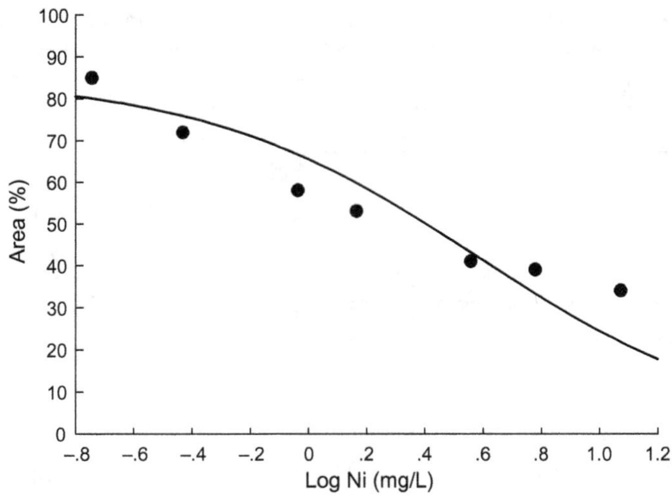

Figure 2. Frond total area of *L. gibba* as a function of Ni concentrations after seven days of exposure.

represented a good predictor of the effects of Ni, as this morphometric character was the most sensitive in detecting the effects on *L. gibba*. A concentration–response curve was constructed using this parameter in order to obtain a spectrum of toxic action for this metal (Fig. 2).

The EC50 -7 days based on frond area estimated from toxicity assays conducted according to the OECD (2006) protocol was 2.39 (2.21–2.79) mg L^{-1} (Martinez, 2019).

The decrease in the values of morphometric characters could be explained by what was observed by Pena *et al.* (2012). These authors suggested that stress first rapidly blocks the progression of the cell cycle, presumably to prevent cells from entering stages where they are particularly vulnerable to damage and allow the cellular defence system to remain activated. This blockage would decrease cell proliferation by negatively regulating cell division stimulators, such as a loss of functionality of a cyclin D-CDK complex (Pena *et al.*, 2012). On the other hand, Chaoui *et al.* (1997) proposed that peroxidases could be involved in lignifying cell walls as a mechanical adaptation to stress conditions, a process that would affect cellular plasticity by limiting proper cell elongation and therefore affecting frond size, for instance, frond area. In accordance with that, it has been suggested that the elastic properties of cell walls may be affected to varying degrees by different metals, where an increase in cell wall rigidity can cause enzymatic disorders in wall constituent biosynthesis, changes in wall polymer assembly, changes in microtubule orientation or an increase in cell adhesion to the middle lamella (Barceló and Poschenrieder, 1992). In this work, they relate an exacerbated activity of the GPX enzyme to a limitation in the growth of *Phaseolus vulgaris.*

Some studies have reported that Ni causes chlorosis and necrosis lesions on leaves as a symptom of exposure to high Ni concentrations, depending on the species, age, pH and duration of exposure (Chen *et al.*, 2002; Cai *et al.*, 2007; Ashraf and Mehmood, 2013; Zhou and Song, 2013). These concentrations have been reported to be as different as 117.38 and 5.86 mg L^{-1} at four and seven days of exposure, respectively. However, macroscopic observations of the plants of *L. gibba* from the above experiment did not reveal such effects. At the end of the exposition time of seven days, all plants had their characteristic green colour and gibbosity, with no evidence of neither dark nor yellow dots or necrosis spots, even at the highest concentration of 11.82 mg L^{-1} (Fig. 3). Besides, no destruction of roots was observed, as was also reported by other authors.

Similar results were found by Appenroth *et al.* (2010) in *Spirodela polyrhiza* and *L. minor* (Fam Araceae, Subfam Lemnoidae, APG IV, 2016) exposed to similar concentrations of Ni for seven days, where they observed that the mother fronds remained relatively unchanged without loss of colour, an effect which they related to not having found any indication of gerontoplast formation, i.e., senescent chloroplast. In accordance with the above results, a macroscopic inspection of the plants of *C. demersum* exposed to 13.69 mg L^{-1} of Ni was performed, observing that the

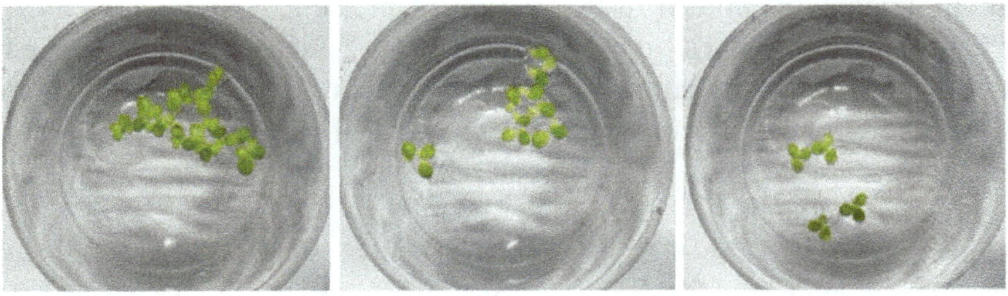

Control **1.46 mg L^{-1}** **11.82 mg L^{-1}**

Figure 3. Plants of *L. gibba* at the end of the Ni toxicity assays.

plants maintained their green colour without signs of chlorosis or areas with necrosis.

Ni can have negative effects on the chlorophyll content of plants. Studies have shown that exposure to high levels of Ni can lead to a decrease in chlorophyll content in plants, which can ultimately affect the plant's ability to carry out photosynthesis (Kumar *et al.*, 2015; Kaur *et al.*, 2017; Li *et al.*, 2020). Although the specific mechanism behind this effect is not yet fully understood, it is believed that Ni may interfere with the synthesis or function of the enzyme involved in chlorophyll production. Some of these events may be attributed to the replacement of Mg^{2+} ions associated with the tetrapyrrole ring of chlorophyll molecules, the inhibition of important enzymes related to the biosynthesis of chlorophyll or the processes of lipid peroxidation of the chloroplast membrane by ROS (Manios *et al.*, 2003; Kumar *et al.*, 2015; Upadhyay *et al.*, 2017). Additionally, Ni may also cause damage to the thylakoid membrane within chloroplasts, which can further impair photosynthesis. Other effects on chloroplasts worth mentioning include disruption of the thylakoid membrane structure, leading to a decrease in photosynthetic efficiency; the inhibition of photosynthetic electron transport, leading to a decrease in ATP and NADPH production that is essential for carbon fixation during photosynthesis; and the damage of chloroplast components, such as lipids, proteins and nucleic acids, by oxidative stress due to an imbalance in the production of ROS. Overall, the impact of Ni on chlorophyll can vary depending on the concentration and duration of exposure, as well as the specific plant species being studied.

Evidence of such effects is found in studies performed on the action of this metal on the main photosynthetic pigments: chlorophyll *a*, chlorophyll *b* and carotenoids of *L. gibba*. They have shown a decrease in the three pigments compared to the control plants (Fig. 4). The content of chlorophyll *a* in plants exposed to 0.92 mg L^{-1} Ni and higher differed significantly from the control plants. The content of chlorophyll *b* was significantly lower in plants treated with 3.61 and 6 mg L^{-1}. In the case of total carotenoid content, a significant decrease was observed in plants exposed to the highest concentration of 6 mg L^{-1} (Fig. 4) (Martinez, 2019; Martinez *et al.*, 2019a; Martinez *et al.*, 2024).

This phenomenon has been observed in species of *S. polyrhiza* and *L. minor* exposed to a range of similar Ni concentrations, as they showed a decrease in photosynthetic pigment content, with chlorophyll *a* being particularly sensitive (Appenroth *et al.*, 2010). These authors reported that

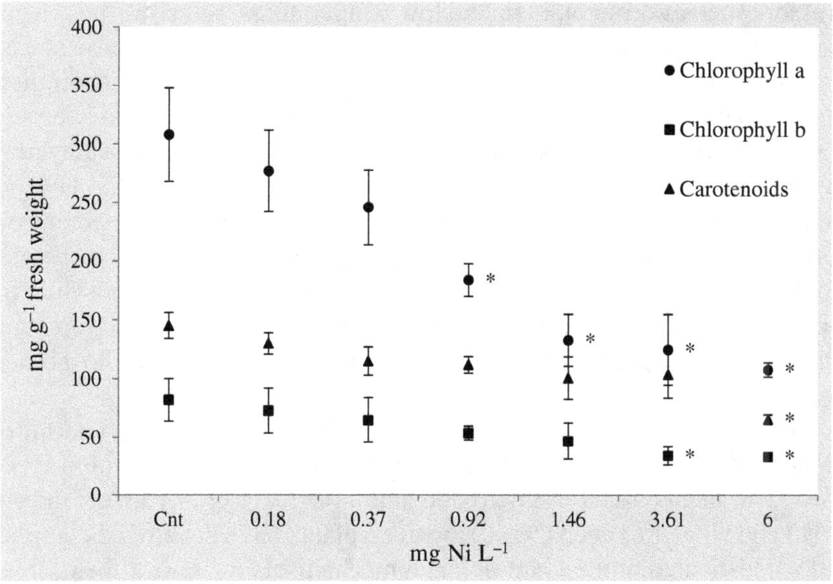

Figure 4. Effects of Ni on photosynthetic pigments of *L. gibba*.

the thylakoid system was greatly affected by treatment with this metal, mainly due to the presence of starch granules in the chloroplasts. They mentioned that there is no record of a Ni treatment generating stimulation in the photosynthesis process; therefore, the accumulation of starch would be a consequence of growth inhibition due to a lower demand for carbohydrates. These authors observed that starch content decreases at high metal concentrations, similar to those mentioned above, which could be due to a significant inhibition of photosynthetic production, leading to a lower accumulation of starch.

An excessive amount of Ni can lead to the induction of oxidative stress in plants through several mechanisms. Oxidative stress is a condition that occurs when there is an imbalance between the production of ROS and the antioxidant defence system of plants. This can result in an excess of ROS that transfer electrons to other molecules, resulting in the production of highly reactive compounds which can cause damage to cellular components, such as proteins, lipids and DNA. These changes affect physiological processes such as photosynthesis, respiration and nutrient uptake. Also, there are a variety of factors, including environmental stressors such as drought, high and low temperatures, salt stress and exposure to pollutants, that can induce oxidative stress in plants. These stressors can cause the plant to produce more ROS than it can neutralise, leading to oxidative stress. On the other hand, ROS are produced naturally in all living cells, being essential for normal plant physiology, as they play an important role in a variety of cellular processes. In particular, H_2O_2 is an important and relatively stable non-radical ROS which, at low concentrations, acts as a signal molecule involved in the regulation of specific biological/physiological processes or functions, such as photosynthetic functions, cell cycles, differentiation, apoptosis, growth and development, defence mechanisms against pathogens and plant responses to abiotic stresses.

ROS are produced primarily in chloroplasts, peroxisomes and mitochondria during photosynthesis and respiration. The most common types of ROS in plants include superoxide anion ($O_2{\cdot}^-$), H_2O_2, hydroxyl radical ($OH{\cdot}$) and singlet oxygen ($^1O_2$). However, plants have developed complex and sophisticated antioxidant defence mechanisms to protect themselves against oxidative stress, including enzymatic antioxidants, such as superoxide dismutase (SOD), catalase (CAT), ascorbate peroxidase (APX), glutathione reductase (GR), monodehydroascorbate reductase (MDHAR), dehydroascorbate reductase (DHAR) and guaiacol peroxidase (GPX), and non-enzymatic low-molecular-weight metabolites, such as ascorbic acid (AsA), glutathione (GSH), carotenoids, flavonoids, α-tocopherol, phenolic compounds and alkaloids. SOD are metalloenzymes that convert $O_2{\cdot}^-$ in H_2O_2 by catalysing its dismutation via a Haber–Weiss reaction, which is considered the primary defence against oxygen radicals. There are three types of SOD in plants that have been identified in the cytosol, mitochondria, peroxisomes and chloroplasts (Fig. 5). CAT efficiently scavenges H_2O_2, without the requirement of a reducing substrate, localised in peroxisomes to scavenge this reactive species produced by glycolate oxidase in the photorespiratory cycle (Fig. 5). APX, GR, MDHAR and DHAR are part of the ascorbate–glutathione cycle (Halliwell–Foyer–Asada cycle) (Fig. 6). APX plays a vital role in scavenging H_2O_2 using ascorbate (ASC) as an electron donor, oxidising it to monodehydroascorbate (MDHA) and dehydroascorbate (DHA). The APX family consists of at least five different isoforms identified in the cytosol (cAPX), mitochondria (mitAPX), membrane bond in the peroxisome (mAPX) and the stroma and thylakoids of the chloroplast (chlAPX) (Fig 5). ASC is regenerated from MDHA and DHA in the reaction catalysed by

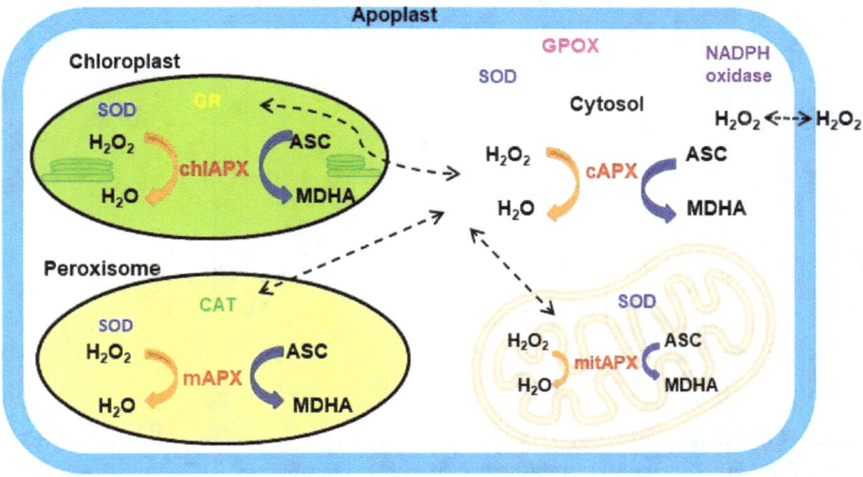

Figure 5. Different subcellular compartments of main antioxidant enzyme (see explanation in the text).

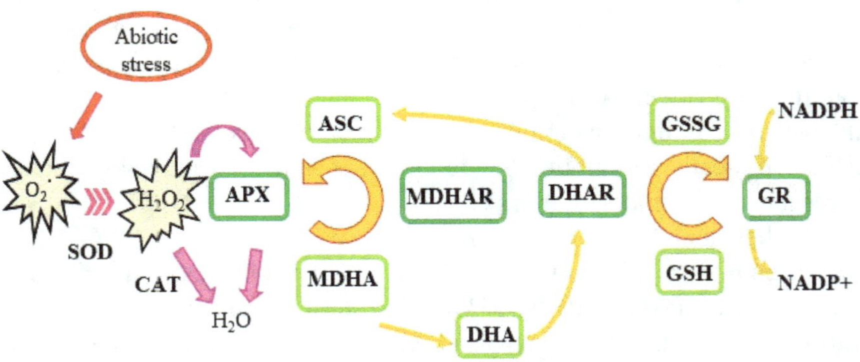

Figure 6. Ascorbate–glutathione cycle (Halliwell–Foyer–Asada cycle) extended with the detoxification of superoxide by SOD and hydrogen peroxide by CAT (see explanation in the text).

NADPH-dependent MDHAR and DHAR. DHAR reduces DHA to ASC at the expense of GSH, oxidising it to GSSG. GSH is regenerated from GSSG by NADPH-dependent GR (Fig. 6).

GPX is an indirect scavenger of H_2O_2 as it has a role in the biosynthesis of lignin in the apoplast, the cell wall and extracellular spaces from phenylpropanoid precursors consuming H_2O_2 (Fig. 5)

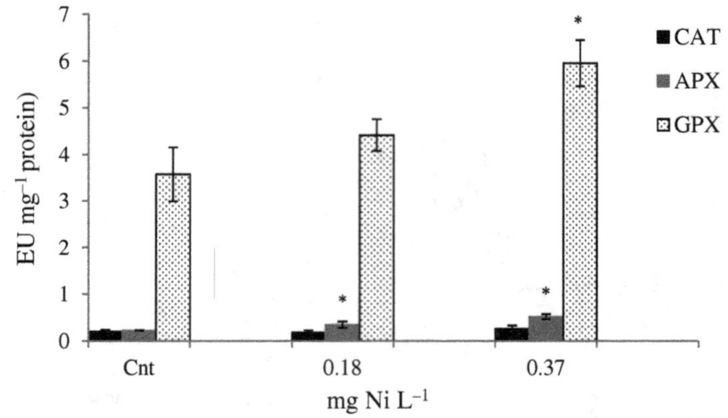

Figure 7. Activities of CAT, APX and GPX in plants of *L. gibba* exposed to Ni (EU: enzymatic units; (*): significant difference from the control at $p < 0.05$).

(Shahid *et al.*, 2014). This mechanism is activated in the presence of pathogen invasion, as ROS is produced at the site of the injury, together with phenolic compounds acting as signal molecules to trigger the biosynthesis of lignin. These responses lead to the mechanical strengthening and rigidification of cell walls, preventing the intrusion of the pathogen. An enhancement in lignification process has been observed as a result of the effects of metals, resembling the defence mechanism displayed by plants in response to a pathogen invasion.

Studies about the effects of Ni on antioxidant enzymes of *L. gibba* have been performed, showing that 0.18 mg L^{-1} of this metal significantly increased the APX activity by 55% with respect to the control. A concentration of 0.37 mg Ni L^{-1} increased APX activity even more, by 130%, with respect to the control plants (Fig. 7). Similar results were reported by Yilmaz and Parlak (2011), as they observed an increase in APX activity when plants of *L. gibba* were exposed to 0.05 mg Ni L^{-1} for seven days.

On the other hand, CAT was the enzyme least activated by this metal. An increase in activity was observed at a concentration of 0.37 mg L^{-1}, but it was not significantly greater than the control values (Fig. 7). Considering the effects of Nion GPX, both concentrations increased this enzyme activity, but only the plants exposed to 0.37 mg L^{-1} showed a significant increase of 66% with respect to the control values (Fig. 7) (Martinez, 2019).

As was said earlier, GPX is very different from CAT and APX in terms of physiological functions, as it has a role in the biosynthesis of lignin by consuming H_2O_2, thus resulting in an indirect or unspecific scavenger of this product.

Similar results were observed when studying the effects of different Ni concentrations on the antioxidant enzyme activity of *C. demersum*. CAT activity was significantly increased at 0.85 mg L^{-1}, while a decrease was observed at the highest concentration of 3.42 mg L^{-1}. The plants of *C. demersum* exposed to the lowest concentration showed an APX activity similar to that of the control; however, an increase was observed in accordance with increasing concentrations, showing the highest activity at 3.42 mg L^{-1}. GPX activity was stimulated after exposure to the two highest Ni concentrations, although a significant difference was found only at 1.71 mg L^{-1} (Fig. 8) (Martinez, 2019). Ni exerted similar effects on antioxidant enzymes of both *L. gibba* and *C. demersum*, with a decrease observed in the activity of CAT, which was the least activated enzyme, while the activities of APX and GPX were increased in both species. Doğanlar (2013) observed similar results when he studied the effects of cadmium on APX and GPX activities on *L. gibba*.

These behaviours could be explained by considering the different affinities of enzymes for the H_2O_2 substrate. Although CAT is an

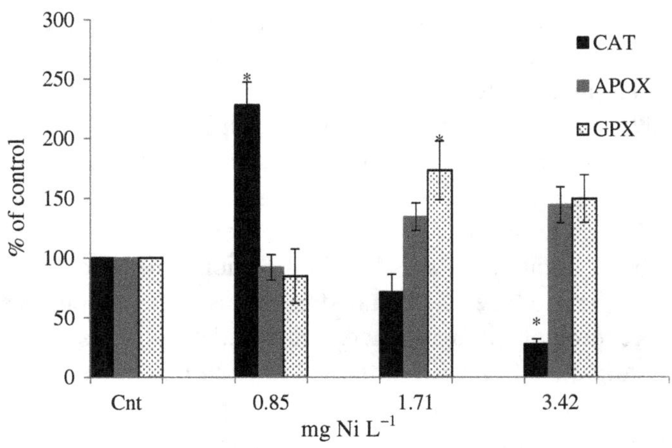

Figure 8. Enzymatic activity expressed as percentage with respect to the control of *C. demersum* exposed to Ni ((*): significant difference with control at $\alpha < 0.05$).

important enzyme for the detoxification of H_2O_2, it may not activate until the concentration of this substrate increases beyond a certain threshold since it has a low affinity for H_2O_2, is mainly located in peroxisomes and is very scarce or inactive in other subcellular compartments (Mizuno *et al.*, 1998). On the contrary, APX can cause a more effective detoxification mechanism of this product, as this enzyme has a higher affinity for H_2O_2, making its reaction possible in response to small amounts of substrate and in more specific locations in several subcellular compartments (Asada, 1992; Gomes-Junior *et al.*, 2006). This could be the reason why the activity of APX was greater than that of CAT in the plants of both *L. gibba* and *C. demersum* exposed to the Ni concentrations tested. These findings mean that the plants are being protected against the attack of free radicals produced by the metal as oxidative stress is generated (Jung *et al.*, 2000). This conclusion supports the proposal that APX is a suitable biomarker for assessing good plant health conditions under metal stress (Martinez *et al.*, 2019b). However, the fact that antioxidant enzymes have different locations and chemical properties makes the antioxidant system more flexible and versatile in controlling ROS at appropriate levels. Thus, a balance is achieved in the enzymes that eliminate or transform these reactive species (Caverzan *et al.*, 2012; Sofo *et al.*, 2015).

Metal toxicity can be mainly caused by an integration of factors, including direct interaction with proteins due to their affinity for thiol, histidine and carboxyl groups, causing the metals to affect structural catalytic and transport sites in the cell; the stimulation of the generation of ROS, causing oxidative stress; and the displacement of essential cations from specific binding sites, causing the collapse of functions (Sharma and Dietz, 2009). Metals tend to bind to the sulphydryl groups of various enzymes, suppressing the functioning of essential biological components, which is evident in the observed toxicity. Additionally, they can produce ionic imbalance due to altered translocation of ions such as Fe^{+2}, Mn^{+2}, Cu^{+2} and Zn^{+2} (Yilma and Parlak, 2011). Toxicity can also be caused by ROS that can be produced by metal stress. This toxicity could be caused by excessive proteolysis associated with oxidative stress promoted by ROS (Palma *et al.*, 2002). Recently, it has been recognised that ROS in plants perform essential functions in various processes, such as growth, development, acclimatisation to the environment and the establishment of resistance to pathogens (Mullineaux *et al.*, 2008). An imbalance in the production or regulation of these chemical species can lead to disorders in the processes mentioned above. Representing up to more than 50% of the dry weight of cells, proteins are biomolecules with many possibilities of reacting with these ROS, so they can suffer oxidative damage (Pena *et al.*, 2012). It has been reported that the amino acids that are most susceptible to attack by oxidants are those capable of delocalising charge, such as those containing aromatic cyclic or thiol side chains, where sulphur increases its valence to stabilise the intermediate radical. Although the main amino acids to undergo oxidation through reactions with radicals appear to be tryptophan, tyrosine, methionine, histidine and cysteine, even the less reactive amino acids are capable of reacting with some radicals, such as hydroxyl, resulting in the oxidation of most of their side chains (Griffiths, 2005). Among the effects of protein oxidation caused by radicals, the introduction of new functional groups in the side chains of amino acids can frequently occur, such as the addition of hydroxyls or carbonyls by hydrogen, contributing to the alteration of functions and their replacement. In addition, after an oxidative attack, oxoacids and aldehydes can be formed (Griffiths, 2005; Pena *et al.*, 2012). Different abiotic stress factors, including metals, cause stomatal closure, leading to a limitation in the supply of carbon dioxide for photosynthesis. Under these conditions, chloroplasts can become a major source of ROS via photorespiration through the recycling process of phosphoglycolate formed by the oxygenation reaction of RuBisCo. This fact leads to a substantial production of H_2O_2 by a glycolate oxidase located in the peroxisome and through the process involving the Mehler reaction, in which molecular oxygen is reduced to $O_2^{\cdot-}$ and H_2O_2 by the electrons of the photosynthetic electron transport chain in photosystem I (Mullineaux *et al.*, 2008).

The antioxidant defence system plays a significant role in providing resistance to plants by protecting labile macromolecules against the attack of free radicals. This ability made possible the use of aquatic plants in technological applications for the restoration of contaminated sediments (Sáenz *et al.*, 2007; Sáenz *et al.*, 2014), solid wastes (Sáenz and Di Marzio, 2011) and organic contaminants (Di Marzio *et al.*, 2005). Accordingly, results have demonstrated that *L. gibba* and *C. demersum* would have the capacity to tolerate and, therefore, resist Ni stress. As previously mentioned, APX would be a more efficient enzyme for the detoxification of H_2O_2 than CAT, as its activity was stimulated at higher exposure concentrations. This means that these plants can counteract the excess of ROS produced by the action of such Ni concentrations through APX activity, as part of the antioxidant defence system. This reinforces the proposal of selecting APX as a good biomarker for the evaluation of the welfare of the plant and growth performance. It is important to keep in mind that plant stress tolerance may be improved by the enhancement of the levels of antioxidant enzymes, in particular those related to the application of aquatic plants in phytoremediation of Ni from wastewaters and industrial effluents.

3. Phytoremediation of Nickel in Liquid Effluents

Phytoremediation is the process of using plants to remove pollutants from the environment. Aquatic plants have been found to be effective in the remediation of heavy metals in contaminated water. The process of phytoextraction involves the uptake and accumulation of metals by plants through their roots and leaf surfaces. Aquatic plants are able to accumulate metals such as arsenic, cadmium, copper, lead, zinc and nickel from water. Once the metals are accumulated in the plant tissues, they can be harvested and removed from the water. Another method of phytoremediation is rhizofiltration, which involves the uptake of metals by the roots of aquatic plants. The metals are then transported to the shoots, where they can be harvested and removed from the water.

Phytoremediation is a sustainable and cost-effective method of remediating metal-contaminated water. Aquatic plants are able to accumulate metals over a period of time, and the plants can be harvested and disposed of safely. Additionally, this process does not require any chemicals or energy to be added to the environment, making it an environmentally friendly solution. Some examples of aquatic plants that have been found to be effective in phytoremediation include water hyacinth, duckweed and water lettuce. However, it's important to note that the effectiveness of phytoremediation depends on a variety of factors, including the type and concentration of the metal contaminant, the type of plant used and the environmental conditions. Moreover, the kinetics of metal removal by aquatic plants during phytoremediation can vary depending on several factors, such as the type of metal, the concentration of the metal in the water, the type of plant used and the environmental conditions. Generally, the removal of metals by aquatic plants follows first-order kinetics. This means that the rate of metal removal is directly proportional to the concentration of the metal in the water. However, as the concentration of the metal decreases in the water, the rate of removal by the plant also decreases. The uptake of metals by plants can occur through several mechanisms, such as adsorption onto the roots or uptake through the plant tissue. The rate of uptake can be influenced by the plant's growth rate, the surface area of the roots and the concentration of the metal in the water. It's important to note that the kinetics of metal removal by aquatic plants can be influenced by the presence of other pollutants in the water, such as hardness, dissolved organic carbon or nutrients. These pollutants can compete with the metal for uptake by the plant and may slow down the rate of removal. Continuous monitoring of the water quality and the rate of metal removal by the plants is essential to determining the efficiency of the remediation process. In general, the initial concentration of metals in the water will be high and will decrease over time as the macrophytes absorb and remove the metals from the water. As the plants continue to grow and develop, their root systems become more established, and their ability to remove pollutants from the water increases. The rate of metal uptake by the macrophytes can vary depending on factors such as the type of plant, the concentration of metals in the water and the hydraulic retention time (HRT) of the treatment system. In some cases, the rate of metal removal may be slower at the beginning of the treatment process and then increase as the plant roots become more established and efficient at absorbing the metals. Over time, the concentration of metals in the water should decrease to a level that is safe for discharge or reuse. The time required to achieve this level will again depend on the initial concentration of metals in the water, the type of plant used and the design of the treatment system. It's important to note that phytoremediation using macrophytes is typically a longer-term treatment process that requires patience and careful monitoring to ensure that the desired level of metal removal is achieved.

3.1. *Kinetics*

The kinetics of Ni removal by aquatic plants can be evaluated by measuring various parameters, such as uptake rate, bioconcentration factor (BCF) and removal efficiency. Here are some examples of these parameters:

Uptake rate: It is the rate at which the plants uptake Ni from the water. This parameter is often measured by monitoring the decrease in Ni concentration in the water over time. The uptake rate can be calculated using the following equation:

$$\text{Uptake rate (mg g}^{-1}\text{ d}^{-1}) = (C_0 - C_t) \times V / (W \times T)$$

where C0 is the initial concentration of Ni in the water (mg L^{-1}), Ct is the concentration of Ni in the water at time t (mg L^{-1}), V is the volume of water (L), W is the weight of the plant biomass (g) and T is the time interval (d).

BCF: It is a measure of the ability of the plant to accumulate Ni from the water. This parameter is calculated as the ratio of the concentration of Ni in the plant tissue to that in the water. The BCF can be calculated using the following equation:

$$BCF = C_{plant}/C_{water}$$

where C_{plant} is the concentration of Ni in the plant tissue (mg g^{-1}) and C_{water} is the concentration of Ni in the water (mg L^{-1}).

Removal efficiency: It refers to the percentage of Ni removed from the water by the plants. This parameter can be calculated using the following equation:

$$\text{Removal efficiency (\%)} = (C_0 - C_T) / C_0 \times 100$$

where C0 is the initial concentration of Ni in the water (mg L^{-1}) and Ct is the concentration of Ni in the water at time T (mg L^{-1}).

These parameters can be used to evaluate the kinetics of Ni removal by aquatic plants in phytoremediation studies. It's important to note that the kinetics of Ni removal can be influenced by several factors, such as the type of plant used, the concentration of Ni in the water and the environmental conditions.

3.2. *Hydraulic retention time*

The HRT and the rate of aquatic phytoremediation are related because the HRT affects the residence time of water in the treatment system and, therefore, the time available for the plants to remove pollutants. In general, a longer HRT will result in a slower velocity of water flow through the treatment system, which can allow for more time for the plants to uptake pollutants from the water. However, the relationship between the HRT and the rate of aquatic phytoremediation can also depend on the type of plant used, the concentration of pollutants in the water, and the hydraulic design of the treatment system. For example, in a horizontal flow-constructed wetland system, the velocity of water flow can be relatively slow, and the HRT can be longer (typically 2–7 days). This allows for a greater opportunity for the plants to remove pollutants from the water. In contrast, in a vertical flow system, the water typically flows more quickly, and the HRT may be shorter (typically 1–2 days). In this case, though the plants may have less time to remove the pollutants, the system can be designed to maximise contact between the plant roots and the water, which can increase the efficiency of pollutant removal. In summary, the relationship between the HRT and the rate of aquatic phytoremediation can vary depending on several factors. A longer HRT can allow for more time for the plants to remove pollutants; however, the design of the treatment system and the type of plant used can also affect the velocity of water flow and the efficiency of pollutant removal.

3.3. *Biomass*

The relationship between the biomass of aquatic plants and the rate of phytoremediation is complex and can depend on various factors, such as the type of plant, the concentration of pollutants in the water, and the hydraulic design of the treatment system. Generally, as the biomass of aquatic plants increases, the rate of phytoremediation may decrease due to the increased resistance of the plant roots and stems to the flow of water. This can result in more time for the water to interact with the plant roots, allowing for more efficient removal of pollutants. However, if the plant biomass becomes too dense, it can restrict water flow and reduce the efficiency of phytoremediation. This is because a high biomass can reduce the amount of light that penetrates the water, which can limit plant growth and photosynthesis. Additionally, a high biomass can create anoxic conditions in the root zone, which can limit the availability of oxygen to the plants and reduce their ability to remove pollutants. Therefore, it's important to balance the biomass of aquatic plants with the rate of phytoremediation to optimise pollutant removal efficiency. The design of the treatment system should consider the type of plant used and the concentration of pollutants in the water to ensure that the biomass and the rate are appropriate for the desired level of pollutant removal.

3.4. *Temperature*

Temperature can have a significant impact on the rate of aquatic phytoremediation. In general, higher temperatures can increase the rate of phytoremediation due to the effects of temperature on plant growth and metabolism, as well as on the rates of biogeochemical processes that occur in the treatment system. As the water temperature increases, the growth rate and photosynthesis of aquatic plants can also increase, which can lead to a higher biomass and greater pollutant uptake. Additionally, higher temperatures can increase the rates of chemical reactions and microbial activity in the water, which can facilitate the breakdown and removal of pollutants.

However, there are limits to the beneficial effects of temperature on phytoremediation. At very high temperatures, the metabolic processes of plants and microbes can become stressed, leading to reduced efficiency of pollutant removal. Moreover, if the water temperature becomes too high, it can become lethal to the plants and other organisms in the treatment system. Therefore, the optimal temperature range for phytoremediation will depend on the specific plant species and the types of pollutants being treated. In general, the temperature should be maintained within a range that supports plant growth and metabolism without causing stress and that also facilitates the biogeochemical processes necessary for pollutant removal.

3.5. *Field experiences*

Most of the published works on the removal of Ni from aquatic plants consisted of exposure to a single concentration of the metal and the evaluation of how it was reduced as a function of time. Among them, Chorom *et al.* (2012), under laboratory conditions, tested a single dose and recorded the bioconcentration over 14 days. With an initial Ni concentration of 6 mg L^{-1} and a density of 6 g wet weight L^{-1}, the plants achieve BCF values of approximately 105 and a maximum BCF of 200 at 4 mg Ni L^{-1}. These exposure conditions are not those expected in the real world. The industries generate a continuous flow of effluent with relatively stable concentrations of the metals associated with their production. In the case of Ni, the main industries are electroplating companies and those related to electronic activities and battery production. In other words, a phytoremediation system must be adapted to the existence of a permanent liquid effluent with variations in flow and metal load.

In order to assess the Ni removal capacity using *C. demersum* exposed to a continuous effluent flow, bioaccumulation tests were carried out according to USEPA (1996, 2003, 2004) and Connel *et al.* (1990). Initial plant material consisted of fresh branches of 15 cm length, presenting an apical meristem on one end (with densities of 7 and 20 g wet weight L^{-1}). Exposures were performed in field conditions in 20 L tanks at an effluent flow rate such as to achieve 12 and 48 hours of hydraulic retention time. Plants were exposed at a fixed concentration of 2.5 mg for a period until the internal metal concentration reached a steady state, in which it remains constant despite continued exposure. The assayed concentration was based on the EC50-14d for Ni, which was 6.51 mg L^{-1}. Plant samples were withdrawn at hours 1, 2, 4 and 8 and at days 1, 2, 4, 7, 9, 11, 14 and 18 for metal tissue determination. The removed samples were immediately immersed in 0.01 M EDTA and then in fresh culture medium without metal.

This procedure removes metals from the surfaces of plants, allowing the determination of internal metal concentration only. After that, they were dried in a vacuum oven at 110 °C and 50 cm Hg for two hours. Dried material was acid digested in a microwave oven in 5 ml of concentrated HNO_3 of analytical grade. The metal concentrations of the digested plant samples were determined using atomic absorption spectroscopy. BCF were determined at the different times, as follows: BCF = concentration in tissue / concentration in external solution (mg kg^{-1}/mg L^{-1}). These were compared with laboratory experiments at unique concentrations, as shown in Fig. 10.

Figure 10 depicts the BCFs obtained under continuous-flow conditions of a constant concentration in a simulated liquid effluent with 2.5 mg L^{-1} and under variable conditions of ambient temperature between 20 and 26 °C and during spring photoperiod. They are three times higher with respect to the data of BCF shown in Fig. 9, and although they do not reach hyperaccumulation values (BCF > 1,000), they can be applied for Ni removal in serial systems. Figures 11 and 12 indicate the bioaccumulation rates (Br) of Ni for different HRTs and biomass over time, until reaching the steady state of bioaccumulation, where plants must be removed.

As can be seen, variable Br is recorded between 0.5 and 0.1 mg Ni g^{-1} dw d^{-1}, with the highest metal retention occurring between 24 and 48 h of exposure. In order to apply these results, sequential or serial field experiments were designed with continuous effluent flow at a nominal Ni concentration of 2.5 mg L^{-1}.

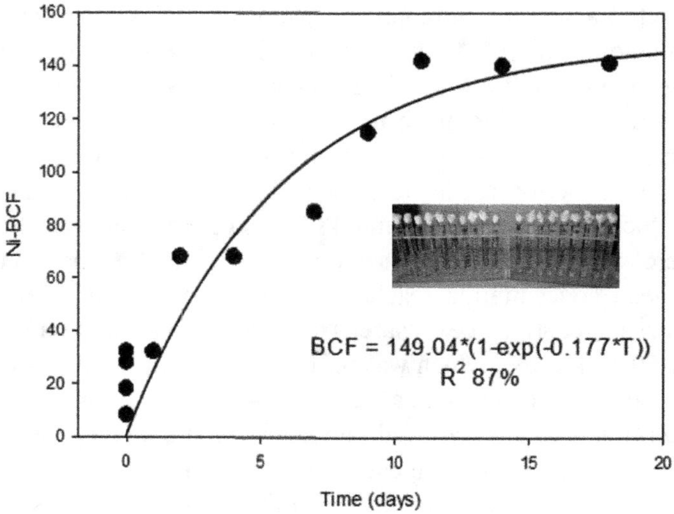

Figure 9. Bioconcentration factor (BCF) for *Ceratophyllum demersum* at a density of 12.85 g ww L^{-1} exposed to 6 mg Ni L^{-1} under laboratory conditions.

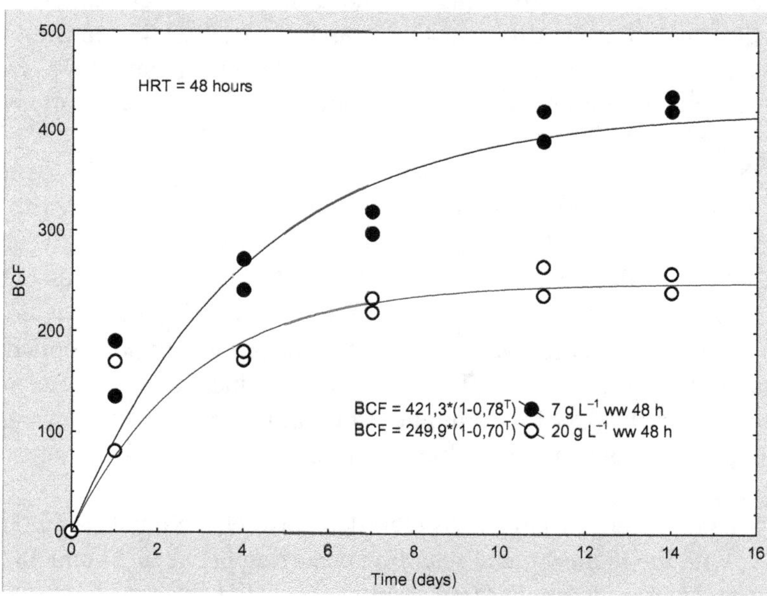

Figure 10. Bioconcentration factor (BCF) for *Ceratophyllum demersum* at a density of 7 and 20 g ww L^{-1} exposed to a continuous flow of 2.5 mg Ni L^{v1} under field conditions.

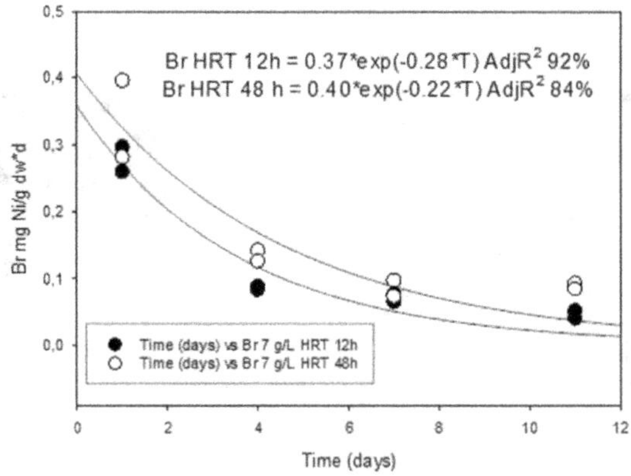

Figure 11. Effects of hydraulic retention time (HRT) on bioaccumulation rate (Br) for 7 g ww L^{-1} of *Ceratophyllum demersum* exposed to continuous effluent with 2.5 mg Ni L^{-1} in semi-field conditions.

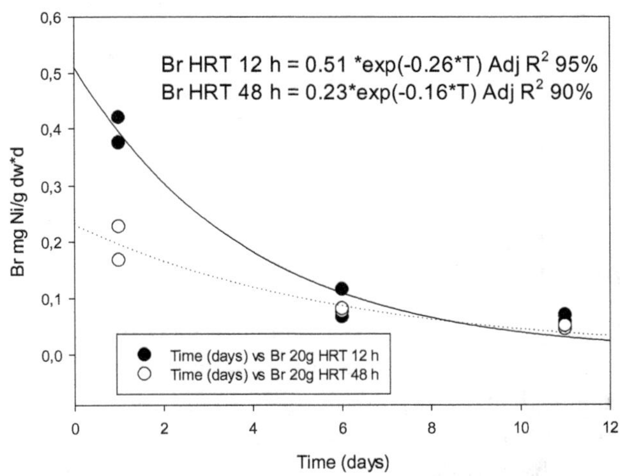

Figure 12. Effects of hydraulic retention time (HRT) on bioaccumulation rate (Br) for 20 g ww L^{-1} of *Ceratophyllum demersum* exposed to continuous effluent with 2.5 mg Ni L^{-1} in semi-field conditions.

The plants were retained in the phytoreactor for 10 days, exploiting their Br to the maximum and in sequence with 100 L volume tanks at a flow of 0.45 L h^{-1} and 20 g ww L^{-1}. In this way, after 44 days of total HRT, removal efficiencies greater than 99% were achieved. The final results are shown in Fig. 13.

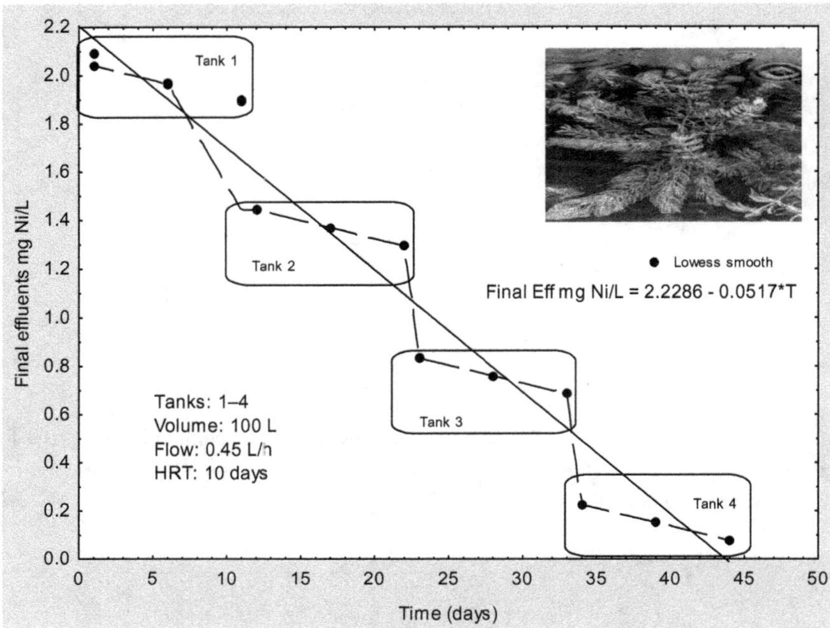

Figure 13. Remotion of initial 2.5 mg Ni L^{-1} under continuous flow, giving an HRT of 10 days and using sequential complete mix phytoreactor during spring season.

4. Conclusions

As reported in this chapter, there are numerous data and information about the incorporation and effects of Ni on aquatic plants. Likewise, a great variety of them demonstrate their ability to accumulate this metal, although they are not considered hyperaccumulators, and without genetically modifying them to increase their Ni retention capacity.

By the other hand, considering aspects related to the design of reactors, such as those used in secondary treatment systems at an industrial level, a configuration with suitable residence times and engineering conditions can be proposed. In this way to achieve the removal of soluble nickel in continuous low-flow industrial effluents with high purification efficiency. Additionally, we can take advantage of the physiological and metabolic information that this metal induces in plants to use it as a means or variable in the process and its control. That allows us to define the best strategies for exposure and removal/replanting of plants at the densities defined in the design of phytoreactors. It is possible to propose the recovery of the Ni retained by the aquatic plants used in the chosen phytoremediation system.

References

Ameen N., Amjad M., Murtaza B., Abbas G., Shahid M., Imran M., Naeem M.A., Niazi N.K. (2019). Biogeochemical behavior of nickel under different abiotic stresses: Toxicity and detoxification mechanisms in plants. *Environ. Sci. Pollut. Res.* 26:10496–10514.

APG (The Angiosperm Phylogeny Group) (2016). An update of the Angiosperm Phylogeny Group classification for the orders and families of flowering plants: APG III. *Bot. J. Linnean Soc.* 161:105–121.

Appenroth K.J., Krech K., Keresztes Á., Fischer W., Koloczek H. (2010). Effects of nickel on the chloroplasts of the duckweeds *Spirodela polyrhiza* and *Lemna minor* and their possible use in biomonitoring and phytoremediation. *Chemosphere* 78:216–223.

Asada K. (1992). Ascorbate peroxidase: A hydrogen peroxide-scavenging enzyme in plants. *Physiol. Plant* 85:235–241.

Ashraf M., Mehmood S. (2013). Response of different tomato cultivars to nickel stress: Growth, leaf pigments and photosynthetic parameters. *Acta Physiol. Planta* 35:2637–2645.

Baran E.J. (2021). Metaloenzimas de plantas. Academia Nacional de Ciencias Exactas, Físicas y Naturales. Serie: Publicaciones científicas N° 17.

Barceló J., Poschenrieder C. (1992). Respuestas de las plantas a la contaminación por metales pesados. *Suelo y Planta* 2:345–361.

Borah P., Rene E.R., Rangan R., Mitra S. (2023). Phytoremediation of nickel and zinc using *Jatropha curcas* and *Pongamia pinnata* from the soils contaminated by municipal solid wastes and paper mill wastes. *Environ. Res.* 219:115055.

Cai Y., Zhang H., Kong X., Xu X., Chen Q. (2007). Nickel indice oxidative damage, antioxidant response and proline metabolism in leaves of tobacco plants. *Biologia Planta* 51:689–694.

Caverzan A., Passaia G., Rosa S.B., Ribeiro C.W., Lazzarotto F., Margis-Pinheiro M. (2012). Plant responses to stresses:role of ascorbate peroxidase in the antioxidant protection. *Genet. Mol. Biol.* 35:1011–1019.

Chaoui A., Mazhoudi S., Ghorbal H., Ferjani E.E. (1997). Cadmium and zinc induction of peroxidation and effects on antioxidant enzyme activities in bean (*Phaseolus vulgaris* L.). *Plant Sci.* 127:139–147.

Chen X., Kumari D., Cao C.J., Plaza G., Achal V. (2019). A review on remediation technologies for nickel-contaminated soil. *Hum. Ecol. Risk Assess.* 26:571–585.

Chen Y.W., Chen Y.X., Li J.L. (2002). Effects of nickel on chlotophyll and photosynthesis in rice leaves. *Plant Sci.* 163:371–376.

Chorom M., Parnian A., Jaafarzadeh N. (2012). Nickel removal by the aquatic plant (*Ceratophyllum Demersum* L.). *Int. J. Environ. Sci. Dev.* 3:372–375.

Dhir B., Sharmila P., Saradhi P.P. (2009). Potential of aquatic macrophytes for removing contaminants from the environment. *Crit. Rev. Environ. Sci. Technol.* 39:754–781.

Di Marzio W.D., Sáenz M.E., Alberdi J.L., Tortorelli M.C., Ambrini G., Vanini P. (2005). Bioaccumulation of Endosulfan from lab-scale contaminated sediment by *Vallisneria spiralis*. *Bull. Environ. Contam. Toxicol.* 74:637–644.

Ding S., Guan D.X., Dai Z.H., Su J., Teng H.H., Ji J., Liu Y., Yang Z., Ma L.Q. (2022). Nickel bioaccessibility in soils with high geochemical background and anthropogenic contamination, *Environ. Pollut.* 310:119914.

Doğanlar Z.B. (2013). Metal accumulation and physiological responses induced by copper and cadmium in *Lemna gibba, L. minor* and *Spirodela polyrhiza*. *Chem. Speciation. Bioavailablity.* 25:79–88.

Elshkaki A., Reck B.K., Graedel T.E. (2017). Anthropogenic nickel supply, demand, and associated energy and water use. *Resour. Conserv. Recycl.* 125:300–307.

Fabiano C.C., Tezotto T., Favarin J.L., Polacco J.C., Mazzafera P. (2015). Essentiality of nickel in plants: A role in plant stresses. *Front. Plant Sci.* 6:754.

Ferreira-Barreto R., Cruz F. (2021). Nickel. In de Mello Prado R (Ed), *Mineral Nutrition of Tropical Plants*, Springer Nature, Switzerland, pp. 253.

Gautam S., Rathoure A.K., Chhabra A., Pandey S.N. (2017). Effects of nickel and zinc on biochemical parameters in plants- a review. *Oct. J. Env. Res.* 5:14–21.

Genchi G., Carocci A., Lauria G., Sinicropi M.S., Catalano A. (2020). Nickel: Human health and environmental toxicology. *Int. J. Environ. Res. Public Health* 17:679.

Ghazanfar S., Komal A., Waseem A., Hassan W., Iqbal R.J., Toor S., Asif M., Saleem I.A., Khan S.U., Tarar Z.H., Nazar S., Rehman H., Ahmed M.I., Rebi A. (2021). Physiological effects of nickel contamination on plant growth. *Nat. Volatiles Essent. Oils* 8:13457–13469.

Gomes-Junior R.A., Moldes C.A., Delite F.S., Pompeu G.B., Grataõ P.L., Mazzafera P., Lea P.J., Azevedo R.A. (2006). Antioxidant metabolism of coffee cell suspension cultures in response to cadmium. *Chemosphere* 65:1330–1337.

Griffiths H.R. (2005). Chemical modifications of biomolecules by oxidants. *Hand Book Environ. Chem.* 2:33–62.

Hao F., Wang X., Chen J. (2006). Involvement of plasma-membrane NADPH oxidase in nickel-induced oxidative stress in roots of wheat seedlings. *Plant Sci.* 170:151–158.

Harasim P., Filipek T. (2015). Nickel in the environment. *J. Elem.* 20:525–534.

Hassan M.U., Chattha M.U., Khan I., Chattha M.B., Aamer M., Nawaz M., Ali A., Khan M.A.U., Khan T.A. (2019). Nickel toxicity in plants: Reasons, toxic effects, tolerance mechanisms, and remediation possibilities — A review. *Environ. Sci. Pollut. Res.* 26:12673–12688.

Held M., Tuchschmid M., Zennegg M., Figi R., Schreiner C., Mellert L.D., Welte U., Kompatscher M., Hermann M., Nachef L. (2022). Thermal runaway and fire of electric vehicle lithium-ion battery and contamination of infrastructure facility. *Renew. Sust. Energy Rev.* 165:112474.

Jain S., Muneerb S., Guerriero G., lieu S., Vishwakarma K., Chauhan D., Dubey N., Tripathia D., Sharma S. (2018). Tracing the role of plant proteins in the response to metal toxicity: A comprehensive review. *Plant Signal Behav.* 13:1507401–150711.

Jung S., Kim J.S., Cho K.Y., Tae G.S., Kang B.G. (2000). Antioxidant responses of cucumber (*Cucumis sativus*) to photo inhibition and oxidative stress induced by norflurazon under high and low PPFDs. *Plant Sci.* 153:145–154.

Kaur R., Yadav S., Kumar N., Singh N. (2017). Effect of nickel on growth, chlorophyll content and photosynthesis of *Ceratophyllum demersum*. *J. Environ. Biol.* 38:801–805.

Kumar N., Singh N., Singh P. (2015). Effect of nickel on biochemical and chlorophyll content of *Ceratophyllum demersum*. *Environ. Sci. Pollut. Res. Internat* 22:9386–9393.

Lešková A., Zvarõk M., Araya T., Giehl, R.F.H. (2020). Nickel toxicity targets cell wall-related processes and PIN2-Mediated auxin transport to inhibit root elongation and gravitropic responses in Arabidopsis. *Plant Cell Physiol.* 61:519–535.

Li Y., Wu S., Liang D., Zhu X., Liao X. (2020). Effects of nickel on growth, photosynthesis and antioxidant system in *Ceratophyllum demersum* L. *Ecotoxicol. Environ. Saf.* 196:110533.

Lyu R., Gu B., Zhang T., Yang Z. (2021). Simultaneous removal of Cd (II), Ni (II), and Pb (II) from water by a submerged macrophyte pondweed (*Potamogeton malaianus*). *Water Environ. Res.* 93:2637–2647.

Maehly A.C., Chance B. (1954). The assay of catalase and peroxidase. In: Glick D. (ed.) *Methods of Biochemical Analysis*, Vol. 1. Intersciences Publishers, Inc., pp. 357.

Maheshwari R., Dubey R.S. (2009). Nickel-induced oxidative stress and the role of antioxidant defence in rice seedlings. *Plant Growth Regul.* 59:37–49.

Malea P., Mylona Z., Panteris E., Kevrekidis D.P., Kevrekidis T. (2021). Nickel uptake kinetics and its structural and physiological impacts in the seagrass *Halophila stipulacea*. *Ecotoxicol. Environ. Saf.* 208:111386.

Manios T., Stentifor E.I., Millner P.A. (2003). The effect of heavy metals accumulation on the chlorophyll concentration of *Typha latifolia* plants, growing in a substrate containing sewage sludge compost and watered with metaliferus water. Ecol. Eng. 20:65–74.

Martinez R.S. (2019). Evaluación ecotoxicológica de metales sobre macrófitas acuáticas. Tesis Doctoral. Facultad de Ciencias Naturales y Museo, Universidad Nacional de La Plata. pp. 375.

Martinez R.S., Alberdi J.L., Di Marzio W.D., Sáenz M.E. (2019a). Efectos del Zn, Ni y Cd sobre parámetros de crecimiento y bioquímicos en *Lemna gibba*. Congreso de Ciencias Ambientales — COPIME. CABA, 9 al 11 Octubre 2019.

Martinez S., Sáenz M.E., Alberdi J.L., Di Marzio W.D. (2019b). Comparative ecotoxicity of single and binary mixtures exposures of nickel and zinc on growth and biomarkers of *Lemna gibba*. *Ecotoxicology* 28:686–697.

Martinez S., Di Marzio W.D., Alberdi J.L., Sáenz M.E. (2024). Efectos del Níquel, Zinc y Cadmio sobre la morfología y el contenido de pigmentos de Lemna gibba L. *Limnetica* 43(2): 2024. doi: 10.23818/limn.43.14

Merlot S. (2020). Understanding nickel responses in plants: More than just an interaction with iron homeostasis. *Plant Cell Physiol.* 61:443–444.

Mishra S., Srivastava S., Tripathi R.D., Kumar R., Seth C.S., Gupta D.K. (2006). Lead detoxification by coontail (*Ceratophyllum demersum* L.) involves induction of phytochelatins and antioxidant system in response to its accumulation. *Chemosphere* 65:1027–1039.

Mizuno M., Kamei M., Tsuchida H. (1998). Ascorbate peroxidase and catalase cooperate for protection against hydrogen peroxide generated in potato tubers during low-temperature storage. *Biochem. Mol. Biol. Int.* 44: 717–726.

Mullineaux M.P., Karpinski S., Creissen G.P. (2008). Integration of signaling in antioxidant defenses. In: Demmig-Adams B. *et al.* (eds.) *Photoprotection, Photoinhibition, Gene Regulation, and Environment*, Springer Nature Switzerland, pp. 223–239.

Myrach T., Zhu A., Witte C.P. (2017). The assembly of the plant urease activation complex and the essential role of the urease accessory protein G (UreG) in delivery of nickel to urease. *J. Biol. Chem.* 292:14556–14565.

Nguyen T.Q., Sesin V., Kisiala A., Emery R.J.N. (2021). Phytohormonal roles in plant responses to heavy metal stress: Implications for using macrophytes in phytoremediation of aquatic ecosystems. *Environ. Toxicol. Chem.* 40: 7–22.

Nishida S., Kato A., Tsuzuki C., Yoshida J., Mizuno T. (2015). Induction of nickel accumulation in response to zinc deficiency in *Arabidopsis thaliana*. *Int. J. Mol. Sci.* 16:9420–9430.

OECD (2006). Guidelines for the testing of chemicals 221 — *Lemna* sp. Growth Inhibition Test. doi:_10.1787/9789264016194.

OECD (2014a). Test No. 238: Sediment-Free *Myriophyllum Spicatum* Toxicity Test, OECD Guidelines for the Testing of Chemicals, Section 2, OECD Publishing, Paris.

OECD (2014b). Guidelines for the testing of chemicals. Water-sediment *Myriophyllum spicatum* toxicity test (TG 239). ENV/JM/WRPR 16. pp. 20.

Palma M., Sandalio L., Corpas F.J., Romero-Puertas M.C., McCarthy I., del Río L.A. (2002). Plant proteases, protein degradation, and oxidative stress: Role of peroxisomes. *Plant Physiol. Biochem.* 40:521–530.

Rathor G., Chopra N., Adhikari T. (2014). Nickel as a pollutant and its management. *Int. Res. J. Environ. Sci.* 3:94–98.

Sáenz M.E., Di Marzio W.D. (2011). Fitorremediación de residuos sólidos utilizando la macrófita acuática *Vallisneria spiralis*. Rev. Ingeniería Sanitaria y Ambiental. *AIDIS*. 114:47–51.

Sáenz M.E., Di Marzio W.D., Alberdi J.L., Tortorelli M.D.C. (2007). Biomarcadores enzimáticos en *Vallisneria spiralis* como indicadores de la calidad de sedimentos. SETAC LA - Uruguay.

Sáenz M.E., Di Marzio W.D., Martinez R.S., Alberdi J.L. (2014). Bioremediation of sediments by rooted aquatic macrophyte *T. domingensis*. SETAC EUROPA, Basilea, Suiza.

Shahzad B., Tanveer M., Rehman A., Cheema S.A., Fahad S., Rehman S., Sharma A. (2018). Nickel; whether toxic or essential for plants and environment — A review. *Plant Physiol. Biochem.* 132:641–651.

Sharma S.S., Dietz K.J. (2009). The relationship between metal toxicity and cellular redox imbalance. *Trend Plant Sci.* 14:43–50.

Shmaefsky B.R. (2020). Principles of phytoremediation. In: Shmaefsky B.R. (ed.) *Phytoremediation. In-situ* applications. Springer Nature, Switzerland, pp. 1–26.

Sofo A., Scopa A., Nuzzaci M., Vitti A. (2015). Ascorbate peroxidase and catalase activities and their genetic regulation in plants subjected to drought and salinity stresses. *Int. J. Mol. Sci.* 16:13561–13578.

Song Y., Zhang L.L., Li J., He X.J., Chen M., Deng Y. (2018). High-potential accumulation and tolerance in the submerged hydrophyte *Hydrilla verticillata* (L.f.) Royle for nickel-contaminated water. *Ecotoxicol. Environ. Saf.* 161:553–562.

Upadhyay R., Sharma Y.K., Srivastava S. (2017). Effects of zinc stress on growth, photosynthetic pigments, antioxidative enzymes and protein content of black gram. *J. Plant Physiol. Pathol.* 5:2.

USEPA (1996). Method 3052. Microwave assisted digestion of siliceous and organically based matrices. In: *Test Methods for Evaluating Solid Wastes: Physical/Chemical Methods*, EPA SW-846, 3rd edn., Vol. 1. Washington, DC.

USEPA (2004). *The Environmental Chemistry of Metals. Risk Assessment Forum.* Washington, DC.

Van der Pas L., Ingle R.A. (2019). Towards an understanding of the molecular basis of nickel hyperaccumulation in plants. *Plants* 8:11.

Yilmaz D.D., Parlak K.U. (2011). Nickel-induced changes in lipid peroxidation, antioxidative enzymes, and metal accumulation in *Lemna Gibba. Int. J. Phytoremediation* 13:805–817.

Zaid A., Mohammad F., Wani S.H., Siddique K.M.H. (2019). Salicylic acid enhances nickel stress tolerance by up-regulating antioxidant defense and glyoxalase systems in mustard plants. *Ecotoxicol. Environ. Saf.* 180:575–587.

Zhou Q., Song Y. (2013). Nickel induced oxidative damage and antioxidant responses in the leaves of wna1, a mutant whit high nickel accumulation capacity. *Biol. Planta* 57:661–666.

https://doi.org/10.1142/9789811283123_0009

Chapter 9

Bioremediation of Lithium and Nickel from Soil and Water

Veysel Turan

Department of Soil Science and Plant Nutrition,
Faculty of Agriculture, Bingöl University, 12000 Bingöl, Türkiye
turan.veysel@gmail.com

Abstract

Human-induced environmental pollution, which has emerged with the increasing population in recent years, has become a global threat. One of the consequences of its rapid and destructive effects is that heavy metals, such as lithium (Li) and nickel (Ni), threaten soil and plant health, and their presence in soil and water poses a risk to both human health and environmental well-being. The purpose of this chapter is to explore the importance of bioremediation in removing Li and Ni from soil and water by investigating various research studies and approaches. Bioremediation is a solution that utilises microorganisms to remove heavy metals by converting them into less toxic forms. Therefore, it is important to carefully select the appropriate microorganisms and optimise the bioremediation conditions to achieve the best results. Moreover, the effectiveness of these techniques depends on various factors, such as the concentration of Li or Ni in the soil, the duration of the remediation process and the environmental conditions.

Keywords: Nickel, lithium, soil and water pollution, bioremediation, heavy metal

1. Introduction

The contamination of soil and water by heavy metals is a primary environmental issue worldwide (Rehman *et al.*, 2023). The occurrence of heavy metals in the environment is mainly attributed to human activities, such as mining, the discharge of industrial effluents and agricultural practices (Sharma and Kumar, 2021; Shaker *et al.*, 2022; Liu *et al.*, 2022). In this context, innovative methods are needed to rehabilitate and recycle natural resources, such as rivers, soil and groundwater, that have been polluted by human activities (Khan *et al.*, 2023; Macedo *et al.*, 2020; Rehman *et al.*, 2022). Heavy metals such as lithium (Li) and nickel (Ni) are particularly toxic, and their accumulation can cause ecological and health problems (Bolan *et al.*, 2021; Genchi *et al.*, 2020). Therefore, the easy accumulation and high toxicity of these metals pose great risks to the ecological environment.

For the removal of heavy metals in soil and water, traditional methods such as soil removal and washing are not cost-effective, and it is clear that new methods should be developed. In this context, bioremediation (BR) has emerged as an efficient and environmentally friendly tool for mitigating the negative effects of various pollutants in soil and water (Naikoo *et al.*, 2021; Urionabarrenetxea *et al.*, 2021). Research has also been conducted with the aim of using microorganisms through BR to reduce the negative effects of heavy metals in both soil and water (Bolan *et al.*, 2021), yielding promising results (Tufail *et al.*, 2022; Schommer *et al.*, 2023). However, it is necessary to consider the aspects of economic effectiveness and environmental friendliness of the methods that will be used to rehabilitate the soil and water contaminated with Li and Ni. In this regard, this chapter aims to explore different approaches to BR in the removal of Li and Ni from soil and water (Fig. 1).

2. Bioremediation of Ni from Soil

Ni is a commonly occurring metal that can be found in soil in discrete forms (Huang *et al.*, 2023). Moreover, Ni is ordinarily used in industrial processes (Wu *et al.*, 2022), such as electroplating (Liu *et al.*, 2023; Wang *et al.*, 2023) and alloy production (Fischer *et al.*, 2023; Gao *et al.*, 2023), and can also be found in pesticides (Egedigwe-Ekeleme *et al.*, 2023), sewage sludge (Ghasemzadeh and Bostani, 2017; Bogusz and Oleszczuk, 2018), agricultural practices (Macedo *et al.*, 2020; Rehman *et al.*, 2022;

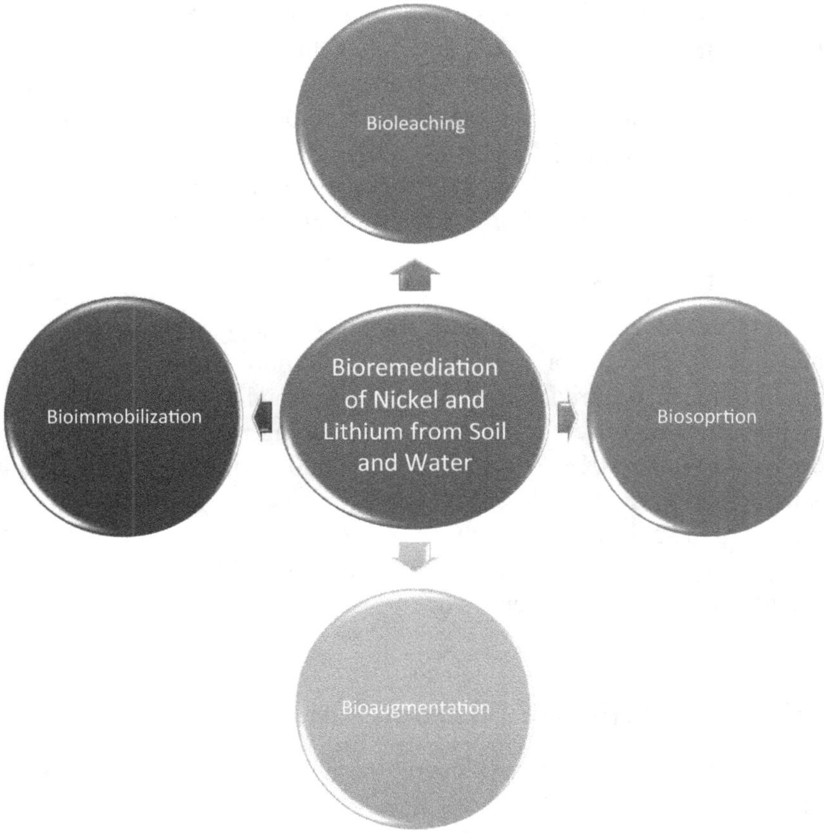

Figure 1. Bioremediation of Ni and Li removal from soil and water.

Iqbal Khan *et al.*, 2023) and as a result of natural weathering of Ni-containing rocks (Ratié *et al.*, 2015; Seiler *et al.*, 2022, 2023). In addition, Ni is a toxic heavy metal that is commonly found in contaminated soil (Borah *et al.*, 2023). Exposure to high levels of Ni in soil can have harmful effects on human health (Genchi *et al.*, 2020) as well as the environment (Ahmad and Ashraf, 2012; Naz *et al.*, 2022). Once Ni enters the soil, it can bind to soil particles and become tightly bound (Begum *et al.*, 2022; Li *et al.*, 2022), making it difficult to remove through conventional remediation methods (Kumar *et al.*, 2021; Rajendran *et al.*, 2022). It is significant to note that the successful application of BR relies upon several factors, including the type and concentration of Ni in the soil, the effectiveness of the BR approach and environmental conditions, such as pH and temperature (Dixit *et al.*, 2015; Genchi *et al.*, 2020).

BR techniques have been developed as a cost-effective and sustainable alternative to traditional remediation methods for the removal of Ni from contaminated soil (Chen *et al.*, 2020; Turan, 2021; Naikoo *et al.*, 2021; Urionabarrenetxea *et al.*, 2021). Furthermore, BR relies on the use of microorganisms (Zhu *et al.*, 2016; Rathour *et al.*, 2022), plants (De Bernardi *et al.*, 2020; Vischetti *et al.*, 2022) or other biological agents (El-Naggar *et al.*, 2021; Hannan *et al.*, 2021) to degrade or remove contaminants from soil. By using BR techniques, it is possible to reduce the concentration of Ni in soil to safe levels while minimising the impact on the environment (Seagren, 2023). Additionally, the BR of Ni from soil can be achieved using different approaches, including microbial remediation (Verma and Kuila, 2019; Nwaehiri *et al.*, 2020; Pande *et al.*, 2022) and bioaugmentation (Ren *et al.*, 2022). For instance, Ni-resistant bacteria can be added to the soil to increase the microbial population that can degrade Ni (Atuchin *et al.*, 2023). In this way, the rate of Ni degradation and removal from soil can be increased.

3. Bioremediation of Ni from Water

Ni contamination in water can occur through a variety of natural and anthropogenic processes, such as atmospheric deposition (Liu *et al.*, 2022), agricultural runoff (Kaur and Sinha, 2019), natural weathering of rocks and minerals (Poznanović Spahić *et al.*, 2019), industrial and mining activities (Guan *et al.*, 2014) and municipal wastewater discharges (Shaker *et al.*, 2022).

Various methods can be used to remove Ni from water, including physical, chemical and biological treatments (Adigun *et al.*, 2020; Nwaehiri *et al.*, 2020; Zhou *et al.*, 2021; Gad, 2023; Abo-Alkasem *et al.*, 2023). BR techniques, such as microbial reduction and biosorption, offer a sustainable and cost-effective approach to removing metals from water (Sharma *et al.*, 2021; Bala *et al.*, 2022; Noman *et al.*, 2022; Sreedevi *et al.*, 2022). As in the case of soil, it is possible to use BR techniques to reduce the concentration of metals in water to safe levels while minimising the impact on the environment (Taştan *et al.*, 2010; Yaashikaa *et al.*, 2022). Microorganisms can be used to convert soluble metals, such as Ni ions, into insoluble forms, which can then be easily removed from the water. Various types of bacteria, fungi and algae have been tested for their ability to bioremediate Ni_2 (Dusengemungu *et al.*, 2021; Ferrier *et al.*, 2021; Carpen and Giese, 2022). Thus, BR can potentially be an efficient and sustainable way of treating polluted water containing Ni.

4. Bioremediation of Li from Soil

Li is a naturally occurring metal that is widely used in various industrial applications, such as battery manufacturing (Mrozik *et al.*, 2021), ceramics production (Aral and Vecchio-Sadus, 2008) and mining (Bolan *et al.*, 2021). Improper disposal of electronic waste (Li and Achal, 2020) can lead to Li contamination in soil. Additionally, Li can contaminate soil through the use of Li-containing fertilisers and the leaching of Li from landfill sites (Robinson *et al.*, 2018; Mrozik *et al.*, 2021). Consequently, Li contamination in soil can have negative impacts on soil quality, plant growth and ecosystem health (Shahzad *et al.*, 2016; Hayyat *et al.*, 2021).

The BR of Li contaminated soil, which uses microorganisms or plants to remove contaminants from the environment, is a potential solution to Li contamination in soil. The microorganisms can be introduced to the soil in high concentrations to enhance their activity and promote the biodegradation of Li (Sedlakova-Kadukova *et al.*, 2020). While BR shows promise for the remediation of Li-contaminated soil, there are also challenges that need to be addressed (Bolan *et al.*, 2021). These challenges stem from the fact that Li is a relatively new, emerging contaminant of concern, and there is limited knowledge about the behaviour and fate of Li in soil, as well as the effectiveness of BR techniques in this regard.

5. Bioremediation of Li from Water

Li can be released into water sources through wastewater discharge, accidental spills and improper disposal of waste materials, leading to environmental pollution (Luo *et al.*, 2015; Zhao *et al.*, 2020; Mrozik *et al.*, 2021). The BR of Li-contaminated water can be achieved through several techniques, including electrochemical microbial remediation and biosorption (Tsuruta, 2005; Dolker and Pant, 2018). In particular, biosorption is a technique that utilises the ability of certain microorganisms or biomaterials to adsorb heavy metals from water (Zinicovscaia *et al.*, 2020; Günan Yücel *et al.*, 2020). Additionally, BR has emerged as a promising approach for the removal of Li from contaminated water, as it offers a sustainable and environmentally friendly resolution (Tsuruta, 2005; Dolker and Pant, 2018). Besides, microbial degradation involves the use of microorganisms that can degrade or transform Li compounds in the water, and several bacterial strains have been isolated that can use Li as a source of energy or carbon, thereby reducing its concentration in the water (Salazar-Ardiles *et al.*, 2020; Sedlakova-Kadukova *et al.*, 2020; Martínez *et al.*, 2021).

6. Conclusion

Heavy metals, such as Li and Ni, pose significant environmental and health risks when they are present in soil and water. Notwithstanding, BR is a promising solution that utilises microorganisms and plants to remove heavy metals by conversion or immobilisation. Moreover, the use of eco-friendly approaches to the removal of heavy metals, such as microbial BR, can provide a sustainable and cost-effective solution to heavy metal pollution. Likewise, BR is an encouraging method for discharging Ni from contaminated soil and water. Nevertheless, the effectiveness of the process can be influenced by several factors, such as the concentration of Ni in the soil or water, the type of microorganisms used and the environmental conditions. Finally, researchers and policymakers need to continue exploring these approaches to ensure that our environment and health are protected.

References

Abo-Alkasem M.I., Hassan N.H., Abo Elsoud M.M. (2023). Microbial bioremediation as a tool for the removal of heavy metals. *Bull. Natl. Res. Cent.* 47:31.

Adigun O.A., Oninla V.O., Babarinde N.A.A., Oyedotun K.O., Manyala N. (2020). Characterization of sugarcane leaf-biomass and investigation of its efficiency in removing Nickel(II), Chromium(III) and Cobalt(II) ions from polluted water. *Surfaces Interfa* 20:100621.

Ahmad M.S.A., Ashraf M. (2012). Essential roles and hazardous effects of nickel in plants. In: Whitacre D.M. (ed.) *Reviews of Environmental Contamination and Toxicology*. Springer, New York, pp. 125–167.

Aral H., Vecchio-Sadus A. (2008). Toxicity of lithium to humans and the environment — A literature review. *Ecotoxicol Environ Saf* 70:349–356.

Atuchin V.V., Asyakina L.K., Serazetdinova Y.R., Frolova A.S., Velichkovich N.S., Prosekov A.Y. (2023). Microorganisms for bioremediation of soils contaminated with heavy metals. *Microorganisms* 11:864.

Bala S., Garg D., Thirumalesh B.V., Sharma M., Sridhar K., Inbaraj B.S., Tripathi M. (2022) Recent strategies for bioremediation of emerging pollutants: A review for a green and sustainable environment. *Toxics* 10:484.

Begum W., Rai S., Banerjee S., Bhattacharjee S., Mondal M.H., Bhattarai A., Saha B. (2022). A comprehensive review on the sources, essentiality and toxicological profile of nickel. *RSC Adv.* 12:9139–9153.

Bogusz A., Oleszczuk P. (2018). Sequential extraction of nickel and zinc in sewage sludge- or biochar/sewage sludge-amended soil. *Sci. Total Environ.* 636:927–935.

Bolan N., Hoang S.A., Tanveer M., Wang L., Bolan S., Sooriyakumar P., Robinson B., Wijesekara H., Wijesooriya M., Keerthanan S., Vithanage M., Markert B., Fränzle S., Wünschmann S., Sarkar B., Vinu A., Kirkham M.B., Siddique K.H.M., Rinklebe J. (2021). From mine to mind and mobiles — Lithium contamination and its risk management. *Environ. Pollut.* 290:118067.

Borah P., Rene E.R., Rangan L., Mitra S. (2023). Phytoremediation of nickel and zinc using Jatropha curcas and Pongamia pinnata from the soils contaminated by municipal solid wastes and paper mill wastes. *Environ. Res.* 219:115055.

Carpen H.L., Giese E.C. (2022). Enhancement of nickel laterite ore bioleaching by *Burkholderia* sp. using a factorial design. *Appl. Water Sci.* 12:181.

Chen X., Kumari D., Cao C.J., Plaza G., Achal V. (2020). A review on remediation technologies for nickel-contaminated soil. *Hum. Ecol. Risk Assess. Int. J.* 26:571–585.

De Bernardi A., Casucci C., Businelli D., D'Amato R., Beone G.M., Fontanella M.C., Vischetti C. (2020). Phytoremediation potential of crop plants in countering nickel contamination in carbonation lime coming from the sugar industry. *Plants* 9:580.

Dixit R., Wasiullah, Malaviya D., Pandiyan K., Singh U., Sahu A., Shukla R., Singh B., Rai J., Sharma P., Lade H., Paul D. (2015). Bioremediation of heavy metals from soil and aquatic environment: An overview of principles and criteria of fundamental processes. *Sustainability* 7:2189–2212.

Dolker T., Pant D. (2018). Bioremediation of metals from lithium-ion battery (LIB) waste. In: Varjani S.J., Gnansounou E., Gurunathan B., Pant D., Zakaria Z.A. (eds.) *Waste Bioremediation, Energy, Environment, and Sustainability*. Springer, Singapore, pp. 265–278.

Dusengemungu L., Kasali G., Gwanama C., Mubemba B. (2021). Overview of fungal bioleaching of metals. *Environ. Adv.* 5:100083.

Egedigwe-Ekeleme C.A., Famurewa A.C., Egedigwe U.O., Onyeabo C., Kanu S.C., Ogunwa S.C., Onuora C.A., Agbo A.O. (2023). Comparative studies on the amino acids, pesticide residue content, biogenic and toxic elements in an underutilized nut: Bambara groundnut (*Vigna subterranea*). *J. Trace Elem. Min.* 3:100051.

El-Naggar A., Ahmed N., Mosa A., Niazi N.K., Yousaf B., Sharma A., Sarkar B., Cai Y., Chang S.X. (2021). Nickel in soil and water: Sources, biogeochemistry, and remediation using biochar. *J. Hazard Mater.* 419:126421.

Ferrier J., Csetenyi L., Gadd G.M. (2021). Selective fungal bioprecipitation of cobalt and nickel for multiple-product metal recovery. *Microb. Biotechnol.* 14:1747–1756.

Fischer L., Volz A., Hagemann U., Ulbricht M. (2023). Polymeric multicomposites with a tailored nickel microenvironment as catalytic flow-through membrane reactors for efficient p-nitrophenol degradation. *Chem. Eng. J.* 463:142437.

Gad S.C. (2023). Nickel and nickel compounds. In: *Reference Module in Biomedical Sciences*. Elsevier. https://doi.org/10.1016/B978-0-12-824315-2.00520-0.

Gao W., Zou Y., Zang Y., Zhao X., Zhou W., Dai Y., Liu H., Wang J.J., Ma Y., Sang Y. (2023). Magnetic-field-regulated Ni-Fe-Mo ternary alloy electrocatalysts with enduring spin polarization enhanced oxygen evolution reaction. *Chem. Eng. J.* 455:140821.

Genchi G., Carocci A., Lauria G., Sinicropi M.S., Catalano A. (2020). Nickel: Human health and environmental toxicology. *IJERPH* 17:679.

Ghasemzadeh P., Bostani A. (2017). The removal of lead and nickel from the composted municipal waste and sewage sludge using nanoscale zero-valent iron fixed on quartz. *Ecotoxicol. Environ. Saf.* 145:483–489.

Guan Y., Shao C., Ju M. (2014). Heavy metal contamination assessment and partition for industrial and mining gathering areas. *IJERPH* 11:7286–7303.

Günan Yücel H., Aksu Z., Yalçınkaya G.B., Karatay S.E., Dönmez G. (2020). A comparative investigation of lithium(I) biosorption properties of *Aspergillus versicolor* and Kluyveromyces marxianus. *Water Sci. Technol.* 81:499–507.

Hannan F., Huang Q., Farooq M.A., Ayyaz A., Ma J., Zhang N., Ali B., Deyett E., Zhou W., Islam F. (2021) Organic and inorganic amendments for the remediation of nickel contaminated soil and its improvement on *Brassica napus* growth and oxidative defense. *J. Hazard Mater.* 416:125921.

Hayyat M.U., Nawaz R., Siddiq Z., Shakoor M.B., Mushtaq M., Ahmad S.R., Ali S., Hussain A., Irshad M.A., Alsahli A.A., Alyemeni M.N. (2021). Investigation of lithium application and effect of organic matter on soil health. *Sustainability* 13:1705.

Huang K., Yang Y., Lu H., Hu S., Chen G., Du Y., Liu T., Li X., Li F. (2023). Transformation kinetics of exogenous nickel in a paddy soil during anoxic-oxic alteration: Roles of organic matter and iron oxides. *J. Hazard Mater.* 452:131246.

Iqbal Khan Z., Ahmad K., Ahmad T., Zafar A., Fahad Alrefaei A., Ashfaq A., Akhtar S., Mahpara S., Mehmood N., Ugulu I. (2023). Evaluation of nickel toxicity and potential health implications of agriculturally diversely irrigated wheat crop varieties. *Arab J. Chem.* 16(8):104934.

Kaur T., Sinha A.K. (2019). Pesticides in agricultural run offs affecting water resources: A study of Punjab (India). *Agric. Sci.* 10:1381–1395.

Kumar A., Jigyasu D.K., Kumar A., Subrahmanyam G., Mondal R., Shabnam A.A., Cabral-Pinto M.M.S., Malyan S.K., Chaturvedi A.K., Gupta D.K., Fagodiya R.K., Khan S.A., Bhatia A. (2021). Nickel in terrestrial biota: Comprehensive review on contamination, toxicity, tolerance and its remediation approaches. *Chemosphere* 275:129996.

Li Q., Wang Y., Li Y., Li L., Tang M., Hu W., Chen L., Ai S. (2022). Speciation of heavy metals in soils and their immobilization at micro-scale interfaces among diverse soil components. *Sci. Total Environ.* 825:153862.

Li W., Achal V. (2020) Environmental and health impacts due to e-waste disposal in China — A review. *Sci. Total Environ.* 737:139745.

Liu L., Xu X., Han J., Zhu J.M., Li S., Liang L., Wu P., Wu Q., Qiu G. (2022). Heavy metal(loid)s in agricultural soils in the world's largest barium-mining area: Pollution characteristics, source apportionment, and health risks using PMF model and Cd isotopes. *Proc. Saf. Environ. Prot.* 166:669–681.

Liu S., Gu A.Y., Zeng Y., Yin L., Dai Y. (2022). Atmospheric deposition of particulate matter and micropollutants as a major mass transport route to surface water in winter: Measurement and modeling in Beijing in 2014 and 2021. *ACS Earth Space Chem.* 6:962–973.

Liu Y., Lian R., Wu Xiaoyun Dai L., Ding J., Wu Xiaoyu Ye X., Chen R., Ding R., Liu J., Van Der Bruggen B. (2023). Nickel recovery from electroplating sludge via bipolar membrane electrodialysis. *J. Coll. Interf. Sci.* 637: 431–440.

Luo X., Guo B., Luo J., Deng F., Zhang S., Luo S., Crittenden J. (2015). Recovery of lithium from wastewater using development of Li ion-imprinted polymers. *ACS Sustainable Chem. Eng.* 3:460–467.

Macedo F.G, Santos E.F., Lavres J. (2020). Agricultural crop influences availability of nickel in the rhizosphere; a study on base cation saturations, Ni dosages and crop succession. *Rhizosphere* 13:100182.

Martínez F.L., Rajal V.B., Irazusta V. (2021). Removal of lithium from aqueous solutions using halotolerant bacteria from El Salar del Hombre Muerto. *J. Environ. Chem. Eng.* 9: 105099.

Mrozik W., Rajaeifar M.A., Heidrich O., Christensen P. (2021). Environmental impacts, pollution sources and pathways of spent lithium-ion batteries. *Energy Environ. Sci.* 14:6099–6121.

Naikoo M.I., Khan F.A., Noureldeen A., Rinklebe J., Sonne C., Rajakaruna N., Ahmad P. (2021). Biotransfer, bioaccumulation and detoxification of nickel along the soil — Faba bean — Aphid — Ladybird food chain. *Sci. Total Environ.* 785:147226.

Naz M., Ghani M.I., Sarraf M., Liu M., Fan X. (2022). Ecotoxicity of nickel and its possible remediation. In: Bhat R.A., Tonelli F.M.P., Dar G.H., Hakeem K. (eds.) *Phytoremediation*. Elsevier, pp. 297–322.

Noman E., Al-Gheethi A., Saphira Radin Mohamed R.M., Al-Sahari M., Hossain M.S., Vo D.V.N., Naushad M. (2022). Sustainable approaches for nickel removal from wastewater using bacterial biomass and nanocomposite adsorbents: A review. *Chemosphere* 291: 132862.

Nwaehiri U.L., Akwukwaegbu P.I., Bright Nwoke B.E. (2020). Bacterial remediation of heavy metal polluted soil and effluent from paper mill industry. *Environ. Anal. Health Toxicol.* 35:e2020009.

Pande V., Pandey S.C., Sati D., Bhatt P., Samant M. (2022). Microbial interventions in bioremediation of heavy metal contaminants in agroecosystem. *Front. Microbiol.* 13:824084.

Poznanović Spahić M.M., Sakan S.M., Glavaš-Trbić B.M., Tančić P.I., Škrivanj S.B., Kovačević J.R., Manojlović D.D. (2019). Natural and anthropogenic sources of chromium, nickel and cobalt in soils impacted by agricultural and industrial activity (Vojvodina, Serbia). *J. Environ. Sci. Health, Part A* 54:219–230.

Rajendran S., Priya T.A.K., Khoo K.S., Hoang T.K.A., Ng H.S., Munawaroh H.S.H., Karaman C., Orooji Y., Show P.L. (2022). A critical review on various remediation approaches for heavy metal contaminants removal from contaminated soils. *Chemosphere* 287: 132369.

Rathour R.K., Sharma D., Sharma N., Bhatt A.K., Singh S.P. (2022) Engineered microorganisms for bioremediation. *Curr. Dev. Biotechnol. Bioeng.* 335–361.

Ratié G., Jouvin D., Garnier J., Rouxel O., Miska S., Guimarães E., Cruz Vieira L., Sivry Y., Zelano I., Montarges-Pelletier E., Thil F., Quantin C. (2015). Nickel isotope fractionation during tropical weathering of ultramafic rocks. *Chem. Geol.* 402:68–76.

Rehman S., Mansoora N., Al-Dhumri S.A., Amjad S.F., Al-Shammari W.B., Almutari M.M., Alhusayni F.S., Al Bakre D.A., Lalarukh I., Alshahri A.H., Poczai P., Galal T.M., Abdelhafez A.A. (2022). Associative effects of activated carbon biochar and arbuscular mycorrhizal fungi on wheat for reducing nickel food chain bioavailability. *Environ. Technol. Innovation* 26:102539.

Rehman Z.U., Junaid M.F., Ijaz N., Khalid U., Ijaz Z. (2023). Remediation methods of heavy metal contaminated soils from environmental and geotechnical standpoints. *Sci. Total Environ.* 867:161468.

Ren W., Liu H., Mao T., Teng Y., Zhao R., Luo Y. (2022). Enhanced remediation of PAHs-contaminated site soil by bioaugmentation with graphene oxide immobilized bacterial pellets. *J. Hazard Mater.* 433:128793.

Robinson B.H., Yalamanchali R., Reiser R., Dickinson N.M. (2018). Lithium as an emerging environmental contaminant: Mobility in the soil-plant system. *Chemosphere* 197:1–6.

Salazar-Ardiles C., Caimanque T., Galetović A., Vilo C., Araya J.E., Flores N., Gómez-Silva B. (2020). Staphylococcus sciuri strain LCHXa is a free-living lithium-tolerant bacterium isolated from Salar de Atacama, Chile. *Microorganisms* 8:668.

Schommer V.A., Vanin A.P., Nazari M.T., Ferrari V., Dettmer A., Colla L.M., Piccin J.S. (2023). Biochar-immobilized Bacillus spp. for heavy metals bioremediation: A review on immobilization techniques, bioremediation mechanisms and effects on soil. *Sci. Total Environ.* 881:163385.

Seagren E.A. (2023). Bioremediation. In: *Reference Module in Biomedical Sciences*. Elsevier, p. B9780128243152005000. https://www.sciencedirect.com/science/article/abs/pii/B9780128243152004139.

Sedlakova-Kadukova J., Marcincakova R., Luptakova A., Vojtko M., Fujda M., Pristas P. (2020). Comparison of three different bioleaching systems for Li recovery from lepidolite. *Sci. Rep.* 10:14594.

Seiler S., Bradshaw P., Klein B. (2023). Awaruite, a new large nickel resource: Activation by ammonium sulfate and thiosulfate for flotation. *Chem. Eng. J. Adv.* 13:100441.

Seiler S., Sánchez G., Teliz E., Díaz V., Bradshaw P., Klein B. (2022). Awaruite (Ni_3Fe), a new large nickel resource: Electrochemical characterization and surface composition under flotation-related conditions. *Min. Eng.* 184:107656.

Shahzad B., Tanveer M., Hassan W., Shah A.N., Anjum S.A., Cheema S.A., Ali I. (2016). Lithium toxicity in plants: Reasons, mechanisms and remediation possibilities — A review. *Plant Physiol. Biochem.* 107:104–115.

Shaker O.A., Safwat S.M., Matta M.E. (2022). Nickel removal from wastewater using electrocoagulation process with zinc electrodes under various operating conditions: Performance investigation, mechanism exploration, and cost analysis. *Environ. Sci. Pollut. Res.* 30:26650–26662.

Sharma P., Kumar S. (2021). Bioremediation of heavy metals from industrial effluents by endophytes and their metabolic activity: Recent advances. *Biores. Technol.* 339:125589.

Sharma P., Kumar S., Pandey A. (2021) Bioremediated techniques for remediation of metal pollutants using metagenomics approaches: A review. *J. Environ. Chem. Eng.* 9:105684.

Sreedevi P.R., Suresh K., Jiang G. (2022). Bacterial bioremediation of heavy metals in wastewater: A review of processes and applications. *J. Water Proc. Eng.* 48:102884.

Taştan B.E., Ertuğrul S., Dönmez G. (2010). Effective bioremoval of reactive dye and heavy metals by Aspergillus versicolor. *Biores. Technol.* 101:870–876.

Tsuruta T. (2005). Removal and recovery of lithium using various microorganisms. *J. Biosci. Bioeng.* 100:562–566.

Tufail M.A., Iltaf J., Zaheer T., Tariq L., Amir M.B., Fatima R., Asbat A., Kabeer T., Fahad M., Naeem H., Shoukat U., Noor H., Awais M., Umar W.,

Ayyub M. (2022). Recent advances in bioremediation of heavy metals and persistent organic pollutants: A review. *Sci. Total Environ.* 850:157961.

Turan V. (2021). Arbuscular mycorrhizal fungi and pistachio husk biochar combination reduces Ni distribution in mungbean plant and improves plant antioxidants and soil enzymes. *Physiol. Planta* 173:418–429.

Urionabarrenetxea E., Garcia-Velasco N., Anza M., Artetxe U., Lacalle R., Garbisu C., Becerril T., Soto M. (2021). Application of *in situ* bioremediation strategies in soils amended with sewage sludges. *Sci. Total Environ.* 766:144099.

Verma S., Kuila A. (2019). Bioremediation of heavy metals by microbial process. *Environ. Technol. Innovation* 14:100369.

Vischetti C., Marini E., Casucci C., De Bernardi A. (2022). Nickel in the environment: Bioremediation techniques for soils with low or moderate contamination in European Union. *Environments* 9:133.

Wang H.Y., Li Y., Jiao S.Q., Chou K.C., Zhang G.H. (2023). Recovery of Ni matte from Ni-bearing electroplating sludge. *J. Environ. Manage.* 326:116744.

Wu X., Jiang W., Yu K., Xu S., Yang H., Wang N., Wei C., Feng C., Sun Y., Xie S. (2022). Coral-inferred historical changes of nickel emissions related to industrial and transportation activities in the Beibu Gulf, northern South China Sea. *J. Hazard Mater.* 424:127422.

Yaashikaa P.R., Kumar P.S., Jeevanantham S., Saravanan R. (2022). A review on bioremediation approach for heavy metal detoxification and accumulation in plants. *Environ. Pollut.* 301:119035.

Zhao C., He M., Cao H., Zheng X., Gao W., Sun Y., Zhao H., Liu D., Zhang Y., Sun Z. (2020). Investigation of solution chemistry to enable efficient lithium recovery from low-concentration lithium-containing wastewater. *Front. Chem. Sci. Eng.* 14:639–650.

Zhou H., Zhao X., Kumar K., Kunetz T., Zhang Y., Gross M., Wen Z. (2021). Removing high concentration of nickel (II) ions from synthetic wastewater by an indigenous microalgae consortium with a Revolving Algal Biofilm (RAB) system. *Algal. Res.* 59:102464.

Zhu X., Li W., Zhan L., Huang M., Zhang Q., Achal V. (2016). The large-scale process of microbial carbonate precipitation for nickel remediation from an industrial soil. *Environ. Pollut.* 219:149–155.

Zinicovscaia I., Yushin N., Pantelica A., Demčák Š., Mitu A., Apostol A.I. (2020). Lithium biosorption by *Arthrospira* (*Spirulina*) platensis biomass. *Ecol. Chem. Eng.* S 27:271–280.

https://doi.org/10.1142/9789811283123_0010

Chapter 10

Occurrence and Speciation of Ni in the Environment and the Health Risk to Living Organisms

Arnab Patra, Nargis Khatun, Trishna Mandal, Sk. Mehebub Rahaman, and Bidyut Saha*

Department of Chemistry, The University of Burdwan, Burdwan 713104, West Bengal, India

**bsaha@chem.buruniv.ac.in*

Abstract

A detailed knowledge and recent advancements on the sources, speciation, toxicity and chemistry of nickel and its different compounds are discussed in this chapter. Nickel is an essential micronutrient for plant growth, and it is also a component of the enzyme urease, which plays a role in nitrogen metabolism in higher plants. Nickel and nickel compounds are also important for several biological processes in animals and soil/water microbes, and they have many industrial and commercial uses. Nickel is a known heavy transition metal and is found at very low levels in the environment. The vast industrial use of nickel leads to widespread environmental pollution. Higher levels of nickel can affect the photosynthetic function and transpiration of higher plants; it reduces seed germination, root and shoot growth, biomass accumulation and final production. Moreover, nickel toxicity leads to chlorosis and necrosis and may cause oxidative damage in plants. In addition, nickel toxicity degrades soil fertility, which may reduce crop production in the near future. The limited knowledge regarding the mechanisms of nickel tolerance in plants further highlights this fact. Furthermore, it causes many chronic diseases in humans, including allergy, cardiovascular and kidney diseases, lung fibrosis and lung and nasal cancers. The molecular mechanisms of nickel-induced toxicity, which cause the above diseases in humans, are still unknown. Mitochondrial dysfunctions and oxidative stress are mainly considered to play a crucial role in inducing toxicity from this metal. Therefore, we should pay attention in future research to find approachable and prominent ways of minimising the entry of nickel into our environment.

Keywords: Nickel, nickel toxicity, chlorosis, necrosis, immunotoxic, carcinogenic

1. Introduction

Nickel is the 28th element in the modern periodic table, and it is a hard, ductile and silvery-white transition metal. It may show several oxidation states (−1 to +4); however, the +2 oxidation state (Ni^{2+}) is the most widespread in the environment and biological systems (Muñoz and Costa, 2012). Nickel is a ferromagnetic element, and it naturally occurs in the Earth's crust, usually in the form of oxides and sulphides. In combination with other elements, nickel may be present in meteorites and soils and emitted from volcanoes. Approximately eight billion tonnes of nickel are present in the sea. Nickel is utilised in modern metallurgy in a wide range of metallurgical processes, including alloy manufacturing, electroplating, the manufacture of nickel–cadmium batteries and as a catalyst in the chemical and food industries due to its special physical and chemical qualities. The wide variety of products containing this metal inevitably causes environmental pollution from nickel and its byproducts during all phases of manufacture, recycling and disposal. Although there is currently no evidence indicating the nutritional value of Ni in humans, various types of microorganisms, plants and animal species have been known to use this metal as an essential nutrient (Song *et al.*, 2017). Ni-containing enzymes or cofactors are still unknown in higher organisms, although Ni-based enzymes are found in archaea, bacteria, algae, primitive eukaryotes and plants (Ragsdale, 2009; Boer *et al.*, 2014; Maroney and Ciurli,

2014; Desguin *et al.*, 2018; Genchi *et al.*, 2020). Nickel plays critical roles in morphological and physiological processes, including seed germination and productivity, and is crucial for the healthy growth and development of plants. Nickel, however, affects the metabolic processes of plants by reducing enzyme activity, photosynthetic electron transport and chlorophyll biosynthesis at high concentrations (Sreekanth *et al.*, 2013). Currently known nickel enzymes include methyl-coenzyme M reductase, urease, CO-dehydrogenase, glyoxalase, Ni-superoxide dismutase, acireductone dioxygenase, prolyl cis-trans isomerase, lactate racemase and [NiFe] hydrogenase (Macomber and Hausinger, 2011; Can *et al.*, 2014). There are also other nickel-dependent enzymes, such as quercitinase from *Streptomyces* sp. FLA and glycerol-1-phosphate dehydrogenase from *Bacillus subtilis* (Guldan *et al.*, 2008; Merkens *et al.*, 2008). The usage and/or generation of gases (CO_2, CO, H_2, CH_4, NH_3 and O_2) that are significant to the global nitrogen, carbon and oxygen cycles are implicated by nickel enzymes (Maroney and Ciurli, 2014; Ragsdale, 2007). Histidine and cysteine residues frequently work together to coordinate the catalytic core of nickel-dependent enzymes, with aspartate and glutamate also contributing (Ragsdale, 2009).

Ni is an immunotoxic and carcinogenic substance that can have a range of negative health effects, including contact dermatitis, lung fibrosis, asthma, cardiovascular disease and respiratory tract cancer, depending on the amount and length of exposure (Genchi *et al.*, 2020). The most common way that nickel causes toxicity in the immune system, lungs and respiratory tract in occupational settings is by inhalation exposure. With a high prevalence of allergic contact dermatitis, non-occupationally exposed individuals who touch stainless-steel- and nickel-plated objects may also be affected through inhalation exposure (Sinicropi *et al.*, 2010a, 2010b). However, since nickel may be a contaminant in food and/or drinking water (Genchi *et al.*, 2020) (Table 1), the major route by which humans are exposed is oral consumption through water and food. Oxidative stress and mitochondrial dysfunction play a significant role in nickel-induced neurotoxicity, despite the fact that the molecular mechanisms are not yet fully understood. The disruption of mitochondrial membrane potential, the decrease in mitochondrial ATP levels and the destruction of mitochondrial DNA can all result in nickel-induced mitochondrial damage. The use of antioxidant compounds, such as L-carnitine, taurine and melatonin, which increase and enhance antioxidant enzyme activity, can protect against nickel-induced neurotoxicity and

Table 1. Nickel-containing foods and items and nickel toxic effects. Adapted from Genchi *et al.* (2020).

Nickel-Containing Foods	Nickel-Containing Items	Nickel Toxic Effects
Cocoa and dark chocolate; hazelnuts; fruits (dates, almonds, pineapple, figs, raspberries, plums); seafood (mussels, shrimps, crab, oysters, salmon); grains (buckwheat, bran, whole grain bread, millet, brown rice, oats, sunflower seeds, sesame seeds); vegetables (savoy cabbage, beans, lettuce, leeks, lentils, spinach, peas, cabbage); soya and soya products; liquorice; peanuts; baking powder.	Inexpensive jewellery; keys; cosmetics; eyeglass frames; cell phones; orthodontic braces; paper clips; nickel plated articles; stainless steel articles; electrical equipment; metallurgical and food processing industries; clothing fasteners (belt buckles, zippers, snap buttons); pigment; alloy; catalysts.	Headaches; gastrointestinal manifestations; nasal cancer; cardiovascular diseases; respiratory manifestations; lung fibrosis; lung cancer; contact dermatitis; epigenetic effects.

carcinogenicity, according to Song *et al.* (2017). Electron respiratory chain interference and reactive oxygen species (ROS) production can both be caused by mitochondrial malfunction. These three antioxidants, which are generated by all mammals, are important for neurotransmission, detoxification and maintaining the balance of mitochondrial energy (Xu *et al.*, 2015). They also lessen oxidative stress and ROS generation.

2. Occurrence and Sources of Nickel

Group 10 of the periodic table is home to nickel (Ni; atomic number: 28, atomic weight: 58.6934, density: 8,908 kg m^{-3}, melting and boiling points: 1,455 and 2,913°C, respectively, and electronic configuration: $[Ar]3d^8 4s^2$), together with palladium, cobalt, platinum, iron and five other elements. The Earth's composition contains 3% nickel (Cempel and Nikel, 2006). Nickel is the fifth most prevalent element in terms of weight after iron, oxygen, magnesium and silicon, and it is the 24th most plentiful element in the Earth's crust. In nature, nickel is primarily found in nickelous form (Ni^{+2}), whereas in soil solution, it is found as $[Ni(H_2O)_6]^{2+}$. While there are mines for Ni minerals all over the world, 20% of the supply comes from Canadian soils. Globally, a variety of products are made with Ni minerals.

It is not easily corroded by air, water or alkalis, although it is easily soluble at pH < 6.5 in dilute oxidising acids. While nickel salts of mild inorganic acids, nickel sulphides and nickel oxides are poorly water-soluble, nickel salts of strong acids (chlorides, nitrates and sulphates) and organic salts are readily water-soluble (Denkhaus and Salnikow, 2002). Nickel nitrate, chloride, sulphate, acetate, carbonate, hydroxide and oxide are substances that are extremely important to industry.

2.1. *Natural sources of Nickel*

2.1.1. *In atmosphere*

Wind-blown dust, which is produced by the weathering of rocks and soils, forest fires and volcanic activity are some of the natural sources of atmospheric nickel. Nickel is also released into the atmosphere as a result of the burning of coal, diesel and fuel oil as well as waste and sewage (Von Burg, 1997).

2.1.2. *In surface and ground water*

Nickel can be found naturally in a variety of aquatic environments. Rocks and soils rich in nickel are subjected to weathering, dissolution and air evaporation to produce the naturally occurring nickel. In aquatic systems, it can also be found on clay particles as soluble salts, organic matter (detritus, algae and bacteria) or humic acid and fulvic acid salts. Because basic bedrock minerals dissolve in rain, surface water may become contaminated with them (Rauch and Pacyna, 2009). After an acid rain, the soil becomes more permeable, increasing the concentration of nickel in the groundwater (Begum *et al.*, 2022). Depending on variables, such as soil type, sample depth and pH, active nickel amounts can be found in both surface water and underground water. Nickel concentrations appear to be more closely related to the element's abundant natural occurrence in soil, minerals, rock and soil dust (Costa, 1995).

2.1.3. *In soil and rocks*

As a result of atmospheric emissions, nickel is primarily released into the soil. The parent rock has a significant impact on the nickel content of soils, although soil formation processes and pollution also have an impact on the amount of nickel present in surface soils. Nickel concentrations in soil can range widely, depending on geology and human activity, although they commonly fall between 3 and 1,000 ppm. Numerous inorganic minerals, such as those that are complexed on inorganic cation exchange surfaces or absorbed on organic cation surfaces, can contribute to the nickel contamination of soils (Gardner *et al.*, 2012). Humic and fulvic acids have been shown to be much more mobile in soil solutions than hydrated divalent cations (Kubicka *et al.*, 2015). These minerals include garnierite, pentlandite, nickel-iron sulphide and nickel-magnesium silicate. Clays, limestones, sandstones and shales have the lowest concentrations, whereas igneous rocks have the highest concentrations (Kachenko *et al.*, 2007).

2.1.4. *In foods*

Nickel is frequently ingested by people due to its tremendous abundance in nature. Due to the great abundance of nickel in food, it is generally difficult for a nickel deficiency to develop in the human body. Alternatively, it is difficult to maintain a diet deficient in nickel. There are many Ni-containing foods, as listed in Table 1.

The amount of nickel ingested through food and drink depends on the type of food consumed, the region and the individual's age and gender. The average daily intake for different categories of people, according to data from studies done in the US, is listed in Table 2. Additionally, cooking tools (such as oven pans and roasting pans) significantly increased the amount of nickel in cooked food. According to Calogiuri *et al.* (2016), stainless-steel pans increase the amount of nickel in foods that are acidic.

2.2. *Nickel from anthropogenic sources*

From anthropogenic sources, nickel is released in a variety of forms, including sulphides, oxides, silicates, soluble compounds and, to a lesser extent, metallic nickel. The ambient air, drinking water, food and cigarette products all contain nickel. A typical cigarette releases 0.04–0.58 μg of nickel. So, if you smoke 40 cigarettes a day, you might breathe in

Table 2. The average daily intake of Ni for different categories of persons in the US (Begum *et al.*, 2022).

Different categories	**Average daily intake of Ni per day**
Adults	101–162 μg
Males	136–140 μg
Females	107–109 μg
Children aged 0–6 months	9 μg
Children aged 7–12 months	39 μg
Children aged 1–3 years	82 μg
Children aged 4–8 years	99 μg
Pregnant women	121 μg
Lactating women	162 μg

2–23 μg of nickel (Poonkothai and Vijayavathi, 2012). The lubricants used in the engines of inefficient vehicles cause nickel to be released during transport. Other uses for nickel include plumbing supplies, cryogenic containers, equipment for cleaning up pollution and oil refining. When pipes and other materials deteriorate, there is a chance that nickel could seep into drinking water, although these leaks are often minor (Begum *et al.*, 2022). Nickel alloys, nickel-plated objects such as steel, coins and jewellery as well as residual nickel that can be found in soaps, oils and fats, are additional sources of nickel exposure.

2.2.1. *Nickel in air from anthropogenic sources*

Even though nickel naturally occurs, concentrations in the environment may also be a result of human activity, such as depositions from the burning of fossil fuels (Poonkothai and Vijayavathi, 2012), energy-producing power plants (petroleum combustion, coal-burning power plants, high-tension lines and nuclear power stations) (Poonkothai and Vijayavathi, 2012), chemical industries (cement manufacturing, pigment manufacturing, plating/metal finishing, etc.) and metallurgical industries (steel manufacturing). As a result of human activity, nickel is primarily emitted into the atmosphere as aerosols (Prueitt *et al.*, 2020; Guo *et al.*, 2019). The emissions produced by people compared to natural emissions is roughly 1.4–1.8 times greater (Begum *et al.*, 2022). Fossil fuel burning was responsible for 62% of all atmospheric nickel emissions during the 1980s. In 1999, 570,000 tonnes of nickel were emitted into the atmosphere due to the burning of fossil fuels. Of this, 326 tonnes were discharged by electric utilities, while 12% went to municipal incineration, 17% went to the refining of nickel metal, 3% went to the fabrication of steel and 2% went to making alloys containing nickel (Cempel and Nikel, 2006).

2.2.2. *Nickel in water from anthropogenic sources*

In addition to the emissions of nickel from industrial processes (such as mining, processing and smelting), home wastewaters, industrial wastewaters and effluents (such as tailing pile runoff) and landfill leachate can also contain nickel. Seawater contains roughly 0.5–2 ppb of nickel, while rivers have about 0.3 ppb of nickel species, according to studies (Begum *et al.*, 2022). The permitted limit for nickel in drinking water is 20 $\mu g\ L^{-1}$ according to the Bureau of Indian Standards (Begum *et al.*, 2022). Up to 980 $\mu g\ L^{-1}$ of dissolved nickel has been detected in groundwater with a $pH < 6.2$. Nickel levels in urban storm runoff water samples have been reported to range from 1 to 87 $\mu g\ L^{-1}$ (Begum *et al.*, 2022; Calogiuri *et al.*, 2016; Poonkothai and Vijayavathi, 2012; Oze *et al.*, 2008).

2.2.3. *Nickel in soil from anthropogenic sources*

Anthropogenic sources that contribute to soil nickel levels include mining, smelting and the waste products of metal manufacturing, commercial waste, fallout and sludge, as well as coal fly ash and coal bottom ash. According to estimates, US manufacturing and processing facilities discharged 82% and 87% of the total amount of nickel that was expected to be released into the environment in 2002, or roughly 5,530 and 14,800 metric tonnes of nickel and its compounds, respectively. Although the nickel issue seems to be exclusive to metropolitan areas, agricultural soils may also experience it as a result of diminished soil liming and acid rain owing to industrialisation (Begum *et al.*, 2022; Cempel and Nikel, 2006).

3. Biological Essentiality of Nickel

Ni is a crucial micronutrient that urease needs in order to hydrolyse urea. On calcareous soil, Ni treatment led to an enhancement in plant development and nitrogen (N) absorption by wheat (*Triticum aestivum* L.) (Singh *et al.*, 1990). In hydroponic investigations, a number of researchers have also demonstrated that Ni has positive impacts on urease activity and enhances plant N consumption efficiency. However, studies on the infrequently observed Ni shortage in soil have shown that it affects plant development, senescence, N metabolism and Fe absorption and may contribute to disease resistance.

Ni is necessary for plants, yet the bulk of plant species have relatively low Ni concentrations (0.05–10 mg kg^{-1} dry weight). Further, as Ni pollution rises, excess Ni, as opposed to a deficit, is more frequently observed in plants. High Ni concentrations in plants have been documented to cause toxic consequences repeatedly. For instance, pigeon pea (*Cajanus cajan* L.) mitotic activity is inhibited, cabbage (*Brassica oleraceae*) plant development is decreased and wheat (*Triticum aestivum* L.) fruit output and quality are negatively impacted. Some farmland is no longer viable for growing crops, fruits or vegetables due to extremely high Ni concentrations in the soil (Duarte *et al.*, 2007).

Since nickel is plentiful in nature, its deficiency is uncommon in people. Additionally, nickel is found in practically all food kinds, which contributes to maintaining a diet rich in nickel. In higher species, nickel deficiency is linked to greater perinatal mortality, altered grooming habits and slower development. When the body lacks nickel, it may also have trouble absorbing iron. Iron, zinc and copper concentrations are similarly shown to be lower in the livers of animals whose diets are deficient in nickel. Abnormal oxidative metabolism, cellular morphology and abnormal lipid levels (increased or decreased) are caused by nickel deficiency (Kumar and Trivedi, 2016). Furthermore, it has been discovered that decreased growth, reduced glucose metabolism and decreased haemoglobin concentrations are true (Kumar and Trivedi, 2016). The specific activity of several enzymes involved in the metabolism of carbohydrates and amino acids is also decreased by a nickel shortage. Similar to nickel exposure, a mildly iron-deficient diet affects serum and hepatic lipids (Min *et al.*, 2015; Kubicka *et al.*, 2015; Guo *et al.*, 2019). Additionally, nickel may be involved in the CO_2-fixation of propinyl CoA to D-methylmalonyl CoA.

Nickel may potentially have a role in the metabolism of lipids, particularly in the formation of phospholipids, as shown by its depletion in the diet. The presence of nickel in three distinct forms in human and rabbit serum has been established: nickel bound to albumin, ultrafilerable ligands and macroglobulins. Albumin serves as the primary protein that transports nickel in human, rat and bovine serums. A metalloprotein called nickeloplasmin has been isolated from the serums of rabbits (α-2 macroglobulin) and humans (α-glycoprotein) (Kachenko *et al.*, 2007).

4. Nickel Exposure

Both the general public and nickel industry personnel are concerned about nickel exposure. It enters the human body through a variety of routes, including food and water consumption, cutaneous contact and inhalation. The specifics of its numerous exposure routes and effects are covered further in the following section.

4.1. *Exposure of the general population*

Nickel may enter the body through ingestion, inhalation or skin absorption. Non-smokers often consume nickel through their food and water. Other exposure mechanisms include percutaneous absorption and ambient air inhalation (Cempel and Nikel, 2006). After being consumed orally, nickel is primarily eliminated through the faeces. Additionally, nickel-related iatrogenic exposures, such as parenteral nickel exposure from nickel alloy jewellery, coins or utensils, as well as cutaneous nickel absorption from nickel-plated items, may be linked to nickel-containing endoprostheses, radiocontrast media, haemodialysis fluids and/or albumin (Ilic *et al.*, 2007).

4.2. *Occupational exposure*

Nickel has long been used economically in a variety of alloy and compound forms (Mondal *et al.*, 2019). Industries that use nickel, such as those that are involved in the manufacturing of alloys and stainless steel, electroplating and electrowinning, welding, grinding and cutting, as well as nickel-producing businesses, such as mines, mills, smelters and refineries, are all sites where people could be exposed to its toxicity through breathing, dermal contact or unintentional ingestion. Because the Mond method is used to purify nickel, it was first believed that the toxic intermediate product, $Ni(CO)_4$, was the only known carcinogen. Yet, the incidence of respiratory malignancies increased in refineries that did not utilise the Mond process.

According to scientific data derived from investigations on occupationally exposed clusters, breathing in nickel and nickel compounds causes a wide range of illnesses, including sinusitis, rhinitis and anosmia. The nasal septum is completely damaged by sustained inhalation exposure. Breathing difficulties, headaches, coughs, nausea, vomiting and diarrhoea are some of the side effects of increased exposure to and consumption of nickel compounds. When it is severe, it becomes lethal and kills both humans and animals. Results were variable and occasionally inconclusive owing to analytic uncertainty regarding precisely which form of nickel was the causal agent. This uncertainty is attributed to the fact that people are co-exposed to several forms of nickel during processing. Researchers found it challenging to distinguish between the effects of metallic nickel and those of people exposed to it in the 1930s and 1970s (Wegman and Fine, 1990). Additionally, the melting of the nickel ores can produce nickel compounds and particles with a range of charges. During the various smelting and refining processes, workers may be exposed to Ni_3S_2, $NiSO_4$, $NiCl_2$, NiO, $NiCO_3$, Ni^0, NiFe oxides and NiCu oxides, and it can be difficult to identify bioactive species. Yet, it was pretty evident that nickel activities with 'low' and 'high' exposure varied in accordance with the risks of developing lung cancer. Recently, 'fingerprints' of workplaces have been created using sophisticated sampling techniques that provide knowledge about Ni species and size categories (Muñoz and Costa, 2012). Currently, further understanding of workplace exposure limits is still required. The idea that Ni may, when combined with other poisons, augment their carcinogenicities was another perplexing aspect. Nickel toxicity (agent code: 50420) was found to be extremely harmful to 50,7681 employees, including 19,673 women, according to the National Occupation Exposure Study (NOES), which was performed in 1981–1983 (Gohre and Paszkowski, 2006). The top six industries that accounted for almost 60% of the exposed workers are as follows: fabrication of metal products (n = 69,984), transportation equipment (n = 44,838) primary metal industries (n = 39,467) and auto repair and garages (n = 27,686) (Begum *et al.*, 2022).

5. Inter-cellular Transportation and Uptake of Nickel by Cells

Determining whether a toxin will have mutagenesis or epigenetic effects on a cell is dependent on how well it integrates into the cell. The method by which such harmful substances are absorbed through cell membranes, distributed inside cells and bound to macromolecules therein determines their bioavailability. We must discover how and why some Ni forms are harmful while others are not in order to comprehend how and why some Ni forms are absorbed by the body while others are not. In the past, the only two primary categories of Ni compounds — soluble and insoluble nickels — were considered to be poisonous, but in modern times, four distinct types of Ni compounds — soluble nickel, oxidic nickel, sulphidic nickel and metallic nickel — are mentioned under observations (Goodman *et al.*, 2009).

There have been several suggestions for categorising nickel compound uptake pathways according to their size, shape and surface properties. To show the uptake and intracellular fates of crystalline nickel sulphide and nickel subsulphide compounds, Max Costa and colleagues employed tissue culture cell systems and a number of different fibroblast cell lines in their research (Costa, 1995; Costa and Mollenhauer, 1980). The presence of the internal acidic pH was discovered to have caused the culture cells to phagocytise the crystalline nickel particles, which ultimately facilitated nickel solubilisation and in fact increased the amounts of soluble nickel inside the cell (Sigel *et al.*, 2013). After that, the DNA and nuclear proteins inside the cell are truly subjected to the carcinogenic consequences of this Ni filling. Amorphous nickel sulphide and water-soluble nickel salts do not absorb as well as crystalline nickel particles, which leads to greater cytosolic and lower nuclear nickel concentrations (Costa, 1995; Costa and Mollenhauer, 1980; Sigel *et al.*, 2013). The impact of surface charge on phagocytosis is seen in Fig. 1. Crystalline nickel sulphide particles tend to have a negative charge on their surfaces, whereas amorphous nickels have a positive charge. Therefore, it appeared that the way in which these various nickel compounds were delivered to cells affected how carcinogenic they were. Following phagocytosis, the particles are taken up in extremely acidic vacuoles, allowing soluble nickel to be released from the particles and raising the quantity of nickel in the cells. Studies show that the nickel ions or complexes that are produced when the soluble nickel particles dissolve are quickly eliminated by ciliary motion. On the other hand, macrophages or epithelial cells have the ability to phagocytose the insoluble particles. These particles steadily dissolve as they move through the cell (Ni_3S_2, for example, dissolves by oxidation), supplying a constant flow of Ni(II) ions (Bal *et al.*, 2000).

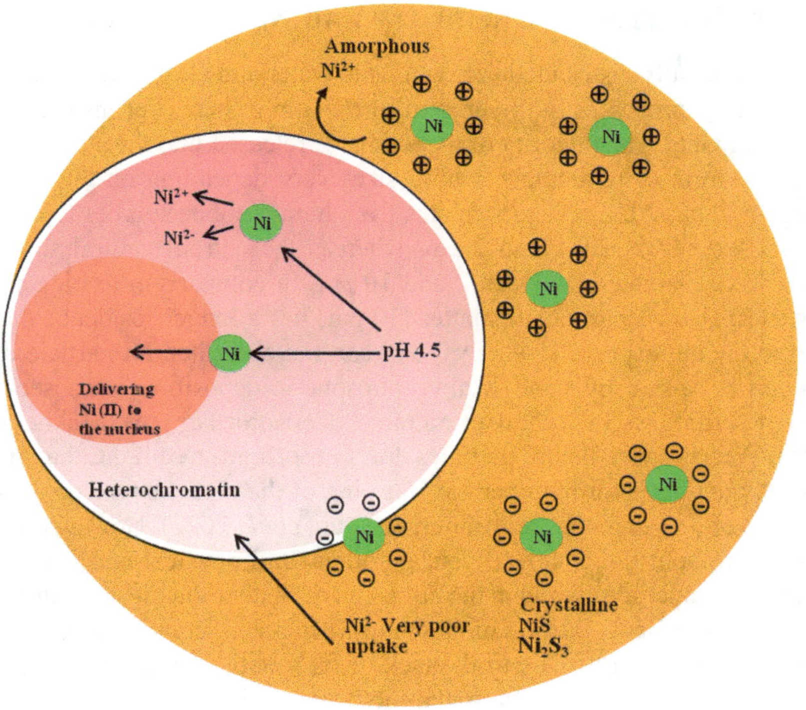

Figure 1. Model of nickel uptake in a living cell. Adapted from Begum *et al.* (2022).

After ingestion, nickel penetrates through the intestinal epithelium, although it is uncertain exactly how it reaches the interstitial regions close to capillaries. According to research, a particular transport pathway might control how nickel absorbs from the lumen and moves to the interstitial space since the junctions of the intestinal epithelium are impermeable and closely packed.

According to several studies, the trans-membrane diffusion process is responsible for nickel absorption and secretion by the rat jejunum (Foulkes and McMullen, 1986). Ca^{2+} channels can also be used to transfer nickel ions. Investigations into the surface binding and uptake of Ni^{2+} in human kidney epithelial cells revealed the importance of Ca^{2+} channels for nickel transport since the calcium ionophore potentiated nickel uptake (Foulkes and McMullen, 1986).

6. Environmental Chemistry and Speciation of Nickel

Ni is continually distributed between land, water and air through a variety of geogenic processes. Ni is present in the atmosphere, but in relatively small amounts. The majority of nickel that is released into the atmosphere is in the form of particulate, whose sizes vary depending on the source (Begum *et al.*, 2022). Ni particulates from human-made sources typically have sizes between 0.1 and 2 μm, whereas those from natural sources often have diameters between 2 and 10 μm. In comparison to bigger Ni particles, fine Ni particulate matter (10 μm) has a longer residence duration in the atmosphere (5–8 days) and can travel farther. Ni particle dispersion is caused by wind, while atmospheric deposition is caused by gravity settling (> 5 μm) or dry and moist depositions (5 μm) (Mukherjee, 1998). Yet, because the Ni particles don't absorb infrared light, they may not be a major factor in either the warming of the planet or the depletion of the ozone layer in the stratosphere (Begum *et al.*, 2022). However, little is known about the chemical configurations and physical and chemical changes it undergoes while in the atmosphere. Rather than its particular or general components, information about the metal content is gathered via analytical procedures. Due to the lack of information on the specifics of different compounds, certain assumptions are made concerning this variation. Generally, it is believed that Ni from anthropogenic sources, notably those derived through burning, is present as nickel oxide, while the wind-blown dust carries the nickel sulphide mineral species. Since nickel oxidises in the presence of sulphur dioxide, nickel sulphate is likely formed in the atmosphere (Begum *et al.*, 2022). The majority of the nickel discharged into the aquatic environment is found as soluble salts combined with organic matter from biological sources, suspended solids or particle matter (Mukherjee, 1998). The redox potential, pH, ionic strength of the water, concentrations of metal and ligand and presence of solid surfaces for adsorption (in particular, hydrous iron and manganese oxides) all affect how Ni behaves and its bioavailability in aquatic systems. These phenomena include adsorption, precipitation, co-precipitation and complexation. For instance, nickel content remains low in anaerobic environments (underground water) due to nickel sulphide precipitation. In reducing circumstances and in the presence of sulphur, some sediments create somewhat insoluble nickel sulphide. In aerobic circumstances and pH < 9, the compounds produced by nickel with naturally occurring organic ligands and hydroxide, sulphate and carbonate are remarkably soluble to keep aqueous Ni^{2+} concentrations above 60 μg L^{-1} (Begum *et al.*, 2022). When the pH of natural waters is 5–9, the divalent ion, which is present as the octahedral hexahydrate ion $[Ni(H_2O)_6]^{2+}$, predominates. Other nickel compounds are dependent on the pH and their inorganic or organic binding partners for their presence, existence and stability (Denkhaus and Salnikow, 2002). The solubility of nickel sulphate hexahydrate and nickel chloride hexahydrate in water is quite high, ranging from 2,400 to 2,500 g L^{-1}. Nickel nitrate and nickel hydroxide, with solubilities of 45 and 0.13 g L^{-1}, respectively, are two nickel compounds with lower aqueous solubilities. Nickel carbonate, with an aqueous solubility of 0.09 g L^{-1}, is the most inferior nickel compound. While nickel forms strong and soluble complexes with SO_4^{2-}, OH^- and HCO_3^-, hydrated Ni^{2+} predominates in groundwater and surface water relative to these species (Begum *et al.*, 2022). Amorphous oxides of Fe and Mn, as well as other soil minerals, significantly adsorb the majority of the Ni that is released into the environment, causing it to concentrate in the soil. In the majority of soils, nickel is specifically adsorbed or co-precipitated with aluminium and iron oxyhydroxides at ion exchange sites. The bulk density, pH, texture, organic matter, type and quantity of clay minerals and specific hydroxides are all soil-specific factors that affect the extent of Ni adsorption. Groundwater flow also has an impact on how much metal is retained and released by the soil. Ni is therefore absorbed by the soil in a site-specific manner. It is more mobile in acidic soil, while its availability and mobility are constrained by irreversible adsorption in alkaline soil (Begum *et al.*, 2022; Harasim and Filipek, 2015). The most prevalent Ni forms have been found to be Ni^{2+}, $NiSO_4$ and $NiHPO_4$ in acidic soil solutions and Ni^{2+} and $Ni(OH)^+$ in alkaline media (Begum *et al.*, 2022). Ni gradually changes in loessal soil from exchangeable and carbonate fractions to reducible oxides and in sandy soil from organic to readily reducible oxide fractions (Rajaie *et al.*, 2008). Xue *et al.* (2001) provided amazing information regarding nickel speciation in lake, river and groundwater samples. For DMG and DPCSV, they used the ligand exchange method. In their study, they came to the conclusion that the dissolved content of nickel in their water samples ranged from 4 nM (in lake water) to 30 nM (in river water), impacted by sewage and agricultural effluents. They found the ligand exchange processes to follow pseudo-first-order kinetics. The study also showed that organic ligands with high binding constant values bound 99.9% of the dissolved nickel. Similar results in several water samples from Mandal and colleagues' study were also validated by another investigation (Mandal *et al.*, 2002). In a recent investigation, Yebra-Biurrun *et al.* attempted to determine the concentration of labile Ni (the real hazardous form of nickel) in samples of river and saltwater (Yebra-Biurrun *et al.*, 2011). The coupled and continuous sonodigestion and solid-phase extraction of Ni using chelating resin constituted its straightforward and efficient instrumentation. In several water samples, they were able to identify both the labile (free) nickel and the organic-bound nickel complexes. The researchers also said that their findings concurred with those of Xue *et al.*, 2001 and Mandal *et al.*, 2002. These studies on nickel chemistry and real-time identification of nickel species in the environment showed that diverse water bodies had dissolved species that are organically bonded to nickel as well as 0.01% free elemental nickel.

7. Nickel Toxicity in Plants

The heavy metal nickel is a necessary microelement for humans, animals and plants; however, it becomes poisonous at levels that are higher than the optimum intake levels. As with other heavy metals, excessive Ni levels in plants lead to chlorosis and necrosis because they interfere with Fe uptake and metabolism (Figure 2) (De kock, 1956; Crooke, 1958). In non-tolerant plants (Robertson and Meakin, 1980), elevated Ni concentrations can prevent cell division at the root meristem and reduce plant development. However, the presence of Ni over the permitted limit alters the activities of numerous cellular parts, which may ultimately lead to cellular damage and plant death (Gohre and Paszkowski, 2006).

7.1. *Effect on plant growth*

The process of plant growth is crucial to sustaining life on earth. The mineral resources in the soil, the air, and the genotype of the plant species are just a few of the many variables that affect how plants grow both internally and externally. The excess Ni in an ecosystem has a negative impact on plant growth and development (Bhalerao *et al.*, 2015).

Ni also inhibits, alters or improves the structure or activity of a broad range of target molecules. According to Ochiai, transition metals, including Ni, are hazardous because of three phenomena (Begum *et al.*, 2022), which are associated with their ability to: remove vital components from

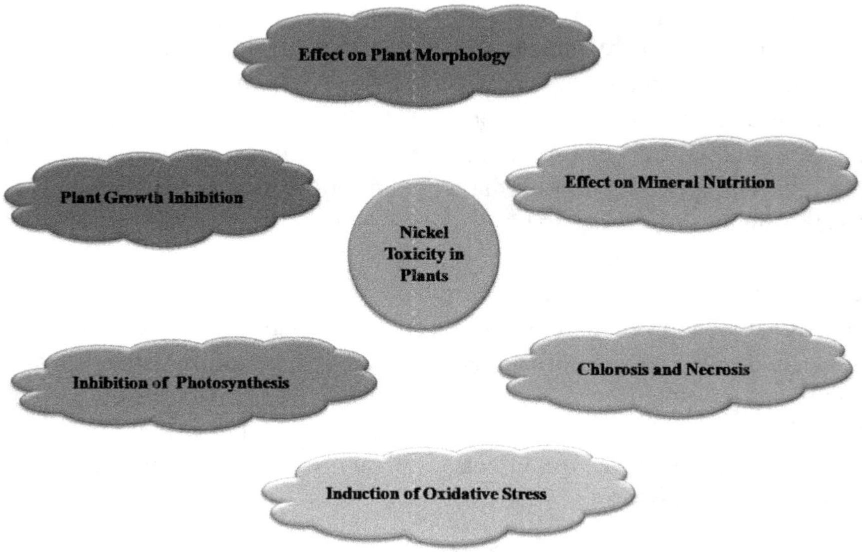

Figure 2. Nickel toxicity in plants shown schematically.

biomolecules; disrupt the action of a vital biological group in the molecules; and modify the structure or operation of enzymes, proteins, plasma membranes and membrane transporters (Begum *et al.*, 2022). Although Ni has oxidative effects, it also limits the growth of roots and shoots and reduces leaf area by interfering with cell division (Robertson and Meakin, 1980).

Inhibition of plant development (Seregin and Ivanov, 2001), a commonly used metric to measure environmental pollution (Seregin and Ivanov, 2001), is the primary manifestation of the harmful effects of nickel and other heavy metals. Higher metal concentrations exacerbate growth inhibition. Root growth is more severely hindered than shoot growth in excluder species, which primarily deposit Ni in their roots (Seregin and Ivanov, 2001; Samantaray *et al.*, 1997). For this reason, the root test is frequently used to determine the toxicity of various agents, including heavy metals (Seregin and Ivanov, 2001; Wang, 1987). A comparison of the root/shoot lengths of the heavy metal-stressed plant and the control plant (Samantaray *et al.*, 1997) was used to calculate the tolerance index, and LC50 (the metal concentration at which root growth is 50% inhibited) was used to measure plant tolerance to heavy metals (Wang, 1987; Bhalerao *et al.*, 2015).

Ni^{2+} inhibits plant growth and development; however, the processes by which it does so are unclear. In addition to causing a general metabolic disorder, heavy metals are known to reduce the plasticity of cell walls, likely through direct binding to pectins and by encouraging the activity of the peroxidases that are necessary for lignification and the linkage of extensin with polysaccharides that contain ferulic acid in the intercellular space (Bhalerao *et al.*, 2015).

To sum up, nickel and other heavy metals impede plant growth by causing a general metabolic disturbance and an immediate reduction in cell division. However, it is unclear if high quantities of Ni enter cell nuclei and, if so, how significant is the initial Ni interaction with DNA and nuclear proteins is. It is also uncertain how Ni^{2+} may influence the formation of phragmoplasts (Bhalerao *et al.*, 2015). We will have a better understanding of nickel's harmful effects on plant growth and morphogenesis if these uncertainties are clarified.

7.2. *Effect on plant morphology*

In addition to impeding plant growth, it also alters the anatomy and morphology of plants by reducing the size of the vascular bundle, the thickness of mesophyll, the plasticity of a cell wall, the content of chlorophyll and many other effects, which impair seed germination and other metabolic processes (Seregin and Kozhevnikova, 2006).

As a result, *Triticum aestivum* (Seregin and Kozhevnikova, 2006) leaves exposed to 1 mM $NiSO_4$ solution had decreased mesophyll thickness, vascular bundle size, vessel diameter in the main and lateral vascular bundles and epidermal cell width, whereas *Brassica oleracea* leaves exposed to $NiSO_4{\cdot}7H_2O$ (10–20 g m^{-3}) had decreased intercellular space volumes and palisade and sponge mesophyll in comparison to the control plants (Molas, 1997). Heavy metals are known to cause general metabolic disorders in addition to reducing the plasticity of cell walls, likely through direct binding to pectins and by enhancing peroxidase activity in the cell walls and intercellular space. For the lignification and linking of extensin to polysaccharides containing ferulic acid, these peroxidases are necessary (Bhalerao *et al.*, 2015).

7.3. *Inhibition of photosynthesis*

The process of photosynthesis in plants is also affected, which is another clear result. Indirectly and directly, nickel interferes with photosynthesis in plants. Its primary effect is to change or displace necessary metal ions, such as Mg, which interfere with the activity of chlorophyll. Disrupted chloroplast structure, hindered chlorophyll synthesis, disordered electron transport, blocked Calvin cycle enzyme activity and a CO_2 shortage caused by stomatal closure are all factors responsible for the reduced rate of photosynthesis (Seregin and Ivanov, 2001).

Brassica oleracea plants grown in agar in the presence of $NiSO_4.7H_2O$ (10–20 g m^{-3}) were found to have chloroplasts of decreased size and number as well as a disorganised chloroplast ultrastructure, including fewer grana and thylakoids, their deformation, the formation of plastoglobuli and changes in the composition of membrane lipids (Bhalerao *et al.*, 2015). Such modifications appeared to be brought on by oxidative stress that led to the peroxidation of membrane lipids or by a fall in cell moisture content driven by Ni (Molas, 1997). By altering the structure of carriers, such as plastoquinone QB, or the reaction centre proteins, it was shown that Ni inhibits electron transfer from pheophytin via plastoquinone QA and Fe to plastoquinone QB (Mohanty *et al.*, 1989; Krupa and Baszynski, 1995). Ni ions also reduced the amounts of ferredoxin, plastocyanin and the cytochromes *b6f* and *B559* in the thylakoids, which led to a decrease in the efficiency of electron transport (Mohanty *et al.*, 1989; Bhalerao *et al.*, 2015). Photosynthesis would be directly inhibited, which would then be amplified by the harmful effects of heavy metals on numerous other metabolic processes. All of these metabolic alterations prevent the growth of plants and interfere with morphogenesis; the resulting occurrences are frequently used to evaluate the phytotoxicity of heavy metals.

7.4. *Effect on mineral nutrition*

Mineral nutrient concentrations in plant organs may rise, fall or remain constant in the presence of nickel. Due to the comparable ionic radii of Ni^{2+} and other cations, one of the likely causes for decreased macro- and micronutrient absorption is the competition for the same binding sites. When the absorption of Mg^{2+} (78 pm), Fe^{2+} (82 pm) and Zn^{2+} (83 pm) is reduced in the presence of Ni^{2+} (78 pm) (ionic radii in parenthesis are from Bhalerao *et al.*, 2015), such a mechanism may be at work (Barsukova and Gamzikova, 1999). It is crucial to emphasise that one of the factors contributing to chlorosis, which is caused by an excess of Ni in the environment, is the decreased intake of Mg and Fe

(Piccini and Malavolta, 1992; Khalid and Tinsley, 1980). Ni-induced metabolic problems that disrupt the structure and enzyme activity of cell membranes (Seregin and Ivanov, 2001) may potentially cause the reduction in nutrition absorption. As a result, Ni^{2+} affects the plasma membrane's sterol and phospholipid composition in *Oryza sativa* shoots, which also caused changes in ATPase activity (Ros *et al.*, 1990). Evidently, these modifications altered the permeability of the membrane, which in turn altered the ion balance in the cytoplasm. The Ni content in the environment influences nutrient uptake in a variety of ways. The ryegrass plant tests showed that Fe content decreased at higher Ni concentrations while increasing at low Ni values in the shoots. The levels of Cu and Mg in the caryopses and those of Mg and Ca in the shoots of *Triticum aestivum* decreased when soil Ni content increased from 50 to 200 mg kg^{-1} soil (Bhalerao *et al.*, 2015).

7.5. *Induction of oxidative stress*

There is mounting evidence that oxidative stress and Ni toxicity in plants are related (MadhavaRao and Sresty, 2000). Excessive Ni causes hydroxyl radicals, superoxide anions, nitric oxide and hydrogen peroxide concentrations to rise noticeably (MadhavaRao and Sresty, 2000; Boominathan and Doran, 2002; Stohs *et al.*, 2001), although Ni cannot produce these ROS directly since it is not a redox-active metal. Moreover, it negatively affects a variety of antioxidant enzymes (Pandey and Sharma, 2002) indirectly, including SOD, CAT, glutathione peroxidase (GSH-Px), glutathione reductase (GR), peroxidase (POD), guaiacol peroxidase (GOPX) and ascorbate peroxidase (APX). It has been demonstrated that exposing plants to Ni at low concentrations (0.05 mM) and/or for short periods of time increases the activities of SOD, POD, GR and GOPX, which in turn enhances the activation of other antioxidant defences and, ultimately, results in the removal (or scavenging) of ROS (Freeman, 2004; Gomes-Juniora, 2006). However, it has been discovered that excess Ni decreases the activity of numerous cellular antioxidant enzymes, both *in vitro* and *in vivo*, as well as a plant's ability to scavenge ROS, which causes ROS to build up and ultimately leads to oxidative stress in plants (Zhou *et al.*, 2009).

8. Nickel Toxicity in Humans and Animals

Any compound's toxicity is determined by its physicochemical characteristics, concentration and exposure route. Studies on occupational exposure to people and animals show that the respiratory tract after oral ingestion, skin and inhalation exposure and the reproductive system in humans after oral exposure are the main targets of the harmful effects. Due to their limited intestinal absorption, almost all nickel compounds, with the exception of nickel carbonyl, are harmless after intake (Begum *et al.*, 2022). Parallel food consumption has a considerable impact on the bioavailability of nickel salt. There is no indication of dose-dependent elimination, and it seems to be excreted quickly through the urine and to follow first-order kinetics. Nickel has teratogenic and embryotoxic effects, as well as the ability to cross the placenta. According to a hen study, exposure to nickel reduced the amounts of magnesium, manganese and zinc in several tissues (Yebra-Biurrun *et al.*, 2011).

Studies have shown that nickel and zinc have adverse impacts on both people and animals. Nickel can cause sensitisation reactions in humans in addition to causing cancer (Fig. 3) (Prueitt *et al.*, 2020; Guo *et al.*, 2019).

8.1. *Nickel carbonyl toxicity*

When workers inhale $Ni(CO)_4$, they frequently experience non-specific, acute symptoms such as nausea, vertigo, headaches, dyspnoea and chest pain (Begum *et al.*, 2022). Both immediate and delayed symptoms

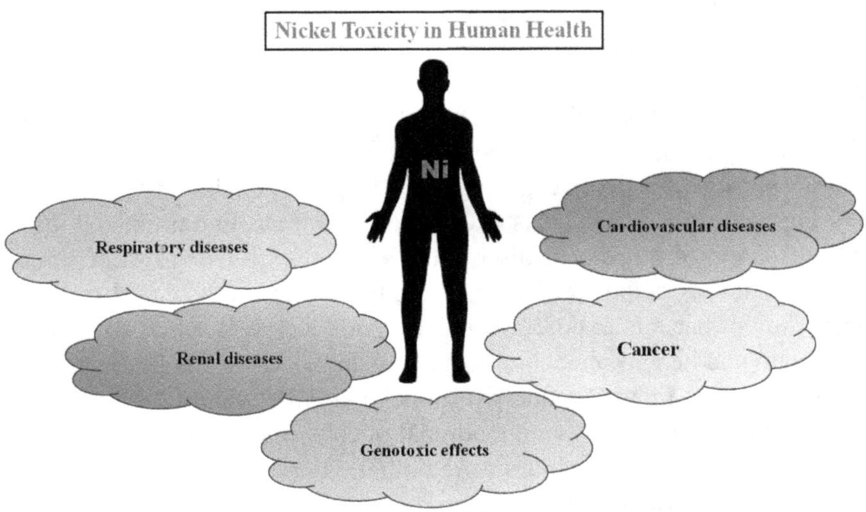

Figure 3. Human health effects of nickel poisoning shown schematically.

manifest in a person exposed to nickel carbonyl (Begum *et al.*, 2022). Initial symptoms usually go away after a few hours; however, in cases of continuous and high exposure lasting between 12 and 36 h, and in some cases up to five days, serious pulmonary symptoms such as coughing, dyspnoea, tachycardia, cyanosis and severe fatigue can develop. After an asymptomatic period, the delayed pulmonary symptoms begin to manifest. Pneumonitis, which is particularly severe, might result in pulmonary haemorrhage, toxic myocarditis, pulmonary oedema and neurasthenic syndrome. Additionally, deaths have reportedly occurred 4–13 days after exposure to high $Ni(CO)_4$ levels. The most frequent causes of death have been cerebral haemorrhage and diffuse interstitial pneumonia (Sunderman, 1997). The EU has categorised $Ni(CO)_4$ as a reproductive toxin (Risk Phrase R61—'may harm an unborn child') (Begum *et al.*, 2022). It has been demonstrated that animals exposed to $Ni(CO)_4$ give birth to offspring with abnormalities (Begum *et al.*, 2022). The acute, subchronic and chronic toxicities of nickel exposure in rats and humans are listed in Table 3 (Begum *et al.*, 2022).

8.2. *Toxicity due to other forms of nickel*

The more soluble forms of nickel compounds exhibit greater toxicity than the less soluble forms of nickel compounds. According to reports, insoluble nickel oxide has the lowest known toxicity, while soluble nickel sulphate has the greatest (Calogiuri *et al.*, 2016). Nickel sulphate showed acute oral LD_{50} values of 39 mg kg^{-1} in female rats and 46 mg kg^{-1} in male rats.

According to research, the LD_{50} values for nickel oxide, less soluble nickel compounds, and sub-sulphide in rats were >3,930, >3,665 and >3,665 mg kg^{-1}, respectively. The oral LD_{50} for nickel acetate in rats is 350 mg kg^{-1}, and in mice, it is 410 mg kg^{-1}. Animals that ultimately died showed signs of diarrhoea, discomfort and lethargy two to three hours after exposure (Cempel and Nikel, 2006; Begum *et al.*, 2022). Also, it has been shown that rats exposed to nickel oxide at low dosages developed malignant and benign pheochromocytomas in the adrenal medulla as well as lung tumours in both genders (Cempel and Nikel, 2006; Begum *et al.*, 2022). The incidences of benign adrenal pheochromocytomas and benign and malignant adrenal tumours were both greater in male rats exposed to metallic nickel. Adrenal cortex tumours in female rats exposed to metallic nickel were more common. Both forms of adrenal tumours exhibited

Table 3. Extreme health consequences from nickel exposure in humans and rats include acute toxicity, sub chronic toxicity and chronic toxicity. Adapted from Begum *et al.* (2022).

Toxicity		**Acute toxicity**	**Sub-chronic toxicity**	**Chronic toxicity**
Exposure duration (day)		1	10–100	>100
Route of exposure	Exposure to nickel dust and nickel fumes, nickel-polluted water and direct inhalation or consumption of nickel compounds		Exposure to nickel fumes and soluble nickel compound	Occupational exposure to nickel dust
Extreme health effects	On rats	Frank haematuria, renal damage,	Kidney and liver failure, ataxia, hyperglycaemia, hypothermia, diarrhoea, lung fibrosis	Emphysema, loss of kidney weight with significant albuminuria
	On humans	Respiratory distress syndrome, cardiac arrest	Tubular dysfunction, visual dysfunction	Bronchitis, asthma, pulmonary and nasal cancers, disruption in oxidative phosphorylation

dose-related reactions. Although metallic nickel inhalation raised blood nickel levels, lung tumour growth was not considerably impacted. Ionic nickel, which is a geno-toxicant and carcinogen, is produced when metallic nickel dissolves in the body. In numerous experiments, nickel subsulphide was also found to induce local sarcomas in mice after numerous injections at various places (Cempel and Nikel, 2006; Begum *et al.*, 2022). Rats exposed to nickel subsulphide also developed liver, spleen and respiratory metastases (Begum *et al.*, 2022).

9. Specific Health Effects of Nickel Toxicity in Humans

9.1. *Effects on the respiratory system*

Humans' acute reactions to nickel inhalation have not been thoroughly researched. The respiratory systems in adults have been found to be fatally affected by 90 min of nickel exposure at a level of 382 mg m^{-3} (Begum *et al.*, 2022). Chronic rhinitis, sinusitis and perforations of the nasal septum are all documented among nickel industry workers. For persons who work with nickel dust or fumes, there have also been instances of lung fibrosis and decreased pulmonary residual capacity. In addition, workers who have been exposed to nickel dust over an extended period of time have shown radiological indications of nickel pneumoconiosis. The pulmonary symptoms, chest radiography and lung ventilation capabilities of 12 steelworkers who stored steel ingots for up to 16 years were investigated (Jones and Warner, 1972). Typical occupational fume concentrations ranged from 1.3 to 294.1 mg m^{-3}, with 98.8% iron oxide and a composition of 6:1:1 for the three oxides of iron, chromium and nickel (fumes from stainless steel and special steel), respectively. These data were used to compute nickel oxide concentrations ranging from 0.15 to 34 mg m^{-3} (Begum *et al.*, 2022). Two workers had significant lung function loss, according to the study, but five men had radiographic evidence of pneumoconiosis (Jones and Warner, 1972).

9.2. *Effects on the renal system*

The symptoms of nephrotoxicity in a worker who consumed toxic water containing nickel sulphate and chloride included oedema, hyperaemia and parenchymal degeneration. Urine albumin concentrations returned to normal on the fifth day after exposure (68, 40 and 27 mg g^{-1} creatinine).

The findings imply that the nephron toxication was minimal and transient. Nickel affects a number of maker enzymes and enzymes involved in the metabolism of ascorbate cholesterol (Das, 2009).

9.3. *Effects on the cardiovascular system*

According to a paper by Sunderman (1997), if a patient has angina pectoris and myocardial infarction, a high serum nickel level may result in coronary vasoconstriction. Although there is no understood mechanism or source for nickel release, some experts advise keeping intravenous nickel concentration to a daily maximum of 5 μg kg^{-1}. According to the study, nickel refinery employees who have worked in the process area for five years are more likely to get lung and nose cancers. While the study predicted 8.5 occurrences of cerebrovascular mortality, 16 cases of the disease were reported (Sunderman, 1997; Das, 2009).

9.4. *Effects on skin*

Nickel ions cannot pass through intact skin, in contrast to nickel absorption through ingesting and inhalation. Nickel metal dust, salts and alloys can all absorb into damaged skin and result in contact dermatitis (Denkhaus and Salnikow, 2002). The most common cause of contact dermatitis in the general population is nickel. People's extended cutaneous exposure is caused by jewellery, buttons and zippers. According to studies, ear piercings in particular are linked to nickel sensitivity, which affects about 10% of females (Kornik and Zug, 2008). *In vitro*, nickel stimulates the proliferation and cytokine secretion of T lymphocytes in nickel-sensitive people. Major histocompatibility complex-encoded molecules may not be necessary for this stimulation, as antibodies to HLA class I and II molecules have little effect on nickel-induced proliferation. Both non-allergic and Ni-sensitised people are stimulated by nickel's immune response. This response is also observed in humans *in vivo.* Contact dermatitis in children and teenagers has been linked to a new source: mobile phones (Wöhrl *et al.*, 2007; Begum *et al.*, 2022).

9.5. *Genotoxic effects*

Nickel sulphides and nickel oxides pose a serious health risk to nickel industry personnel engaged in the processes of crushing, roasting and smelting. Workers in the nickel-electrolysis department are often exposed to nickel chloride and nickel sulphate. Both categories of workers have alarmingly elevated levels of chromosomal abnormalities, primarily gaps, according to cytogenetic investigations. Surprisingly, a similar finding was also discovered among former nickel workers. The report determined that those retired workers had plasma nickel levels of 2 $\mu g\ L^{-1}$ (Wöhrl *et al.*, 2007). Specific chromosomal damage has been demonstrated to be caused by nickel compositions, which are weakly clastogenic, especially in the heterochromatic regions. Insoluble Ni_3S_2 and soluble $NiCl_2$ have a noticeable impact on the heterochromatic long arm of the X chromosome in the Chinese hamster, which is commonly decondensed and where aberrations are generated (Begum *et al.*, 2022).

9.6. *Carcinogenicity of nickel* (*nickel cancer*)

In 2008, nickel received the rather awful label 'Allergen of the Year'. The human respiratory system is also affected by nickel aerosols produced in nickel electrorefining plants, such as Ni(II) sulphate. Previously, only the water-insoluble nickel compounds or dust, such as Ni_3S_2, and NiO, were believed to be carcinogenic. However, epidemiological data from recent studies have countered this belief. The order of nickel particles' carcinogenic activity against rats following intramuscular injection is Ni_3S_2~ βNiS (nickel(II) sulphide) > NiO (nickel(II) oxide) >> Ni^0 >>> NiS (amorphous) (Prueitt *et al.* 2020; Ilic *et al.*, 2007; Kornik and Zug, 2008; Wöhrl *et al.*, 2007). Effluent byproducts of nickel refining, such as nickel sulphate, sulphides and oxides, are carcinogenic to people exposed to them. Of the 100 instances of sino-nasal cancer among nickel refinery employees, a case study revealed that 48% were squamous cell carcinomas, 39% were anaplastic and undifferentiated carcinomas, 6% were adenocarcinomas, 3% were transitional cell carcinomas, 15% were anaplastic small cell and oat cell carcinomas, 3% were large cell carcinomas, 1% were other malignant tumours and 6% were unspecified cancers. In fact, this study gives us compelling proof that occupational nickel inhalation causes squamous cell carcinomas (Ilic *et al.*, 2007). The primary cause of sporadic local tumours in humans, aside from occupational exposures, has been hypothesised, though not verified, to be nickel found in endoprostheses, bone-fixing plates, screws and other medical devices (Ilic *et al.*, 2007).

10. Conclusions

Due to its increasing anthropogenic impact on the environment, nickel has drawn significant attention as a cause of dangerous heavy metal pollution. Despite being a heavy metal, it is necessary for plants, animals and people to have healthy lives. On the other hand, necrosis of leaves, reduced development and inhibition of photosynthesis are all effects of nickel poisoning in plants, and in people, it results in dermatitis, cardiovascular diseases and nasal and lung disorders. The thorough analysis presented in this chapter has focused on the toxicity and carcinogenicity of nickel in plants and people. Being exposed to low levels of nickel is inevitable since nickel is so common in the environment. Although it is biologically necessary for plants, its relevance to higher animals, including humans, is still up for dispute. Nickel and its derivatives have become unavoidable causes of environmental contamination across all phases of their manufacturing, disposal and recycling due to the extraordinary growth in the consumption of nickel and nickel-containing goods in modern times. Ni is continually distributed between land, water and air through a variety of geogenic processes. According to scientific data derived from investigations of occupationally exposed groups, breathing in nickel and nickel compounds causes a variety of illnesses, including sinusitis, rhinitis and anosmia. Breathing difficulties, headaches, coughs, nausea, vomiting and diarrhoea are some of the side effects of increased exposure to and consumption of nickel compounds. When it is severe, it becomes lethal and kills both humans and animals.

References

Bal W., Kozłowski H., Kasprzak S.K. (2000). Molecular models in nickel carcinogenesis. *J. Inorg. Biochem.* 79:213–218.

Barsukova V.S., Gamzikova O.I. (1999). Effects of nickel surplus on the element content in wheat varieties contrasting in Ni resistance. *Agrokhimiya* 1:80–85.

Begum W., Rai S., Banerjee S., Bhattacharjee S., Mondal M.H., Bhattarai A., Saha B. (2022). A comprehensive review on the sources, essentiality and toxicological profile of nickel. *RSC Adv.* 12:9139–9153.

Bhalerao S.A., Sharma A.S., Poojari A.C. (2015). Toxicity of nickel in plants. *Int. J. Pure Appl. Biosci.* 3:345–355.

Boer J.L., Mulrooney S.B., Hausinger R.P. (2014). Nickel-dependent metalloenzymes. *Arch. Biochem. Biophys.* 544:142–152.

Boominathan R., Doran P.M. (2002). Ni-induced oxidative stress in roots of the Ni hyperaccumulator, *Alyssum bertolonii*. *New Phytol.* 156:205–215.

Calogiuri G.F., Bonamonte D., Foti C., Al-Sowaidi S. (2016). Nickel hypersensitivity: A general review on clinical aspects and potential co-morbidities. *J. Allergy Ther.* 7:243.

Can M., Armstrong F.A., Ragsdale S.W. (2014). Structure, function, and mechanism of the nickel metalloenzymes, CO dehydrogenase, and acetyl-CoA synthase. *Chem. Rev.* 114:4149–4174.

Cempel M., Nikel G. (2006). Nickel: A review of its sources and environmental toxicology *Pol. J. Environ. Stud.* 15:375–382.

Costa M (1995). Molecular mechanisms of nickel carcinogenesis. Toxicol Environ Chem 49:145–148.

Costa M., Mollenhauer H.H. (1980). Phagocytosis of nickel subsulfide particles during the early stages of neoplastic transformation in tissue culture. *Cancer Res.* 40:2688–2694.

Crooke W.M. (1958). Effect of heavy metal toxicity on the cation exchange capacity of plant roots. *Soil Sci.* 86:231–241.

Das K.K. (2009). A comprehensive review on nickel (II) and chromium VI toxicities–possible antioxidant (*Allium Sativum* Linn) defenses. *Al Ameen J. Med. Sci.* 2:43–50.

De kock P.C. (1956). Heavy metal toxicity and iron chlorosis. *Ann. Bot.* 20:133–141.

Denkhaus E., Salnikow K. (2002). Nickel essentiality, toxicity, and carcinogenicity. *Crit. Rev. Oncol. Hematol.* 42:35–56.

Desguin B., Fellner M., Riant O., Hu J., Hausinger R.P., Hols P., Soumillion P. (2018). Biosynthesis of the nickel-pincer nucleotide cofactor of lactate racemase requires a CTP-dependent cyclometallase. *J. Biol. Chem.* 293: 12303–12317.

Duarte B., Delgado M., Caador I. (2007). The role of citric acid in cadmium and nickel uptake and translocation, in halimioneportulacoides. *Chemosphere* 69:836–840.

Foulkes E.C., McMullen D.M. (1986). On the mechanism of nickel absorption in the rat jejunum. *Toxicology* 38:35–42.

Freeman J.L. (2004). Increased glutathione biosynthesis plays a role in nickel tolerance in Thlaspi nickel hyperaccumulators. *Plant Cell* 16: 2176–2191.

Gardner M., Comber S., Scrimshaw M., Cartmell E., Lester J., Ellor B. (2012). The significance of hazardous chemicals in wastewater treatment works effluents. *Sci. Total Environ.* 437:363–372.

Genchi G., Carocci A., Lauria G., Maria S.S., Catalano A. (2020). Nickel: Human health and environmental toxicology. *Int. J. Environ. Res. Public Health* 17:679.

Gohre V., Paszkowski U. (2006). Contribution of arbuscular mycorrhizal symbiosis to heavy metal phytoremediation. *Planta* 223:1115–1122.

Goodman J.E., Prueitt R.L., Dodge D.G., Thakali S. (2009). Carcinogenicity assessment of water-soluble nickel compounds. *Crit. Rev. Toxicol.* 39: 365–417.

Gomes-Juniora R.A. (2006). Nickel elicits a fast antioxidant response in *Coffea arabica* cells. *Plant Physiol. Biochem.* 44:420–429.

Guldan H., Sterner R., Babinger P. (2008). Identification and characterization of a bacterial glycerol-1-phosphate dehydrogenase: Ni(2+)-dependent AraM from *Bacillus subtilis*. *Biochemistry* 47:376–384.

Guo H., Liu H., Wu H., Cui H., Fang J., Zuo Z., Deng J., Li Y., Wang X., Zhao L. (2019). nickel carcinogenesis mechanism: DNA damage. *Int. J. Mol. Sci.* 20:4690.

Harasim P., Filipek T. (2015). Nickel in the environment. *J. Elem.* 20:525–534.

Ilic V., Bojanic V., Jovic B. (2007). Epidemiological and pathogenetic aspects of nickel poisoning. *Acta Medaca Medianae* 46:37–44.

Jones J.G., Warner C.G. (1972). Chronic exposure to iron oxide, chromium oxide, and nickel oxide fumes of metal dressers in a steel works. *Br. J. Ind. Med.* 29:168–177.

Kachenko A.G., Singh B., Bhatia N.P. (2007). Heavy metal tolerance in common fern species. *Aust. J. Bot.* 55:63–73.

Khalid B.Y., Tinsley J. (1980). Some effects of nickel toxicity on rye grass. *Plant Soil* 55:139–144.

Kornik R., Zug A.K. (2008). Nickel. *Dermatitis* 19:3–8.

Krupa Z., Baszynski T. (1995). Some aspects of heavy metals toxicity towards photosynthetic apparatus–direct and indirect effects on light and dark reactions. *Acta Physiol. Plant* 17:177–190.

Kubicka K., Samecka-Cymerman A., Kolon K., Kosiba P., Kempers A.J. (2015). Chromium and nickel in *Pteridium aquilinum* from environments with various levels of these metals. *Environ. Sci. Pollut. Res.* 22:527–534.

Kumar S., Trivedi A.V. (2016). A review on role of nickel in the biological system. *Int. J. Curr. Microbiol. Appl. Sci.* 5:719–727.

Macomber L., Hausinger R.P. (2011). Mechanisms of nickel toxicity in microorganisms. *Metallomics* 3:1153–1162.

MadhavaRao K.V., Sresty T.V. (2000). Antioxidative parameters in the seedlings of pigeon pea (*Cajanus cajan* (L.) Millspaugh) in response to Zn and Ni stresses. *Plant Sci.* 157:113–128.

Mandal R., Hassan N.M., Murimboh J., Chakrabarti C.L., Back M.H. (2002). Chemical speciation and toxicity of nickel species in natural waters from the Sudbury area (Canada). *Environ. Sci. Technol.* 36:1477–1484.

Maroney M.J., Ciurli S. (2014). Nonredox nickel enzymes. *Chem. Rev.* 114:4206–4228.

Merkens H., Kappl R., Jakob R.P., Schmid F.X., Fetzner S. (2008). Quercetinase QueD of Streptomyces sp. FLA, a monocupin dioxygenase with a preference for nickel and cobalt. *Biochemistry* 47:12185–12196.

Min E.Y., Cha Y.J., Kang J.C. (2015). Effects of waterborne nickel on the physiological and immunological parameters of the Pacific abalone Haliotis discus hannai during thermal stress. *Environ. Sci. Pollut. Res.* 22:13546–13555.

Mohanty N., Vaas I., Demeter S. (1989). Impairment of photosystem 2 activity at the level of secondary quinone electron acceptor in chloroplasts treated with cobalt, nickel and zinc ions. *Physiol. Plant* 76:386–390.

Molas J. (1997). Changes in morphological and anatomical structure of cabbage (*Brassica oleracea* L.) outer leaves and in ultrastructure of their chloroplasts caused by an *in vitro* excess of nickel. *Photosynthetica* 34:513–522.

Mondal M.H., Ali M.A., Pal A., Saha B. (2019). A review on micellar catalysed oxidation reactions of organic functional groups in aqueous medium using various transition metals. *Tenside Surfactants Deterg.* 56:516–525.

Mukherjee A.B. (1998). Nickel: A review of occurrence, uses, emissions, and concentration in the environment in Finland. *Environ. Rev.* 6:173–187.

Muñoz A., Costa M. (2012). Elucidating the mechanisms of nickel compound uptake: A review of particulate and nano-nickel endocytosis and toxicity. *Toxicol. Appl. Pharmacol.* 260:1–16.

Oze C., Skinner C., Schroth A.W., Coleman R.G. (2008). Growing up green on serpentine soils: Biogeochemistry of serpentine vegetation in the Central Coast Range of California. *Appl. Geochem.* 23:3391–3403.

Pandey N., Sharma C.P. (2002). Effect of heavy metals Co^{2+}, $Ni^{2+}$and Cd^{2+}on growth and metabolism of cabbage. *Plant Sci.* 163:753–758.

Piccini D.F., Malavolta E. (1992). Effect of nickel on two common bean cultivars. *J. Plant Nutr.* 15:2343–2350.

Poonkothai M., Vijayavathi B.S. (2012). Nickel as an essential element and a toxicant. *Int. J. Environ. Sci.* 1:285–288.

Prueitt R.L., Li W., Chang Y.C., Boffetta P., Goodman J.E. (2020). Systematic review of the potential respiratory carcinogenicity of metallic nickel in humans. *Crit. Rev. Toxicol.* 50:605–639.

Ragsdale S.W. (2007). Nickel and the carbon cycle. *J. Inorg. Biochem.* 101:1657–1666.

Ragsdale S.W. (2009). Nickel-based enzyme systems. *J. Biol. Chem.* 284:18571–18575.

Rajaie M., Karimian N., Yasrebi J. (2008). Nickel transformation in two calcareous soil textural classes as affected by applied nickel sulfate. *Geoderma* 144:344–351.

Rauch J.N., Pacyna J.M. (2009). Earth's global Ag, Al, Cr, Cu, Fe, Ni, Pb, and Zn cycles. *Global Biogeochem. Cycles* 23:1–16.

Robertson A.I., Meakin M.E.R. (1980). The effect of nickel on cell division and growth of *Brachystegias piciformis* seedlings. *J. Bot. Zimb.* 12:115–125.

Ros R., Cooke D.T., Burden, R.S., James C.S. (1990). Effects of the herbicide MCPA, and the heavy metals, cadmium and nickel on the lipid composition, Mg^{2+} -ATPase activity and fluidity of plasma membranes from rice, *Oryza sativa* (cv. Bahia) shoots. *J. Exp. Bot.* 41:457–462.

Samantaray S., Rout G.R., Das P. (1997). Tolerance of rice to nickel in nutrient solution. *Biol. Plant* 40:295–298.

Seregin I.V., Ivanov V.B. (2001). Physiological aspects of cadmium and lead toxic effects on higher plants. *Russ. J. Plant Physiol.* 48:523–544.

Seregin I.V., Kozhevnikova A.D. (2006). Physiological role of nickel and its toxic effects on higher plants. *Fiziol. Rast.* 53:285–308.

Singh B., Dang Y.P., Mehta S.C. (1990). Influence of nitrogen on the behavior of nickel in wheat. *Plant Soil* 127:213–218.

Sigel A., Sigel H., Sigel R.K.O. (2013). Interrelations between essential metal ions and human diseases. *Met. Ions Life Sci.* 13:321–357.

Sinicropi M.S., Amantea D., Caruso A., Saturnino C. (2010a). Chemical and biological properties of toxic metals and use of chelating agents for the pharmacological treatment of metal poisoning. *Arch. Toxicol.* 84:501–520.

Sinicropi M.S., Caruso A., Capasso A., Palladino C., Panno A., Saturnino C. (2010b). Heavy metals: Toxicity and carcinogenicity. *Pharmacologyonline* 2:329–333.

Song X., Kenston S.S.F., Kong L., Zhao J. (2017). Molecular mechanisms of nickel induced neurotoxicity and chemoprevention. *Toxicology* 392: 47–54.

Sreekanth T.V.M., Nagajyothi P.C., Lee K.D., Prasad T.N.V.K.V. (2013). Occurrence, physiological responses and toxicity of nickel in plants. *Int. J. Environ. Sci. Technol.* 10:1129–1140.

Stohs S.J., Bagchi D., Hassoun E., Bagchi M. (2001). Oxidative mechanisms in the toxicity of chromium and cadmium ions. *J. Environ. Pathol Toxicol. Oncol.* 20:77–88.

Sunderman F.W. (1997). A review of the metabolism and toxicology of nickel. *J. Ann. Clin. Lab. Sci.* 7:377–398.

Von Burg R. (1997). Toxicology update. *J. Appl. Toxicol.* 17:425–431.

Wang W. (1987). Root elongation method for toxicity testing of organic and inorganic pollutants. *Environ. Toxicol. Chem.* 6:409–414.

Wegman D.H., Fine L.J. (1990). Occupational Health in the 1990s. *Ann. Rev. Public Health* 11:89–103.

Wöhrl S., Jandl T., Stingl G., Kinaciyan T. (2007). Mobile phone as new source of nickel dermatitis. *Contact Dermatitis* 56:113.

Xu S., He M., Zhong M., Li L., Lu Y., Zhang Y., Zhang L., Yu Z., Zhou Z. (2015). The neuroprotective effects of taurine against nickel by reducing oxidative stress and maintaining mitochondrial function in cortical neurons. *Neurosci. Lett.* 590:52–57.

Xue H.B., Jansen S., Prasch A., Sigg L. (2001). Nickel speciation and complexation kinetics in freshwater by ligand exchange and DPCSV. *Environ. Sci. Technol.* 35:539–546.

Yebra-Biurrun M.C., Castro-Romero J.M. (2011). Speciation of dissolved trace nickel in environmental waters by on-line sonodigestion-flow injection solid phase extraction coupled to flame atomic absorption spectrometry. *Am. J. Anal. Chem.* 2:116–125.

Zhou X., Li Q., Arita A., Sun H., Costa M. (2009). Effects of nickel, chromate, and arsenite on histone 3 lysine methylation. *Toxicol. Appl. Pharmacol.* 236:78–84.

https://doi.org/10.1142/9789811283123_0011

Chapter 11

Health Risks and the Regulations Linked to the Li/Ni Toxicity in Food and the Environment

Jozef Sabol[*,‡] **and Pavel Otřísal**[†]

[*]*Faculty of Security Management,*
Police Academy of the Czech Republic in Prague,
Lhotecká 559/7, 14301 Prague, Czech Republic

[†]*Faculty of Physical Culture, Palacky University Olomouc,*
Tr. Míru 117, 771 11 Olomouc, Czech Republic

[‡]*sabol@polac.cz*

Abstract

The chapter summarises some essential features and findings related to the toxicities of lithium (Li) and nickel (Ni), which are always present in trace concentrations in food and the environment. These metal elements occur almost everywhere in nature, including in soils, rocks and the atmosphere. Therefore, due attention should be paid to their presence in various substances, including food, water and air. These materials can enter the human body through ingestion of food, drinking water and other routes, such as inhalation and skin penetration. The occurrence of Li/Ni in the environment has increased tremendously recently due to some human activities, especially specific technological processes, which have resulted in an increased presence of these elements in various products and goods. Recently, Li and Ni are gradually being used as main components in Li-ion and NiCd batteries. The first type of battery is mainly used in electric cars, while the less powerful NiCd batteries are primarily used in portable electronic devices and toys.

In addition, the chapter discusses the specific forms of toxicity of these metal elements and their occurrence in some foods. The processes and pathways by which these elements can enter the human organism are also described. They can have severely harmful effects at certain excessive concentrations and amounts. Because of these detrimental impacts, Li and Ni levels should be monitored to enable intervention when their presence exceeds regulatory limits or recommendations. Based on the results of such measurements, effective control mechanisms can then be implemented to minimise their effects on human health, including introducing appropriate measures to protect workers who come into contact with these elements in the workplace. This chapter also describes the actions necessary to regulate these elements' presence and impose adequate controlling mechanisms and preparations, including those to respond effectively to possible emergencies.

Keywords: Lithium, nickel, toxicity, health effects, occurrence in nature, applications, contamination, detection and identification, emergency response

1. Introduction

Chemical elements are the building blocks of life. They make up the amazing variety of molecules that combine to form DNA, cells, tissues and organs. In order to discuss the effects of *lithium* and *nickel* on human health, it is desirable to recapitulate their roles in the human body compared to other chemical elements. Some of them play an essential role in guaranteeing the proper functioning of all vital organs and tissues through the unique metabolic processes in which they are involved. The body is an extraordinarily complex mechanism and requires many parts to work together in intricate interactions, from the micro- to the macromolecular levels. The organisation of the building blocks that make up the sum of these parts, such as proteins and nucleic acids, is determined by the ratio and interaction of the chemical elements.

On average, *the human body consists of about 99% of only six elements*, namely, oxygen (O), hydrogen (H), nitrogen (N), carbon (C), calcium (Ca) and phosphorus (P). Another five elements, sulphur, potassium, sodium, chlorine and magnesium, comprise about 0.85% of the remaining mass. All of these 11 elements are essential for life; the remaining 0.15% of the human body consists of *trace elements*, some of which also play a significant role.

In general, the human body needs about 20 vital elements for its normal activity, 10 of which are *metallic elements*. The body cannot do without some of them. Until 1950, little attention was paid to the so-called 'inorganic elements', while earlier, high priority was given to the study of the 'organic elements' (C, N, O and H) and organic compounds, and the study of the essential inorganic elements was to a certain extent neglected. Today, we know that metals such as Na, K, Mg, Ca, Fe, Mn, Co, Cu, Zn and Mo are vital elements that must be available to our bodies in specific quantities.

On the other hand, there are a number of unique trace elements that are present in our bodies. Under normal circumstances, their mass fractions are several orders of magnitude less than those of the main components contained in living organs. The influence and role of these elements have not yet been thoroughly researched.

An overview of the elements essential to the human body expressed as a percentage of body weight is shown in Fig. 1. The following elements have the highest mass percentage in the human body: oxygen (65%),

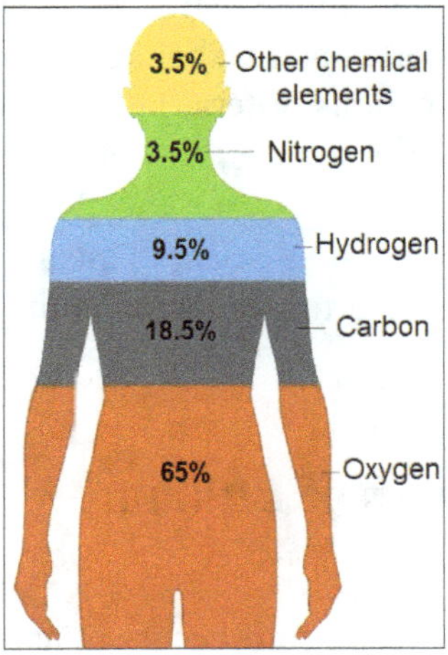

(a)

Element	Mass(%)
Selenium (Se)	19×10^{-3}
Iron (Fe)	6×10^{-3}
Zinc (Zn)	3.2×10^{-3}
Cobalt (Co)	100×10^{-6}
Copper (Cu)	100×10^{-6}
Barium (Ba)	31×10^{-6}
Manganese (Mn)	17×10^{-6}
Nickel (Ni)	14×10^{-6}
Molybdenum (Mo)	13×10^{-6}
Lithium (Li)	3.1×10^{-6}
Chromium (Cr)	2.4×10^{-6}

(b)

Figure 1. Some of the essential elements in the human body: (a) five of the most abundant elements, (b) heavy metal elements significant for physiological and biochemical functions in the human body. Based on (Underutilization of Lithium, 2018).

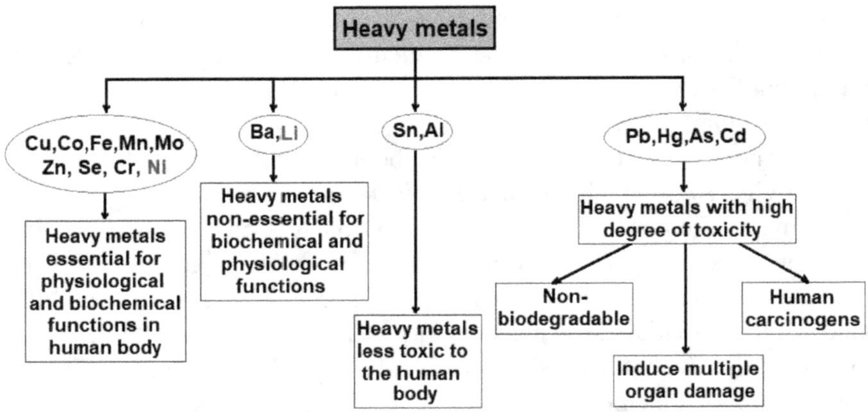

Figure 2. Illustration of the classification of heavy metal elements by their function in the human body.

carbon (18.5%), hydrogen (9.5%) and nitrogen (3.5%). The remaining elements make up less than 3% of the human mass. In addition to calcium (1.5%), phosphorus (1.0%), potassium (0.4%), sulphur (0.3%), sodium (0.2%), chlorine (0.2%) and magnesium (0.1%), several trace elements are also included, namely iron, zinc, copper, chromium, cobalt, vanadium, molybdenum, manganese and some others, as well as lithium and nickel.

Any substance that can cause a negative biological effect can be considered a toxicant. There are several different and elaborate classification systems for toxic substances based on various criteria depending on the desired use. These classifications include pesticides, solvents, food additives, metals and war gases. Critical pollutants that can damage both human health and the ecological balance are non-essential metals. A classification system for heavy metals according to their roles in the body is shown in Fig. 2 (Filipoiu *et al.*, 2022).

2. Some Specific Characteristics of Lithium and Nickel

Lithium is a silvery and highly reactive metal, so it occurs in nature only in compounds. It is the lightest of all metals and very soft but forms strong alloys with other metals. Lithium is the first of the alkalis in the periodic table (Fig. 3). Some of its characteristic properties can be derived from its

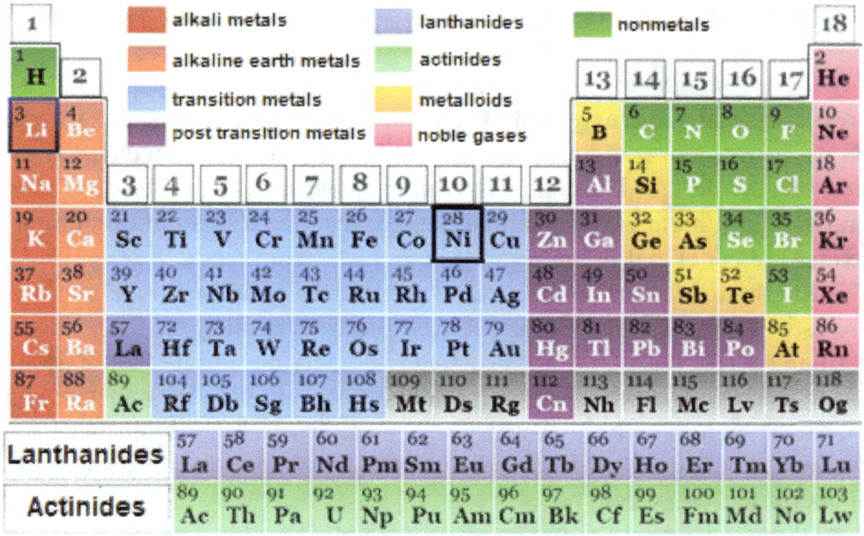

Figure 3. Positions of lithium and nickel in the periodic table of the elements. Based on Bashyal (2022).

position in the periodic table. Many of its physical and chemical properties are more similar to those of the alkaline earth metals than to those of its own group's elements. In nature, it is a mixture of the isotopes Li-6 and Li-7. In addition to being soft and the lightest solid metal, lithium has a low melting point and is reactive.

The most important properties of lithium include its high specific heat (heat capacity), wide temperature interval in the liquid state, high thermal conductivity, low viscosity and very low density. Metallic lithium is soluble in short-chain aliphatic amines, such as ethylamine. It is insoluble in hydrocarbons.

The most significant potential for the use of lithium (mainly in the form of compounds) is in the manufacture of *rechargeable batteries.* They can be found primarily in portable electronics (mobile phones, tablets, laptops, notebooks, cameras, medical devices, etc.). Recently, their importance has increased due to their use in batteries for electric cars and large-capacity batteries for energy storage, where consumption is growing the fastest and currently accounts for about one-third of total consumption. Only a slightly smaller share is accounted for by using lithium oxides as fluxes (lowering the firing temperature, which brings about great energy savings) in the ceramics and glass industries. Lithium metal is also used in alloys (mainly with Al and Mg).

Lithium metal alloys (mainly with Al and Mg) improve mechanical properties and reduce weight while maintaining the strength of many products. Lithium compounds are also utilised as lubricants in the production of polymers or aluminium and in the pharmaceutical industry. It is known that lithium citrate can treat bipolar disorders. In addition, lithium oxides are beneficial in the ceramics and glass industries.

Although *lithium markets* vary by location, global end-use markets are estimated as follows: batteries, 74%; ceramics and glass, 14%; lubricating greases, 3%; continuous casting, 2%; polymer production, 2%; air treatment, 1%; and other uses, 4%. Lithium consumption for batteries has increased significantly in recent years, as rechargeable lithium batteries are extensively used in the growing market for electric vehicles and portable electronic devices.

Global estimated lithium reserves were identified in such countries as Bolivia, 21 million tonnes; Argentina, 19 million tonnes; Chile, 9.8 million tonnes; Australia, 7.3 million tonnes; China, 5.1 million tonnes; Congo (Kinshasa), 3 million tonnes; Canada, 2.9 million tonnes; Germany, 2.7 million tonnes; Mexico, 1.7 million tonnes; the Czech Republic, 1.3 million tonnes; Serbia, 1.2 million tonnes; Russia, 1 million tonnes; Peru, 880,000 tonnes; Mali, 700,000 tonnes; Zimbabwe, 500,000 tonnes; Brazil, 470,000 tonnes; Spain, 300,000 tonnes; Portugal, 27,0000 tonnes; Ghana, 130,000 tonnes; Austria, 60,000 tonnes; and Finland, Kazakhstan and Namibia, 50,000 tonnes each (USGS, 2022).

The most recent information on mined lithium production in tonnes is shown in Table 1 (Wikipedia, 2023). It can be seen that production is increasing in most of these countries due to increased demand.

Table 1. Global lithium mine production by top countries (tonnes) in the past six years. Based on (Wikipedia, 2022).

Rank	Country	2018	2019	2020	2021	2022
1	Australia	51,000	45,000	39,700	55,000	61,000
2	Chile	16,000	19,300	21,500	26,000	39,000
3	China	8,000	10,800	13,300	14,000	19,000
4	Argentina	6,200	6,300	5,900	6,200	6,200

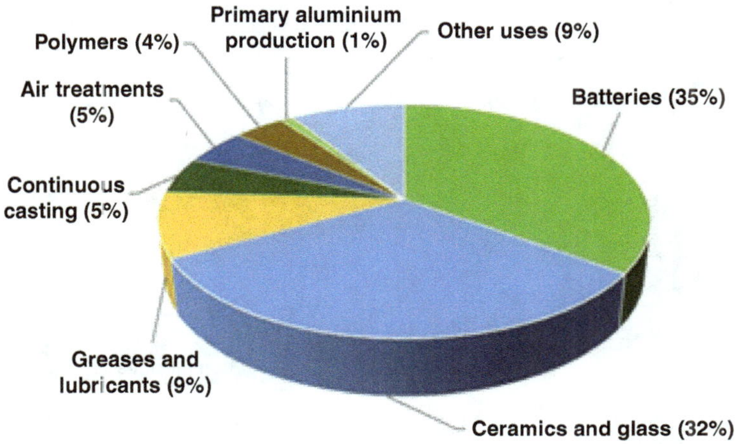

Figure 4. Estimated global use of lithium by individual applications. Based on Guiral (2018).

Figure 4 provides an overview of the wide range of applications of lithium as of the situation in 2016 (Gujral, 2018), where the share of batteries is now estimated to contribute up to 50% of the total use of lithium.

Lithium can always be found in the environment, where its level is increased in areas with mining activities. Its occurrence in drinking water is usually 0.2–2 μg L^{-1}. While exceeding concentrations of Li in water are found to be damaging to health, its low levels may also be damaging, as illustrated in Fig. 5, which shows that a shortage in intake of Li may increase the probability of suicide (Ando *et al.*, 2022).

Nickel is a silvery-white, hard, malleable and ductile metal. It belongs to the iron group and is highly lustrous. It is a relatively good conductor of heat and electricity. In the known compounds, nickel is divalent but also takes on other valences. It also forms a number of complex compounds. Most nickel compounds appear blue or green. Nickel dissolves slowly in dilute acids but, like iron, becomes passive when treated with nitric acid. Finely divided nickel adsorbs hydrogen. The chemical properties of nickel can also be deduced from its position in the periodic table (Fig. 3). It belongs to the transition metals and is hard and malleable. It occurs in combination with sulphur in the mineral millerite, arsenic in the mineral niccolite, and arsenic and sulphur in the mineral nickel glance. It is the 28th element in the periodic table. Nickel may exist in several

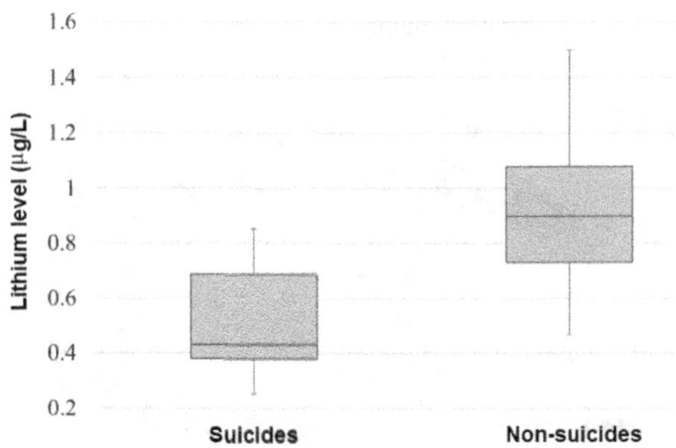

Figure 5. Comparison of lithium levels between suicide and non-suicide fatalities: a cross-national study. Based on Anke *et al.* (2003).

oxidative states (from −1 to +4); nevertheless, the +2 oxidation state (Ni^{2+}) is the most widespread in the environment and biological systems.

Because of its stability in air and its inertness against oxidation, nickel is used in smaller coins, for plating iron, brass, etc., for chemical instruments and in some alloys, such as nickel silver. It is magnetic and is found very often together with cobalt. Both elements occur in meteoric iron.

Most of the mined nickel is used for the production of thousands of different alloys (with iron, chromium, aluminium, zinc, etc.); up to 65% of the world's nickel production is then used for the production of stainless steel. As part of many enzymes and proteins, nickel is essential in small amounts for bacteria, plants and animals. However, it is toxic in excessive amounts. In the case of skin contact, it can cause a local allergic reaction. Ingestion or inhalation causes damage to the liver, kidneys, lungs and vascular and nervous systems. Nickel is considered a suspected human carcinogen.

Nickel usually occurs together with sulphur, arsenic and antimony in minerals, for example in gersdorffite (NiAsS), nickeline (NiAs) and millerite (NiS). The anthropogenic source of nickel is primarily wastewater from the surface treatment of metals, where it is mostly complexly bound, and also wastewater from non-ferrous metallurgy. It is also used in the ceramic and glass industries and some chemical syntheses as catalysts. Another source can be nickel-plated parts of equipment that come into contact with water, for example in distribution networks. Nickel is present in water in oxidation state II and mostly in inorganic forms. The solubility of nickel in water is mainly limited by either $NiCO_3$ or $Ni(OH)_2$. In the presence of sulphides, NiS also comes into consideration.

Nickel is most useful in alloy form, especially high-alloy alloys. The production of nickel is quite complex and depends on the ores used. However, nickel oxide (NiO) is usually obtained in the final stage, which is further reduced with coke to form elemental nickel. Most of the nickel is used to produce alloys, the most important of which are the metal monel (68% Ni + 32% Cu + traces of manganese and iron), which is extremely rust-resistant and even resistant to fluorine, and nichrome (60% Ni + 40% Cr), which is mainly used in electrical engineering. The physical and mechanical properties of nickel also make it ideal for the production of special refractory alloys and intermetallic alloys (Ni_3AL and NiAl). Pure nickel is mainly used in the food industry as a catalyst for solidifying fats, as well as in batteries and electroplating. Stainless steel contains about 8–12% nickel.

Lithium was known as early as 1817, when it was discovered by Johan August Arfvedson (1792–1841) in Stockholm while he was examining petalite, one of the first lithium minerals to be located. Nickel was discovered in 1751 by Alex Frederik Cronstedt, who was also working in Stockholm. First, he thought that the mineral he was investigating might contain copper; however, what he extracted was a new metal, which, in 1754, he named nickel. Both scientists were Swedish chemists and mineralogists. The basic chemical and physical properties of lithium and nickel are presented in Table 2 (Periodic Elements, 2022).

Table 2. Summary of main parameters characterising lithium and nickel. Based on (Periodic Elements, 2022).

	Element	
Parameter	**Lithium**	**Nickel**
Atomic number	3	28
Atomic mass	6.941 g mol^{-1}	58.71 g mol^{-1}
Density	0.53 g cm^{-3} at 20°C	8.9 g cm^{-3} at 20°C
Melting point	180.5°C	1453°C
Boiling point	1,342°C	20,913°C
Isotopes	2	10

3. Lithium and Nickel in Food and Water

3.1. *Lithium*

Lithium can be a dietary supplement and is commonly found in drinking water and especially in many foods, where the primary sources of Li in the diet are cereals, potatoes, tomatoes, vegetables, mustard, kelp, pistachios, dairy, fish and cabbage. Lithium can also be found in trace amounts in many other foodstuffs. Of course, the concentration of lithium in various foods may vary significantly depending on the geographic position of the country or area where the food is grown. The time of year may also play a certain role in this variation. For example, overall lithium intake in Europe is higher in the winter than in the summer (Szklarska and Rzymski, 2019).

According to Biological Trace Element Research (Corleone, 2019), some of the major sources of lithium of natural origin and the corresponding amount of the element present in units of μg per gram of dry weight are as follows:

- Nuts: 8.8,
- Cereals: 4.4,
- Fish: 3.1,
- Vegetables: 2.3,
- Mushrooms: 0.19,
- Dairy: 0.5,
- Meats: 0.012.

Some more information about the typical ranges of lithium concentration found in common foodstuffs is given in Table 3 (Greenblatt, 2023).

Although lithium is not considered an essential nutrient currently, some researchers (Chandra, 2021) have proposed that lithium should be regarded as a micronutrient since there is some evidence about the beneficial effects of trace lithium levels on the brain.

However, based on some other studies, nutritional studies (Anke *et al.*, 2003), lithium can be considered an essential trace element with a recommended dietary allowance of 1 mg d^{-1}. Lithium is often used to treat patients with depressive episodes with therapeutic doses. The natural lithium intake doses are much lower than those used for the treatment of patients with psychiatric disorders, but even very low lithium levels

Table 3. The typical ranges of lithium levels in some common foodstuffs. Based on Anke *et al.* (2003).

Foodstuff	Li level (mg kg^{-1})
Nuts	8.8
Cerial grains	0.11–4.4
Vegetables	0.8–4.6
Fruits	0.08–1.3
Meat	0.012–3.2
Diary	0.09–3.2

ingested by the routine consumption of lithium from water and food may have anti-suicidal effects. Food consumption can naturally increase lithium intake to levels that are unlikely to cause negative side effects. Nutritional lithium, consumed through drinking water, plant-based foods or low-dose supplementation, is a well-tolerated naturally occurring nutrient. The World Health Organization recognises lithium as a nutritionally essential trace element, and the role of trace concentrations of this element in the human body is compelling.

The lithium content of *foods and beverages* depends on many factors, with the main one being the occurrence of this element in soil and water, where, for example, water from swamps and bogs contains up to eight times more lithium than water from diluvial sands. There is a relatively substantial difference in how different plants react to the concentration of Li in the soil and how they absorb it. Cereals, and thus flour, contain relatively little lithium. Honey and sugar are also enormously low in lithium. Richer in lithium in this direction are ready-made soups with meat and eggs, while various puddings and pasta usually contain < 1 mg kg^{-1} of lithium. The data in Table 4 illustrate this fact. Bread, cake and pastries are generally poor sources of lithium. On average, they contain less lithium than wheat flour. Adding sugar has been shown to reduce the lithium content of bread, cakes and pastries.

A peeled potato has a lower lithium content than an unpeeled one because the skin of the potato stores more lithium than the inner part of the potato.

As a rule, fruits contain less lithium than the vegetative parts of plants (vegetables). Lemons and apples contain significantly more lithium, with

Table 4. The lithium content in several groups of foodstuffs. Based on Anke *et al.*, (2003).

Foodstuffs	Content (mg kg^{-1})	Dispersion
Beverages	0.192	0.018–0.329
Sugar, honey	0.363	0.199–0.527
Bread, cake	0.6121	0.317–0.935
Cereal products	0.888	0.538–1.391
Spices	1.850	1.046–2.316
Fruits	2.846	0.383–6.707
Meat	3.217	2.379–3844
Milk, dairy products	3.626	1.070–7.533

about 1.4 mg kg^{-1}, than peas and beans, which, like various types of cereal grains, are extremely poor in lithium. Due to the small amounts used in their application, spices do not contribute much to the lithium diet. Surprisingly, mustard is relatively rich in lithium, with 3.4 mg kg^{-1} dry weight, while mustard seed contains extremely little lithium. The lithium content in chocolate, chocolate candies and sweets is around 0.5 mg kg^{-1}. Cocoa is slightly richer in lithium. Adding sugar to chocolate reduces its lithium content. On average, eggs, meat, sausages and fish show considerably more lithium per kg of dry substance than most cereal foods. Eggs, liver and kidneys of cattle have a typical lithium content of 5 mg kg^{-1}. Beef and mutton contain more lithium than poultry. Green fodder and silage consumed by cattle and sheep are much richer in lithium than the grains that comprise the more significant part of poultry feed. Sausage and fish contain comparable amounts of lithium to meat. However, considerable lithium content in milk, which includes a relatively high lithium concentration of up to 10 mg kg^{-1}, was also noted. Similar to margarine made from vegetable fat, butter is also low in lithium (around 1.2 mg kg^{-1}).

The Earth's crust contains 50–65 mg Li kg^{-1}. The geological origin of the soil is reflected in its lithium content. The presence of Li in the human body is influenced primarily by the intake of this element. It is estimated that the daily oral intake of Li is in the range of 0.5–3 mg. A typical daily intake of Li is around 1 mg day^{-1}. The content of Li in some foodstuffs, which affects the amount of Li in the body, is illustrated in Fig. 5 (Anke *et al.*, 2003).

Other than drinking water, food can also be a source of minerals. However, growing practices combined with local soil levels influence the lithium content of food crops significantly. In addition, if soil levels of lithium get too high, it can have growth-inhibiting effects on crops and other plants.

Lithium levels in foods are somewhat variable in the limited published data available. Studies often found that the same foodstuff could have undetectable levels of up to a few milligrams per kilogram of lithium. The primary lithium sources in food are nuts, legumes, cereal grains and vegetables. Of the vegetables, leafy greens may have higher levels; however, the data are too sparse to draw solid conclusions. The data on many foods are derived from one or two sources, making this information difficult to generalise, especially considering the differences among published reports. In Spain, for example, significantly higher levels were found in meat, poultry and dairy products — different from other locations.

The main source of lithium in food is mainly some types of vegetables, but its content varies greatly depending on its content in the soil. High contents of lithium can be found in foods such as cucumbers, cabbages and tomatoes. Lithium is also found in smaller concentrations in fruits (for example, apples and bananas), cereals, pasta and all types of baked goods. In order to meet the daily requirements for lithium, reliable sources of the mineral are needed. Water is often a source of lithium; however, concentrations can vary dramatically from location to location.

The level of Li in water commonly ranges from 2 to 70 $\mu g\ L^{-1}$. It is believed that the incidence of suicide, rape and violence was significantly lower in counties with lithium levels in drinking water ranging from 70 to 160 $\mu g\ L^{-1}$ compared to those with 0–12 $\mu g\ L^{-1}$ (Anke *et al.*, 2003). This is why lithium is intentionally added to drinking water to decrease the suicide rate. Since not all the details behind this are clear, it is important to continue researching this area. Lithium dampens aggression and self-destructive tendencies. Deficiency causes depression, rage, manic-depressive psychoses, suicidal tendencies, obsessions and partner and child abuse. Conversely, excess lithium causes tremors, confusion, dizziness and psychomotor retardation.

3.2. *Nickel*

Nickel is almost absent from meat and dairy-based foods, with no quantified concentrations reported in beef rump steak, butter and cheese.

Plant-based foods were essentially all positive for quantified nickel concentrations, with nut-based foods, in particular, having steadily higher concentrations (peanuts and peanut butter: 0.94–10.2 mg kg^{-1}). The reported concentrations ranged from 0.0006 mg kg^{-1} in tap water to 10.2 mg kg^{-1} in whole peanuts (Anke *et al.*, 2003).

If you are exposed to nickel, many factors will determine whether you will be harmed. These factors include the dose (how much), the duration (how long) and how you come into contact with it. You must also consider any other chemicals you are exposed to and your age, gender, diet, family traits, lifestyle and state of health. Some foods known for an increased concentration of nickel are illustrated in Fig. 6. Individuals allergic to nickel can reduce the possibility of allergic dermatitis by avoiding foods high in nickel:

- Foods high in nickel (>0.5 mg kg^{-1}) include mussels, nuts, seeds, beans, peas, soy flour, oatmeal, wheat bran, buckwheat, millet, dark chocolate, cocoa powder and liquorice.
- Moderate amounts of nickel (0.1–0.5 mg kg^{-1}) are found in oysters, milk chocolate, rice, rye, barley, kale, parsnips, garlic and raspberries.

While kale, soy, peanut butter, lima beans, canned pork and beans, chocolate wheat germ, spirulina, oat ring cereal, canned refried beans and

Figure 6. Some foods with higher contents of nickel. Based on (What is Nickel, 2015; Schrauzer, 2002).

Table 5. Mean Li content in dry food stuffs in μg^{-1}. Based on Anke *et al.* (2003).

Li content in some vegetables	Average concentration (μg^{-1})
Cereals	4.4
Fish	3.1
Mushrooms	0.19
Vegetables	2.3
Meat	0.012
Dairy products	0.5
Nuts	8.8

sunflower seeds are characterised by high nickel content, there are other foods with a much lower content of nickel. To this category belong mushrooms, cucumbers, bell peppers, celery, broccoli, olive oil, salmon, chicken breast, Swiss cheese, eggs, white rice, steak and common fruits, such as apples, oranges and bananas. Limiting the consumption of drinks, vitamin supplements with nickel and canned foods is also recommended. Other foods that may trigger nickel allergy include margarine, acidic foods such as stewed fruits or rhubarb cooked in stainless-steel utensils, tea from drink dispensers, beer, red wine, herring, mackerel, tuna, tomatoes, onions, carrots, apples, citrus fruits, dates, figs, pineapples and prunes.

Another illustration of recommendations for a low-nickel diet is given in Table 6 (Sheman *et al.*, 2008). These recommendations apply mainly to nickel-sensitive individuals.

4. Some Specific Health Effects of Lithium and Nickel

In the vast majority of cases, an excess of minerals is not suitable for the human body. With the element *lithium*, an excess can manifest in impaired memory, hair loss, swelling, tremors, convulsions, nausea, vomiting, diarrhoea and, subsequently, changes in the kidneys and liver. Since it is relatively the most toxic element of all metals, any synthetic intake must be managed exclusively by a doctor.

Table 6. General guidelines for a low-nickel diet. Based on Genchi *et al.* (2020).

Diet	Recommendations	More information
Grains	Avoid	Whole wheat and multigrain bread, multigrain cereals, wheat bran, whole wheat pasta, oats
	May eat	Rice, corn, rye
	May eat in moderation	White flour, white bread, white pasta
Vegetables	Avoid	Beans, lentils, peas, soy products, kale, lettuce, spinach, vegetable juices, canned vegetables
	May eat	Other fresh or frozen vegetables
Fruits	Avoid	Dates, figs, pineapples, plums, raspberries, canned fruits
	May eat	Other fresh or frozen fruits
Dairy	Avoid	Chocolate milk
	May eat	Milk, yoghurt, cheese
Meat	Avoid	Shellfish, processed meats with filters or coatings, canned meats or fish
	May eat	Beef, chicken, fish, turkey, eggs
Other sources of dietary nickel to avoid		Chocolate or cocoa powder, all nuts, all seeds, black tea, commercial salad dressings, vitamins containing nickel

Certain lithium compounds, also known as lithium salts, are used as psychiatric medications, primarily for bipolar and major depressive disorders. Lithium is taken orally. Common side effects include increased urination, shakiness of the hands and increased thirst.

Lithium is used to control and prevent manic and depressive episodes, acts as a mood stabiliser in people with bipolar disorders and is also used for adjunctive therapy in patients with major depressive disorders and schizoaffective disorders. Lithium has been reported to be associated with a lower risk of suicide and all-cause mortality in people with mood disorders. In addition, lithium is suggested to have a neuroprotective effect and a role in reducing aggression and impulsivity (Greenblatt, 2023). Although suicide is considered the leading cause of preventable mortality worldwide, we do not have effective strategies to prevent it. Lithium has been consistently associated with reducing the risk of suicide. This effect could occur at very low concentrations, such as trace amounts of lithium in tap water. Several ecological studies and recent meta-analyses have suggested an inverse relationship between lithium in water and suicide in the general population, although knowledge about clinically significant side effects is lacking.

The extraction of lithium carbonate from natural minerals generally involves pressurised carbon dioxide dissolving lithium carbonate. Releasing the pressure then allows lithium carbonate to precipitate out of the solution. The best-known uses of lithium carbonate in health supplements are for managing moods, supporting metabolic syndromes, retaining minerals and managing occasional headaches (Fig. 7) (Health Benefits and Uses of Lithium Carbonate, 2023).

Lithium is not defined as an essential element, so it is not an essential, important or main element for the human organism. Its physiological effect as a mood stabiliser (fluctuations between mania and melancholy) in humans is still only used by the pharmaceutical industry to suppress the symptoms of bipolar disorders. However, people can also take lithium without a prescription in certain waters or herbs. Some groundwater has a naturally elevated lithium content because it does not go through treatment technologies to treat drinking water. However, the lithium content of these waters is only in the tens of micrograms per litre. The usual daily dose of lithium carbonate for symptoms of the disease is 600–1,200 mg,

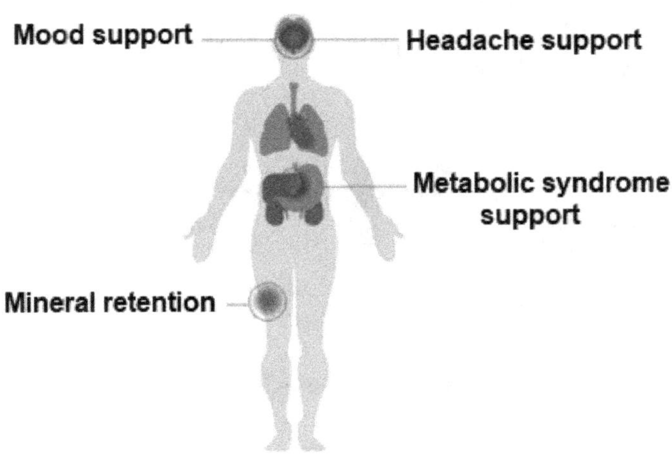

Figure 7. Health effects of lithium carbonate on the human body. Based on Thyssen *et al.* (2011).

i.e., significantly higher. However, such waters are not recommended by doctors for a person's regular drinking regime, and they only recommend their use for a certain period of time. Water containing lithium and other elements can be purchased in many general stores.

Interestingly, prescription lithium salts are recommended for some people for the rest of their lives. It is an open secret that mental illness can never be proven to be properly diagnosed or even cured. In the case of mental illnesses, no precise procedures or laboratory tests would accurately determine these illnesses. It is based only on the listed symptoms of the patient and the manifested symptoms of the disease during a visit to the doctor. However, the consequences of the adverse effects of lithium ingestion are already irreversible and destructive for some internal organs. An illustration of a balance between the causes of the negative perception of lithium and the countermeasures for its challenging and optimising long-term lithium administration is summarised in Fig. 8.

Although lithium is not an essential nutrient for humans, low Li intakes are associated with increased suicide and homicide rates, aggressive behaviours, unipolar/bipolar disorders, acute mania, etc. On the other hand, Li is one of the most effective psychopharmacological agents used for the treatment of these psycho-behavioural disorders. The beneficial normothymic effect of Li could be achieved at lower doses. Therefore, modern psychiatry has been called upon to consider Li biofortification of foods to improve dietary intake. The concept of agronomic biofortification of crops with Li is juvenile, and there exist a limited number of studies, mainly focused on vegetables or mushrooms.

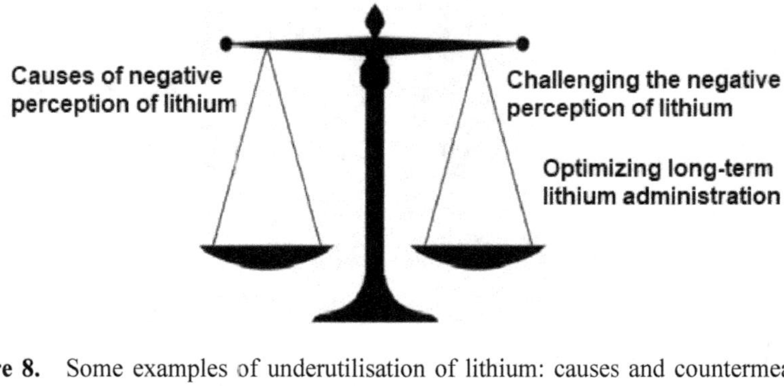

Figure 8. Some examples of underutilisation of lithium: causes and countermeasures. Based on (Underutilization of Lithium, 2018).

Like fluoride, lithium is another element with beneficial pharmacological properties; it has been used effectively in treating manic-depressive states and psychosis. Lithium deficiency (<1.5 mg kg^{-1}) is reported to result in reduced fertility, birth weight, lifespan and altered activity in several liver and blood enzymes in goats. In rats, lithium deficiency (5–15 mg kg^{-1}) results in reduced fertility, birth weight, litter size and weaning weight. Additionally, these lithium-deficient rats had reduced lithium content in the testes, seminal vesicles and epididymis, but had relatively high lithium concentrations in the pituitary and adrenal glands even after two generations on a low-lithium diet. These findings suggest that lithium may have a role in the regulation of some endocrine functions. Other biochemical *in vitro* actions providing indirect support for lithium's essentiality include growth stimulation of some cultured cells and insulinomimetic action. Its ability to affect mental function may explain the report that the incidence of violent crime is lower in areas with drinking water containing relatively high amounts of lithium.

Even though two research groups using two animal species reported similar signs of lithium deficiency (reduced growth and fertility), the symptoms could have been more pronounced and specific; this has hindered widespread recognition that lithium is possibly necessary. Moreover, the large difference between the apparently insufficient intake (based on rat experiments) and the typical daily intake prevents the assumption that lithium could be of practical nutritional importance. Thus, although more credibility should probably be given to the essentiality of lithium, it most likely will not occur until a specific biochemical function is identified for this element.

Inhalation exposure in occupational contexts is the main route for nickel-induced toxicity in the respiratory tract, lungs and immune system. This route may also affect non-occupationally exposed individuals, mainly those who handle stainless-steel- and nickel-plated articles, with a high prevalence of allergic contact dermatitis. However, the exposure of human beings mainly concerns oral ingestion through water and food, as nickel may be a contaminant in drinking water and/or food (Table 7) (Genchi *et al.*, 2020).

Although the molecular mechanisms of nickel-induced neurotoxicity are not yet clear, an important role is played by oxidative stress and mitochondrial dysfunction. Nickel-induced mitochondrial damage can occur due to impairment of mitochondrial membrane potential,

Table 7. Nickel-containing items and nickel related toxic effects. Based on Genchi *et al.* (2020).

Nickel in food, some items and effects	Food containing nickel and its toxic effects
Nickel-containing food	Hazelnuts, cocoa and dark chocolate, fruits; grains, seafood, vegetables, tea from drinks dispensers, soya and soya products, peanut, licorice, baking powder
Nickel-containing items	Inexpensive jewellery, cosmetics, keys, cell phones, eyeglass frames, paper clips, orthodontic braces, stainless steel articles, nickel plated articles, clothing fasteners, electrical equipment, armaments, alloy, metallurgical and food processing industries, pigments, catalysts
Nickel's toxic effects	Contact dermatitis, headaches, gastrointestinal manifestations, respiratory manifestations, lung fibrosis, cardiovascular diseases, lung cancer, nasal cancer, epigenetic effects.

reduction of mitochondrial ATP concentration and destruction of mitochondrial DNA.

More women are sensitive to nickel than men. This difference between men and women is thought to be a result of greater exposure of women to nickel through jewellery and other metal items containing nickel. Two cases of the effects of nickel allergy are shown in Fig. 9 (Nickel Health Effect Asian Metals, 2022). Once a person is sensitised to nickel, further contact with the metal may cause a reaction. The most common response is a skin rash at the site of contact. For some sensitised people, dermatitis (a type of skin rash) may develop in an area of the skin that is away from the site of contact. Some workers exposed to nickel by inhalation can become sensitised and have asthma attacks, although this is rare. People sensitive to nickel react when nickel comes into prolonged contact with the skin. Some sensitised individuals react when they eat nickel in food or water or breathe dust containing nickel. Jewellery allergy or nickel allergy are terms used to denote skin reactions arising due to jewellery or nickel. Suppose the skin gets exposed to allergens or irritants present in jewellery or nickel. In that case, there is every chance one's skin will feel itching or irritation, accompanied by the appearance of rashes or inflammation.

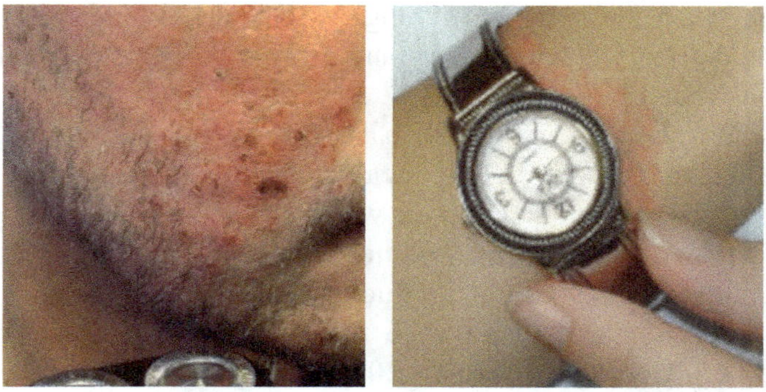

Figure 9. Examples of the effects of nickel as one of the most common allergic metals. Based on (Nickel: Health Effects, 2022; Thyssen *et al.*, 2011).

Lithium has adverse effects on the kidneys, thyroid gland and parathyroid glands, necessitating monitoring of these organ functions through periodic blood tests.

5. Lithium and Nickel in the Environment

5.1. *Lithium*

Lithium is expected to play an undesirable role in the environment, especially due to its mining, production and ever-increasing use in car batteries.

Lithium is extensively used in Li-ion batteries (LIBs), which also contain metals such as cobalt, nickel and manganese, which are toxic and can contaminate water supplies and ecosystems if they leach out of landfills. Additionally, fires in landfills or battery-recycling facilities have been attributed to the inappropriate disposal of LIBs. Despite the environmental cost of improper disposal of LIBs, the rate of recycling is still relatively low, as recycling processes remain costly and immature (Wikipedia, 2023; Mrozik *et al.*, 2021).

The demand for lithium for electric vehicle (EV) batteries is driving a mining boom in the arid Andes regions of Argentina, Chile and Bolivia, home to half the world's reserves. As with all mining, there are concerns about lithium mines; however, some experts overstate the potential environmental cost while neglecting to mention a big advantage: mining for lithium is much cleaner than mining for coal. Hydrologists warn that mines could drain vital ecosystems and deprive Indigenous communities of precious water.

The environmental impact of LiBs begins with its mining, which leads to the contamination of the local environment, where, through the consumption of the contaminated water by people and animals, the final effects become harmful. Another contribution due to LiBs comes from lithium supply, lithium production, use of Li products, battery leakage, recycling and disposal to the appropriate landfill (Fig. 10) (Lithium U.S. Geological Survey, 2022).

The impact of lithium on the environment is mainly related to its extraction. The extraction of raw materials for batteries often results in harmful consequences for the environment. The mining and refining of high-purity metals will continue to increase the complexity of processing techniques in the future. Graphite-containing dust particles are a severe concern in mining and processing operations, which are covered with large amounts of graphite-containing dust, posing a serious health risk to workers.

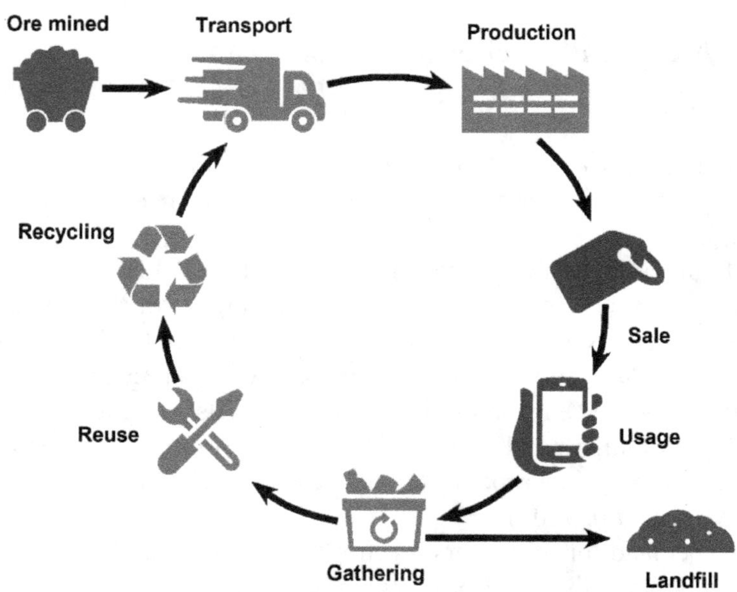

Figure 10. The main stages of the Li life cycle, from the mining of ore containing this element up to its final disposal. Based on (Lithium-ion Used Batteries, 2023).

In addition to exposure by inhalation of contaminated dust particles, mining raw materials for Li presents additional operational risks. Common hazards include fires and explosions, working from heights and in confined spaces, chemical exposure and noise-related injuries. Lithium mining can lead to stream contamination, resulting in livestock and soil contamination. Mined mountains in many parts of the world demonstrate the irreversible impact of lithium mining on natural landscapes and ecosystems.

Lithium prices have doubled in recent years due to a steady increase in demand. During this period, the International Energy Agency (IEA) predicted that the number of EVs would reach 125 million worldwide by 2030. We now have approximately 5 million electric cars. Using LiBs in electric cars introduces the paradox of 'clean' versus 'dirty' lithium mining. Even though electric vehicles are emission-free, if they are at least powered by electricity from renewable sources, they still have a climate impact that results from manufacturing both cars and batteries. Mining and refining battery materials and manufacturing cells, modules and kits require significant amounts of energy that could generate greenhouse gas emissions.

The benefits of driving electric cars in cities will be immediate: reduction of noise and air pollution and elimination of greenhouse gases and toxins such as NOX. However, LIB recycling is still not economically viable. As more old batteries become available, the price will go down. Unfortunately, the client has to pay if he wants to recycle the old car battery.

Storing energy in LIBs offers a set of advantages that can help us achieve sustainability goals considering energy use: for instance, allowing us to ease our reliance on fossil fuels in favour of renewable energy resources and LIBs. However, with these advantages come drawbacks throughout the battery's supply chain. In the following, we outline some of these advantages and disadvantages.

Some obvious benefits of using *electric cars*:

- *Renewable energy sources*: LIBs can store energy from renewable resources, such as solar, wind, tidal currents, bio-fuels and hydropower.
- *Resilience*: Stored energy, as with battery systems, can be beneficial in the case of disasters that threaten the power supply.

- *Long service life*: LIBs have a long service life and can be recharged thousands of times. While the recycling of LIBs is not yet optimised, the long-term use of batteries and products can result in reduced consumption and electronic waste.
- *Smarter energy use*: We can use devices powered by LIBs to monitor and manage our energy usage and efficiency.

In addition, as previously mentioned, there are also some environmental drawbacks, including intensive extraction (open-pit mining and brine extraction) and problems with the disposal of LIBs (they become electronic waste, and much of this waste is not disposed of properly).

There is a growing demand for LIBs for electric transportation and to support the application of renewable energies by auxiliary energy storage systems. This surge in demand requires a concomitant increase in production and, down the line, leads to large numbers of spent LIBs. The ever-increasing battery waste needs to be managed accordingly. Currently, there are no universal or unified standards for waste disposal of LIBs around the globe. Each country uses one or a combination of practices such as landfilling, incineration and complete or partial recycling, depending on the number of batteries leaving the market, current legislation and infrastructures (The Environmental Impact of Lithium Batteries, 2020; Nickel Mining, 2017).

5.2. *Nickel*

The mining and processing of nickel-rich ores consistently generate vast amounts of dust in the air. The dust is contaminated by high concentrations of potentially toxic metals, such as nickel, cobalt, copper and chromium. The costs associated with the environmental impacts of nickel mining, which is the main component of batteries for electric cars, are underestimated by the pioneers of emission-free vehicles. In many countries where nickel is mined, they have to deal with the consequences of mining, which is very unfriendly to the environment and causes serious health problems for people around the mines. Nickel, which is mainly mined in Australia, Canada, Indonesia, Russia and the Philippines, is not as environmentally friendly as it would seem from the use of its final products. The percentage of some countries where most of the world's nickel is mined is shown in Fig. 11 (Nickel: Supply, Risk and ESG Issues, 2023).

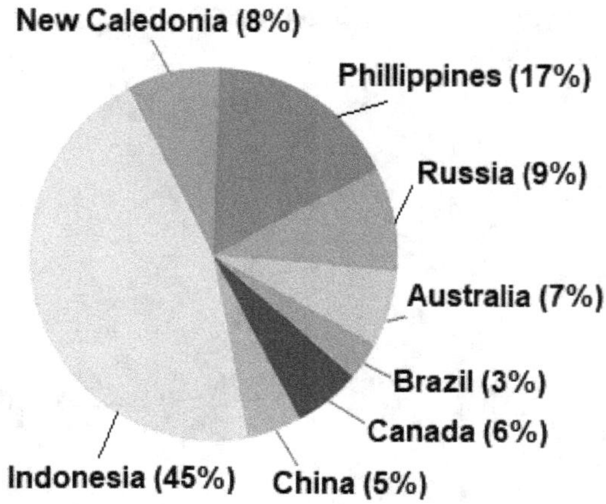

Figure 11. The percentage of some countries where most of the world's nickel is mined. Based on (Nickel: Supply, Risk and ESG Issues, 2023).

The advent of the era of electric cars with rechargeable batteries seems inevitable. Supporters of this idea often emphasise the destructive potential of internal combustion engines that use fossil fuels. Less is said about the ecological impact of mining precious metals, which are needed to produce accumulators.

Since almost 50% of nickel is mined in developing countries, where there are not sufficient regulations for the protection of the surrounding environment, the countryside is heavily affected. This is illustrated in Fig. 12, showing the nickel mine in Indonesia (Investment Minister Denies Indonesia Nickel Reserve, 2023).

While stainless steel will continue to be the main first use for nickel, the major engine of demand growth over the next two decades will be batteries. From only 7% of the total market in 2021, we expect battery use to grow to 40% of nickel consumption by 2040. That will push nickel demand to double in size to six million tonnes per year (Fig. 13, based on Mitchell and Pickens, 2022). It is estimated that nickel use in batteries has increased by 0.9 megatonnes in the past six months, mainly due to more significant net zero commitments from governments and automakers. The increasing importance of energy storage to enable more extensive use of renewables will also be an essential factor in driving demand.

Most countries are trying to replace petrol and diesel cars with batteries, which would require not only lithium but also particularly nickel.

Figure 12. A view on one of many nickel mines in Indonesia. Based on (Investment Minister Denies Indonesia Nickel Reserves, 2023).

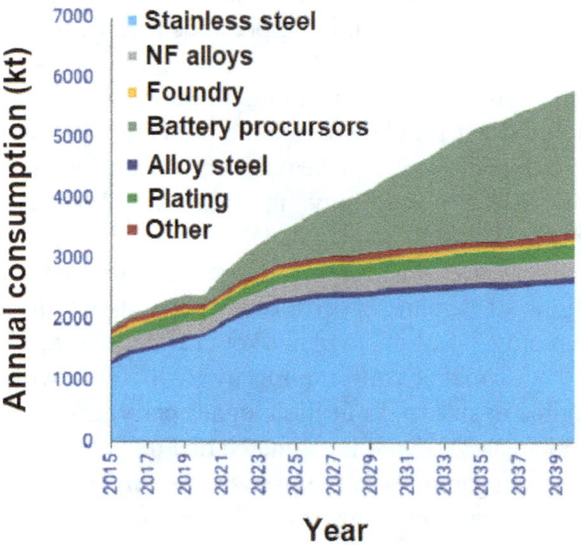

Figure 13. Use in batteries is supposed to double global demand by 2040. Based on Mitchell *et al.* (2023).

Emissions from the smelting and refining of nickel-rich ores can release harmful substances into the air, including sulphur dioxide, nitrogen oxides and particulate matter. These pollutants can have serious health consequences for nearby communities and contribute to global climate change. It may have associated negative impacts on the environment, including deforestation, erosion, contamination and alteration of soil profiles, contamination of local streams and wetlands and increased noise level, dust and pollution. Mine exploration, construction, operation and maintenance may alter land use.

Nickel is among the most significantly toxic heavy metals. Nickel present in the air can reach the soil or water through atmospheric deposition. It then binds to particles containing iron and manganese, which are often found in soil and sediments. This is why most of the nickel in the environment is found here. The toxicity of nickel for some aquatic organisms is relatively high. Therefore, its permissible concentration in water supply streams is limited more strictly than in drinking water. Plants receive nickel from the soil mainly through the roots, and they are able to accumulate it. Lowering the pH increases the mobility of nickel and thus its uptake by plants.

Often, the most important environmental impact of mining activities is the discharge of metals and other elements into the soil, surface water and groundwater. The most common and significant environmental impact involves discharges to soil and watercourses. The implications of discharges differ from element to element. When released in effluent, nickel is a toxic element that often occurs in higher concentrations than normal background levels, posing a severe threat to ecosystems.

Nickel is becoming a toxic pollutant in agricultural environments due to its increasing concentrations through anthropogenic activities. Different activities are responsible for Ni release into the environment, such as mining, smelting, burning of fossil fuels and industrial wastes. Some negative impacts of mining are loss of vegetation cover, mass destruction of water bodies, loss of biodiversity, land-use changes, food insecurity, increased social vices and conflicts, a high cost of living and air pollution.

6. Conclusion

Lithium and nickel metals have recently experienced a certain renaissance, mainly due to their increased use in automotive and other batteries.

The environmental consequences of their extraction, use, recycling and disposal have not yet received due attention. This is mainly due to the fact that the countries where these activities are performed do not have enough resources to properly ensure environmental protection, including the impact on food contamination and the regulation of the application of these metals not only in industry but also in their use in consumer goods. It is also related to the fact that most of these metals are mined in third-world countries where adequate attention is not paid to protecting the environment against excessive Li and Ni pollution.

In many countries worldwide, the necessary legislation to protect health, food, water and the environment from the harmful effects of Li and Ni still needs to be included. Sufficient attention is also not paid to monitoring food and the environment in relation to the levels of content and concentration of these metals, including their movement in the environment.

In connection with the enormous looming consumption of Li and Ni in the automotive industry, an even more significant increase in the extraction of these metals can be expected soon, for which especially developing countries are not yet prepared. There is also a lack of more detailed and uniform international standards in this area.

The real positive impact of both metals concerning their unique applications is often overestimated because the negative impacts of their extraction and processing are not sufficiently considered. In any case, the use of both Li and Ni can be expected to grow further. It will require adopting more effective mechanisms to protect society and the environment from some obvious negative impacts.

References

Ando S., *et al.* (2022). Comparison of lithium levels between suicide and non-suicide fatalities: Cross-sectional study. *Transl. Psychiatry* 12:466–471.

Anke M., *et al.* (2003). Lithium — History, nature and function. Friedrich-Schiller-Universitat, Jena, Germany, Elsevier Science Ltd. (20 February 2023), https://www.sciencedirect.com/referencework/9780122270550/encyclopedia–of-food-sciences-and-nutrition#book-description.

Bashyal J. (2022). Periodic table of elements. https://thechemistrynotes.com/periodic-table-of-elements/.

Corleone J. (2019). Food high in lithium. Biological Trace Elements Research. Online (21 April 2023), https://www.livestrong.com/article/519437-what-foods-contain-lithium/.

Chandra S. (2021). Is lithium an essential mineral for your mental health? Psychiatry and Cognition. (21 April 2023), https://chandramd.com/lithium-mental-health/.

Filipoiu D.C., *et al.* (2022). Characterization of the toxicological impact of heavy metals on human health in conjunction with modern analytical methods. *Toxics* 10:716–734.

Genchi G, *et al.* (2020). Nickel: Human health and environmental toxicology. *Int. J. Environ. Res.* 17:679.

Greenblatt J. (2023). Lithium as a nutrient. Psychiatry Redefined. (21 April 2023), https://www.psychiatryredefined.org/lithium-as-a-nutrient/.

Guiral J.S. (2018). Textural and mineralogical characterization of Li-pegmatite deposit: Using microanalytical and image analysis to link micro and macro properties of Spodumene in drill cores, Keliber Lithium Project, Finland. M. Sc. Thesis, August 2018. doi: 10.13140/RG.2.2.20597.22243.

Health Benefits and Uses of Lithium Carbonate (2023). Xtend-Life, Christchurch, New Zeeland. (21 February 2023), https://www.xtend-life.com/blogs/supplement-ingredients/lithium.

Investment Minister Denies Indonesia Nickel Reserves (2023). Tempo, Co. (30 April 2023), https://en.tempo.co/read/1692486/investment-minister-denies-indonesia-nickel-reserves-wont-last-long.

Lithium (2022). What is lithium? Lithium in the periodic table. (20 September 2023), https://medium.com/periodic-table-elements/what-is-lithium-e9bb42 77e6bd.

Lithium U.S. Geological Survey (2022). U.S. Geological Survey Mineral commodities summaries data 2022 release. (20 September 2023), https://www.sciencebase.gov/catalog/item/6197ccbed34eb622f692ee1c.

Lithium-ion Used Batteries (2023). US Environmental Agency. (21 February 2023): https://www.epa.gov/recycle/used-lithium-ion-batteries.

List of Countries by Lithium Production (2022). Wikipedia. (20 September 2023), https://en.wikipedia.org/wiki/List_of_countries_by_lithium_production.

Mrozik W., *et al.* (2021). Environmental impacts, pollution sources and pathways of spent lithium-ion batteries. *Energy Environ. Sci.* 14:6099–6121.

Mitchell A. Pickens N. (2022). Nickel and copper: Building blocks for a greener future. Wood Mackenzie. (20 September, 2023), https://www.woodmac.com/news/opinion/nickel-and-copper-building-blocks-for-a-greener-future/.

Nickel (2022). (20 September 2023), https://www.nutrientsreview.com/minerals/nickel.html.

Nickel: Supply, Risk and ESG Issues (2023). (20 September 2023), https://insights.issgovernance.com/posts/nickel-supply-risks-and-esg-issues/.

Nickel: Health Effects (2022). Asian Metal (2022). Beijing, China. Online (27 February 2023), http://metalpedia.asianmetal.com/metal/nickel/health.shtml.

Nickel Mining (2017). The hidden in the environmental cost of electric cars. Business and Human Rights, New York, USA. (23 April 2023), https://www.business-humanrights.org/en/latest-news/nickel-mining-the-hidden-environmental-cost-of-electric-cars-incl-company-comments/.

Periodic Elements (2022). Lenntech, USA, 2022. (12 February 2023), https://www.lenntech.com/periodic/elements/ni.htm#ixzz7t7acBCSw.

Schrauzer G.N. (2002). Lithium: Occurrence, dietary intakes, nutritional essentiality. *J. Am. Coll. Nutr.* 2:14–21.

Sheman A.J. *et al.* (2008). Contact allergy: Alternative for the 2007 North American contact dermatitis group (NACDG) standard screening tray. *Dis. Mon.* 54:7–156.

Szklarska D., Rzymski P. (2019). Is lithium a micronutritient? From biological activity and epidemiological observation to food fortification. *Biol. Trace Elem. Res.* 189:18–27.

The Environmental Impact of Lithium Batteries (2020). Institute for Energy Research, Washington, USA. (23 March 2023), https://www.instituteforenergyresearch.org/renewable/the-environmental-impact-of-lithium-batteries/.

Thyssen J.P., *et al.* (2011). Allergic nickel dermatitis caused by shaving: Case report and assessment of nickel release from an electric shaver. *Acta Derm. Venereol.* 92:95–96.

Underutilization of Lithium (2018). Challenging the negative perception of lithium and optimizing its long-term administration. (21 February 2023), https://www.researchgate.net/publication/328024889_Challenging_the_Negative_Perception_of_Lithium_and_Optimizing_Its_Long-Term_Administration/figures?lo=1.

Wikipedia (2023). The free encyclopaedia, composition of human body. (20 January 2023), https://en.wikipedia.org/wiki/Composition_of_the_human_body.

https://doi.org/10.1142/9789811283123_bmatter

Index

www.ingramcontent.com/pod-product-compliance
Lightning Source LLC
LaVergne TN
LVHW010332220926
848484LV00004BB/885